ECE/STEEL/88

ECONOMIC COMMISSION FOR EUROPE
Geneva

STRUCTURAL CHANGES IN CONSUMPTION AND TRADE IN STEEL

UNITED NATIONS
New York and Geneva, 1994

NOTE

The designations employed and the presentation of the material in this publication do not imply the expression of any opinion whatsoever on the part of the Secretariat of the United Nations concerning the legal status of any country, territory, city or area, or of its authorities, or concerning the delimitation of its frontiers or boundaries.

ECE/STEEL/88

UNITED NATIONS PUBLICATION

Sales No. E.94.II.E.42

ISBN 92-1-116 619-5
ISSN 1014-9163

TABLE OF CONTENTS

TABLES

FIGURES

ANNEX TABLES

SYMBOLS EMPLOYED

... = figure not available
- = nil or negligible quantity
Mt = millions of tonnes

Unless otherwise stated, tonnages are in metric tonnes.

PREFATORY NOTE

At its second session, held in October 1992, the Working Party on Steel of the Economic Commission for Europe decided to conduct a study on recent developments in the consumption and trade in steel in ECE member countries. The study attempts to identify and explain important changes and their consequences for consumption and trade in steel from 1980 to 1993. It is a follow-up to two former studies carried out by the Steel Committee of the Economic Commission for Europe, the current Working Party on Steel: The Evolution of the Specific Consumption of Steel (1984) and Structural Changes in International Steel Trade (1987). The present study covers more recent data from 1980 to 1992 and in some cases to 1993, and attempts to use those data available with the maximum efficiency, focusing on data analysis and its explanation.

The study consists of four chapters. Chapter 1 provides an overview and analysis of the relationship between steel consumption, production and trade. The structure and balance of world steel consumption, production and trade are examined using data from 1980 to 1993. Chapter 2 analyses structural changes in world steel consumption. In the first part of chapter 2, growth patterns in steel consumption in major regions and countries are analysed in comparison with GDP growth, showing the importance of steel in the total economy. In the second part, the steel intensity or the specific consumption of steel in selected countries is analysed, based on data obtained by a questionnaire. The steel intensity or the specific consumption of steel is the relation between steel consumption and measures of the activity of a production sector or an end use. The importance of steel in several industrial sectors such as manufacturing and construction is discussed on a country-by-country basis. The last part of chapter 2 describes in detail the situation and important factors for steel consumption in three major regions and countries; the EC, the United States and Japan. Chapter 3 concentrates on international trade in steel products. The structure of exports, imports and trade balances of regions and countries are examined and a product-based analysis is provided. In chapter 4, the major findings and conclusions of the study are summarized. In the text, only the summary tables and figures appear. However, all the basic data discussed in the text appear in annex tables for the convenience of readers.

For the sake of more effective analysis, a clear distinction is drawn between crude steel equivalent and finished steel. In chapter 1, all analysis is made on the basis of crude steel equivalent, because this is the most efficient means of comparing three different phases: consumption, production and trade. However, in chapter 2 and chapter 3, finished steel figures are used to reflect a more realistic picture of steel consumption and trade.

Detailed analysis is focused on ECE member States and other major steel producing countries. Groupings of countries based on geographic and economic regions are made for convenience of analysis. Africa, Middle East, Far East, North America, Other America, Oceania, EC, EFTA, Eastern Europe, former USSR and Other Europe are these groupings. Countries listed in the basic tables in the annex are limited to steel producers with more than 2 million tonnes of crude steel production annually or ECE member countries. The rest are grouped in regions. The reader should be aware of the following factors concerning the EC, the former USSR and the former Yugoslavia. Because the purpose of this study is to analyse developments from

1980 to 1993, consistency in the grouping of countries is necessary. Even though the European Community, the current European Union, increased its membership during the 1980s, the region referred to as the EC in this study includes all 12 current member States for all periods which the study covers. In the same manner, even though the three Baltic States became independent in 1991 and the former USSR was dissolved after that, the grouping of the former USSR is used even for data for 1991 and 1992. This also applies to the former Yugoslavia.

All the data used in the study come from either the Economic Commission for Europe (ECE) or the International Iron and Steel Institute (IISI) except when specified otherwise. In general, data for Europe, the United States, Canada and Japan are from the ECE/ITD database or other information made available to the ECE and the United Nations, while data for other countries are from the IISI.

This publication was written by Nobuhisa Iwase and Alexander Cavic of the ECE Industry and Technology Division. The secretariat wishes to thank all those who responded to the questionnaire for the study and made valuable comments.

INTRODUCTION

The 1980s was a period in which the world political and economic scene showed a dramatic change. The Berlin Wall fell in November 1989 and the former CMEA countries started a transition phase to market economies, after experiencing difficulties in economic growth under the old communist regime. The United States adopted new supply-side economic policies called Reaganomics in the first half of the 1980s, which first resulted in the appreciation of the key currency of the world, the dollar, and then in its depreciation after 1985. Western Europe moved towards a "unified Europe" by strengthening ties inside the European Community and EFTA.

With these dramatic changes in the political and economic environment, the world steel industry faced a challenging period. Moreover, the steel industry itself brought about an internal evolution through continual technological innovation. Increasing yield ratios and cost reductions were enhanced by maximizing the use of continuous casters and other process technologies. In many countries, advanced coating technologies for higher value-added steel sheets were introduced. With well distributed technological information, developing steel producing countries began to get easier access to modern technologies and know-how. Thin-slab casting technology was also introduced to commercial production and with improvements in electric arc furnace technologies this innovation decreased the initial cost of establishing flat-rolled product plants, and made it possible for small developing countries and small investors to challenge established flat product markets. With the threat of material substitution from steel to other products and the increasing needs of steel users for greater varieties, higher added-value and higher cost performance of steel products, world steel producers were forced to tackle difficult tasks. The trend toward creating regional economic blocs and continual disputes about unfair trade practices encouraged domestic steel producers to become global, developing joint activities with other producers in different countries. The world steel industry saw significant changes in its structure in terms of consumption, production, trade and technologies throughout the 1980s.

The exchange of steel products can be divided into two categories. One is the steel consumption of each country, which is the demand side and in most cases is the major delivery orientation of steel products by domestic producers. Domestic consumption of steel also covers the demand for indirect exports of steel products such as automobiles and machinery produced inside the country and exported to other countries. By knowing the general consumption situation and the specific orientation of steel products, the flow of steel products inside the country can be understood as well as the potential of indirect exports of steel. The other category is international trade which meets the shortage in the steel supply of importing countries. Exporting countries can achieve higher scale economies with larger production capacities than merely by fulfilling domestic demand. In other cases, countries can concentrate on certain steel products and exchange those products to effectively utilize their competitive advantages. The study presents figures for the exchange of steel products for both consumption and international trade.

CHAPTER 1
OVERVIEW OF WORLD STEEL CONSUMPTION, PRODUCTION AND TRADE

1.1. General evolution of consumption, production and net exports

World crude steel production in the 1980s showed the same cyclical pattern as that of world economic growth (Figure 1-1). However, unlike the constant growth in the world economy, the growth rate of crude steel production was quite modest and on occasion even recorded sharp declines. As a result, average crude steel production in the 1980s stayed at 717 million tonnes, which was only a 7.6 per cent increase from the average of the 1970s, even though the highest ever volume of crude steel production, 785 million tonnes, was recorded in 1989.

The same 10-year average growth rates from 1950 to 1960 and from 1960 to 1970 were 74.5 per cent and 52.9 per cent respectively, with average crude steel production of 436 million tonnes in the 1960s and 667 million tonnes in the 1970s. Because these figures are for crude steel equivalent and do not reflect exactly the total volume of final steel products, however, it can be said that in general growth in steel production has been slowing down.

Even though the growth rate of world crude steel production as a whole has been slowing down, there are large differences in changes in crude steel production by regions (Figure 1-2). The Far East, excluding Japan, recorded a surprisingly big increase of 84.9 million tonnes over the period from 1980-1982 to 1991-1993. Such developing regions as Other America (Latin America), the Middle East, Other Europe (Turkey) and Africa had stable increases in production.

On the other hand, most developed economies faced declines. Three major steel producing countries and regions, the European Community, Japan and the United States recorded declines of 1.5 per cent, 1.7 per cent and 8.6 per cent during the same period, respectively. Furthermore, the so-called transition economies, the former USSR and eastern Europe, faced the largest declines of 21.3 per cent and 40.1 per cent.

Of the 19 largest crude steel producers in 1980 and 1993 (Figure 1-3), three characteristic features are apparent. First of all, such emerging steel producing countries as the Republic of Korea, Brazil and India entered the top ten list in 1993. At the same time, China, in fifth position in 1980, advanced to third in 1993, more than doubling its crude steel production to 90 million tonnes. Secondly, mainly due to increases by these newly emerging producers, total production by the second largest group of countries, from the fourth largest to the nineteenth largest on the list, increased from 285 million tonnes in 1980 to 353 million tonnes in 1993.

Figure 1-1 World crude steel production and the growth rate of the world economy a/

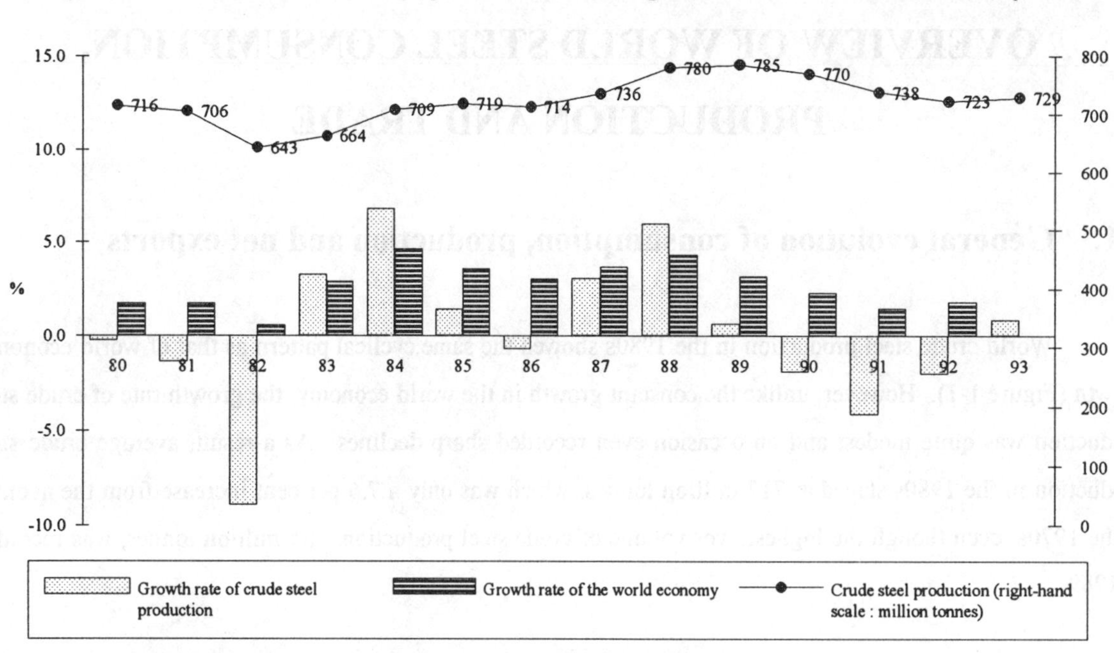

a/ The growth rate is the percentage change from the previous year.

Figure 1-2 Changes in crude steel production from 1980 to 1993 a/

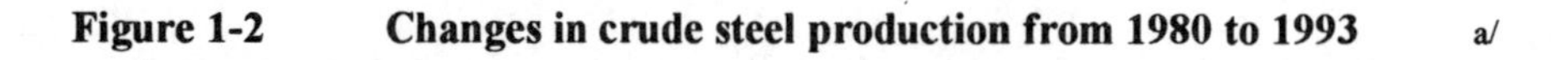

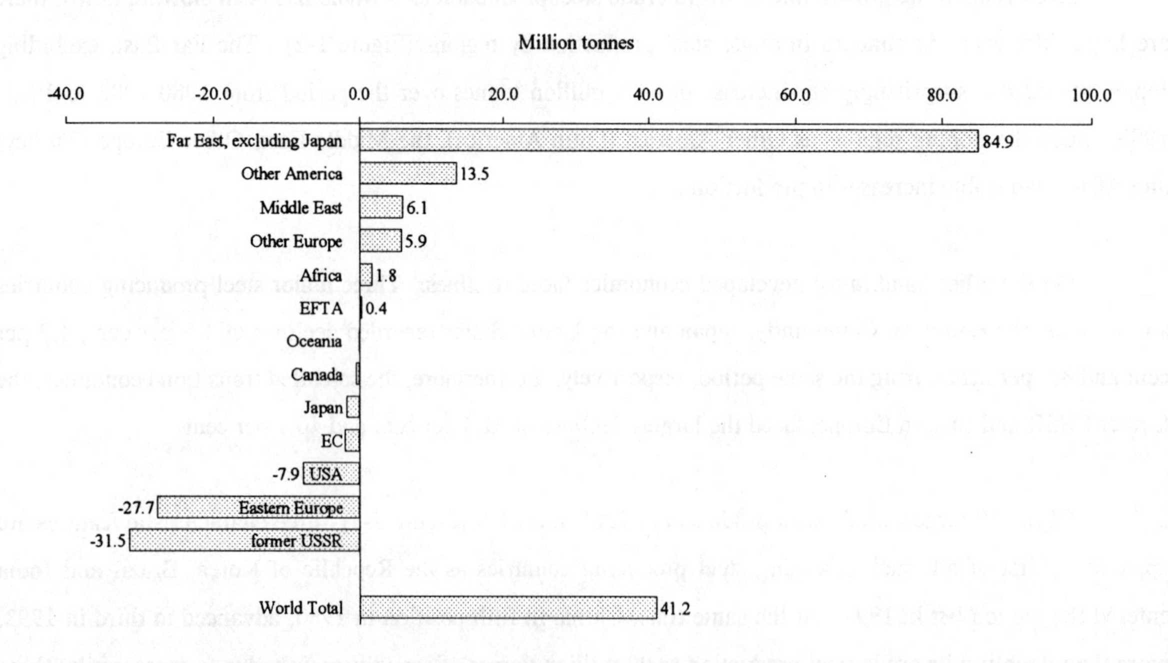

a/ Average change from 1980-1982 to 1991-1993.

Figure 1-3 Largest crude steel producers in 1980 and 1993

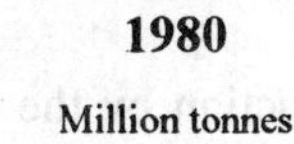

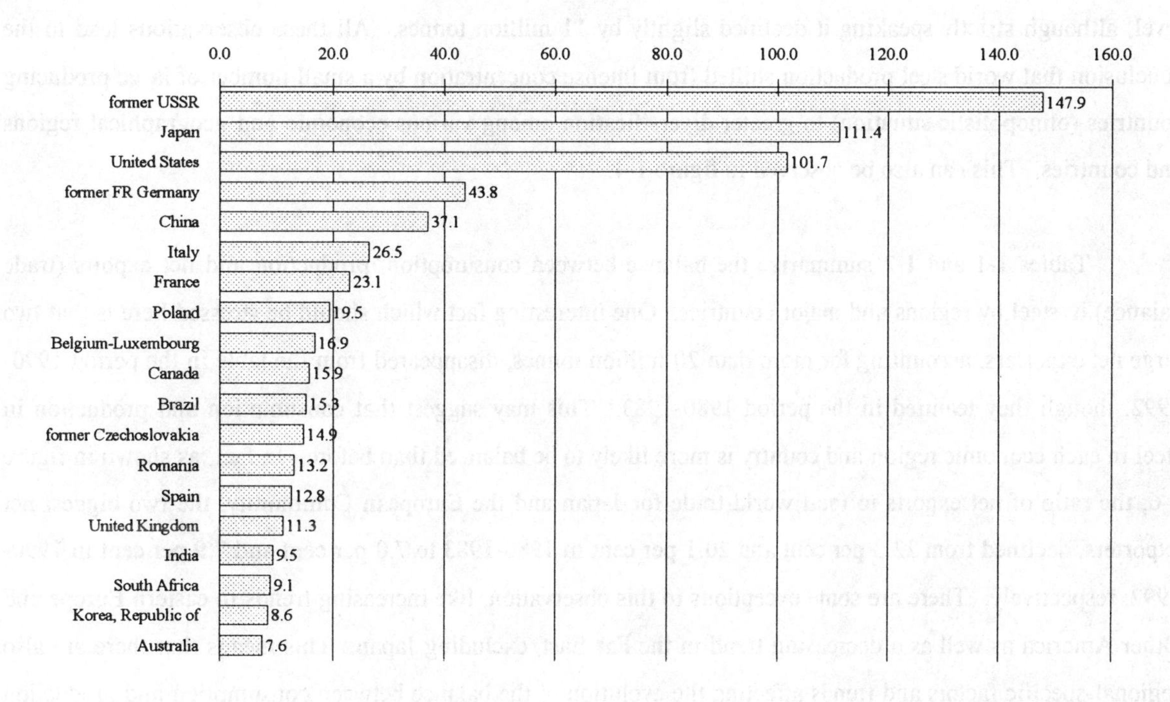
1980
Million tonnes
0.0
20.0
40.0
60.0
80.0
100.0
120.0
140.0
160.0
former USSR
147.9
Japan
111.4
United States
101.7
former FR Germany
43.8
China
37.1
Italy
26.5
France
23.1
Poland
19.5
Belgium-Luxembourg
16.9
Canada
15.9
Brazil
15.3
former Czechoslovakia
14.9
Romania
13.2
Spain
12.8
United Kingdom
11.3
India
9.5
South Africa
9.1
Korea, Republic of
8.6
Australia
7.6

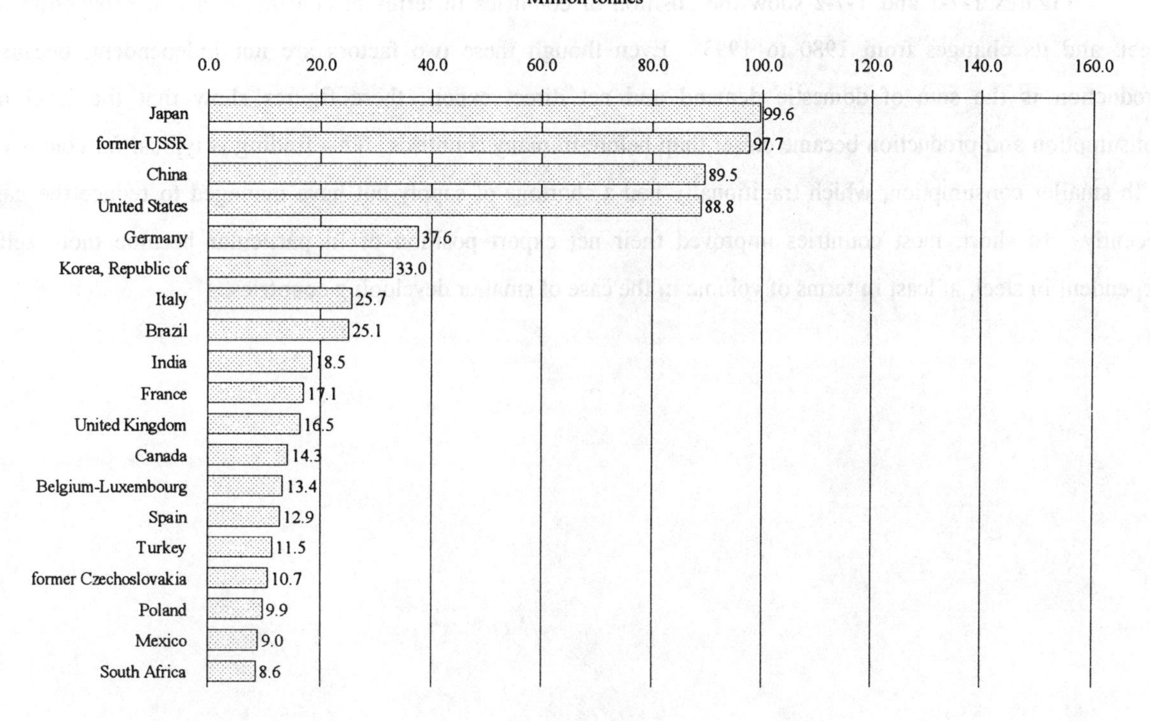
1993
Million tonnes
0.0
20.0
40.0
60.0
80.0
100.0
120.0
140.0
160.0
Japan
99.6
former USSR
97.7
China
89.5
United States
88.8
Germany
37.6
Korea, Republic of
33.0
Italy
25.7
Brazil
25.1
India
18.5
France
17.1
United Kingdom
16.5
Canada
14.3
Belgium-Luxembourg
13.4
Spain
12.9
Turkey
11.5
former Czechoslovakia
10.7
Poland
9.9
Mexico
9.0
South Africa
8.6

Thirdly and finally, even though the top three producers in 1980, the former USSR,[1/] Japan and the United States remained in almost the same positions in 1993, their combined production levels decreased by 75 million tonnes.

As a result of these changes, total production by the top 19 producers remained at almost the same level, although strictly speaking it declined slightly by 11 million tonnes. All these observations lead to the conclusion that world steel production shifted from intense concentration by a small number of large producing countries (oligopolistic situation) to greater diversification among various economic and geographical regions and countries. This can also be observed in figure 1-4.

Tables 1-1 and 1-2 summarize the balance between consumption, production and net exports (trade balance) in steel by regions and major countries. One interesting fact which should be stressed here is that two large net exporters, accounting for more than 20 million tonnes, disappeared from the table in the period 1990-1992, though they featured in the period 1980-1983. This may suggest that consumption and production in steel in each economic region and country is more likely to be balanced than before. In fact, as shown in figure 1-6, the ratio of net exports to total world trade for Japan and the European Community, the two biggest net exporters, declined from 22.3 per cent and 20.1 per cent in 1980-1983 to 7.0 per cent and 7.9 per cent in 1990-1992, respectively. There are some exceptions to this observation, like increasing trends in eastern Europe and Other America as well as a decreasing trend in the Far East, excluding Japan. This means that there are also regional-specific factors and trends affecting the evolution of the balance between consumption and production in steel.

Figures 1-7-1 and 1-7-2 show the position of countries in terms of consumption and production in steel, and its changes from 1980 to 1993. Even though these two factors are not independent, because production is the sum of domestic demand and net direct export, these figures show that the level of consumption and production became closer than before in many countries. This finding is typical for countries with smaller consumption, which traditionally had a shortage of supply but have managed to reduce the gap recently. In short, most countries improved their net export position or in particular became more self-dependent in steel, at least in terms of volume in the case of smaller developing countries.

[1/] As described in the prefatory note, even though the former USSR dissolved into many independent States and did not exist as a single country in 1993, this study uses the name, the former USSR, as a grouping of those independent States for the sake of medium-term comparison and analysis.

Figure 1-4 Share of crude steel production in 1980-1983 and 1990-1992 a/

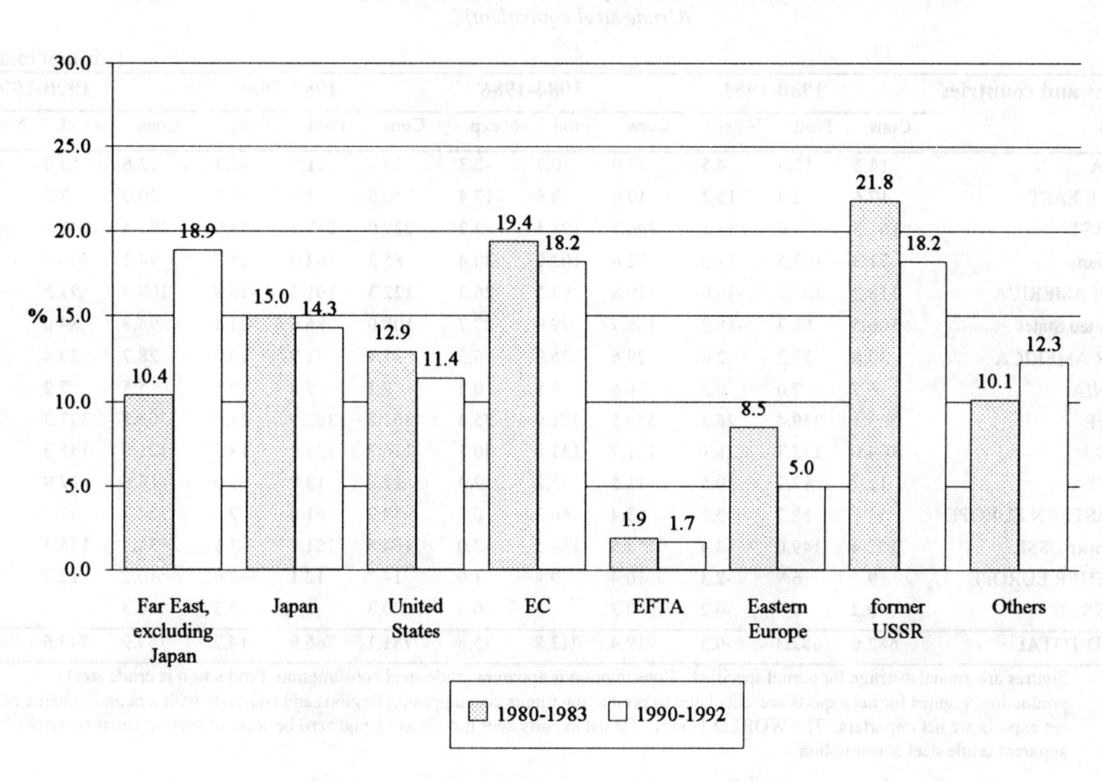

a/ Figures are annual average for period specified.

Figure 1-5 Share of apparent crude steel consumption in 1980-1983 and 1990-1992 a/

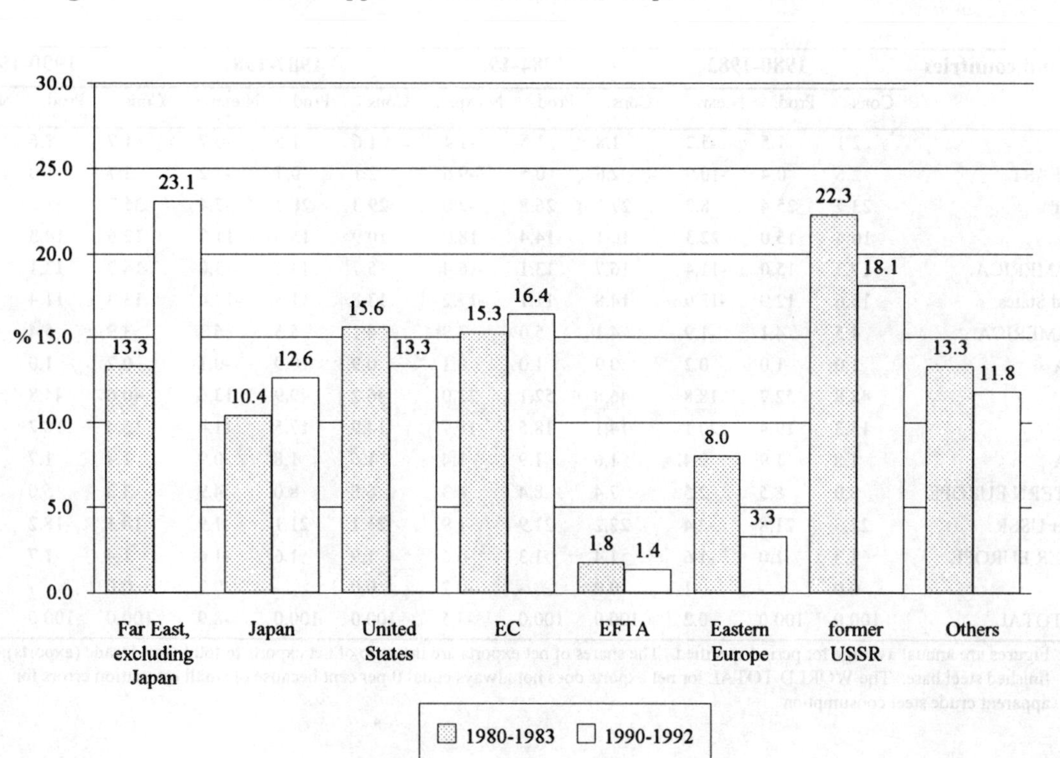

a/ Figures are annual average for period specified.

Table 1-1 Consumption, production and net exports a/
(Crude steel equivalent)

(Million tonnes)

Regions and countries	1980-1983			1984-1986			1987-1989			1990-1992		
	Cons.	Prod.	N.exp.	Cons.	Prod.	N.exp.	Cons.	Prod.	N.exp.	Cons.	Prod.	N.exp.
AFRICA	14.5	10.0	-4.5	13.0	10.7	-2.3	12.6	11.5	-1.1	12.6	12.0	-0.6
MIDDLE EAST	17.6	2.4	-15.2	19.0	3.6	-15.4	16.6	5.1	-11.5	20.0	7.2	-12.8
FAR EAST	161.5	173.0	11.5	194.7	191.5	-3.2	229.0	217.1	-11.9	265.4	246.6	-18.8
Japan	71.3	102.5	31.2	72.6	103.0	30.4	85.3	104.0	18.7	94.1	106.0	11.9
NORTH AMERICA	118.2	102.2	-16.0	119.8	93.5	-26.3	122.3	101.5	-20.8	109.4	97.5	-11.9
United States	106.5	88.3	-18.2	106.7	79.0	-27.7	108.0	86.6	-21.4	98.8	84.6	-14.2
OTHER AMERICA	30.8	28.2	-2.6	29.6	35.9	6.3	31.9	41.9	10.0	28.7	39.8	11.1
OCEANIA	6.7	7.0	0.3	6.6	6.8	0.2	7.1	7.0	-0.1	5.5	7.2	1.7
EUROPE	333.1	359.4	26.3	336.5	371.9	35.4	361.2	382.9	21.7	302.1	333.3	31.2
EC	104.6	132.7	28.1	101.7	131.8	30.1	116.3	134.5	18.2	121.9	135.3	13.4
EFTA	12.2	12.7	0.5	11.5	13.8	2.3	12.2	13.7	1.5	10.6	12.9	2.3
EASTERN EUROPE	54.7	58.2	3.5	53.4	60.3	6.9	53.2	61.0	7.8	24.7	37.2	12.5
former USSR	152.4	149.0	-3.4	159.5	156.5	-3.0	164.8	161.7	-3.1	134.7	135.3	0.6
OTHER EUROPE	9.2	6.9	-2.3	10.4	9.4	-1.0	14.7	12.1	-2.6	10.2	12.7	2.5
OTHERS	0.2	-	-0.2	0.3	-	-0.3	0.3	-	-0.3	0.3	-	-0.3
WORLD TOTAL	682.6	682.3	-0.3	719.4	713.8	-5.6	781.1	766.9	-14.2	743.9	743.6	-0.3

a/ Figures are annual average for period specified. Consumption is apparent crude steel consumption. Production is crude steel production. Figures for net exports are calculated as production minus consumption. Regions and countries with a negative value of net exports are net importers. The WORLD TOTAL for net exports does not always equal zero because of small estimation errors for apparent crude steel consumption.

Table 1-2 Regional share of consumption, production and net exports a/
(Crude steel equivalent)

(%)

Regions and countries	1980-1983			1984-1986			1987-1989			1990-1992		
	Cons.	Prod.	N.exp.	Cons.	Prod.	N.exp.	Cons.	Prod.	N.exp.	Cons.	Prod.	N.exp.
AFRICA	2.1	1.5	-3.2	1.8	1.5	-1.4	1.6	1.5	-0.7	1.7	1.6	-0.4
MIDDLE EAST	2.6	0.4	-10.9	2.6	0.5	-9.6	2.1	0.7	-7.2	2.7	1.0	-7.5
FAR EAST	23.7	25.4	8.2	27.1	26.8	-2.0	29.3	28.3	-7.4	35.7	33.2	-11.0
Japan	10.4	15.0	22.3	10.1	14.4	18.9	10.9	13.6	11.7	12.6	14.3	7.0
NORTH AMERICA	17.3	15.0	-11.4	16.7	13.1	-16.4	15.7	13.2	-13.0	14.7	13.1	-7.0
United States	15.6	12.9	-13.0	14.8	11.1	-17.2	13.8	11.3	-13.4	13.3	11.4	-8.3
OTHER AMERICA	4.5	4.1	-1.9	4.1	5.0	3.9	4.1	5.5	6.3	3.9	5.4	6.5
OCEANIA	1.0	1.0	0.2	0.9	1.0	0.1	0.9	0.9	-0.1	0.7	1.0	1.0
EUROPE	48.8	52.7	18.8	46.8	52.1	22.0	46.2	49.9	13.6	40.6	44.8	18.3
EC	15.3	19.4	20.1	14.1	18.5	18.7	14.9	17.5	11.4	16.4	18.2	7.9
EFTA	1.8	1.9	0.4	1.6	1.9	1.4	1.6	1.8	0.9	1.4	1.7	1.3
EASTERN EUROPE	8.0	8.5	2.5	7.4	8.4	4.3	6.8	8.0	4.9	3.3	5.0	7.3
former USSR	22.3	21.8	-2.4	22.2	21.9	-1.9	21.1	21.1	-1.9	18.1	18.2	0.4
OTHER EUROPE	1.3	1.0	-1.6	1.4	1.3	-0.6	1.9	1.6	-1.6	1.4	1.7	1.5
OTHERS	0.0	-	-0.1	0.0	-	-0.2	0.0	-	-0.2	0.0	-	-0.2
WORLD TOTAL	100.0	100.0	-0.2	100.0	100.0	-3.5	100.0	100.0	-8.9	100.0	100.0	-0.2

a/ Figures are annual average for period specified. The shares of net exports are the ratio of net exports to total world trade (exports): finished steel base. The WORLD TOTAL for net exports does not always equal 0 per cent because of small estimation errors for apparent crude steel consumption.

Figure 1-6 Ratio of net exports to total world trade a/

Net exporters

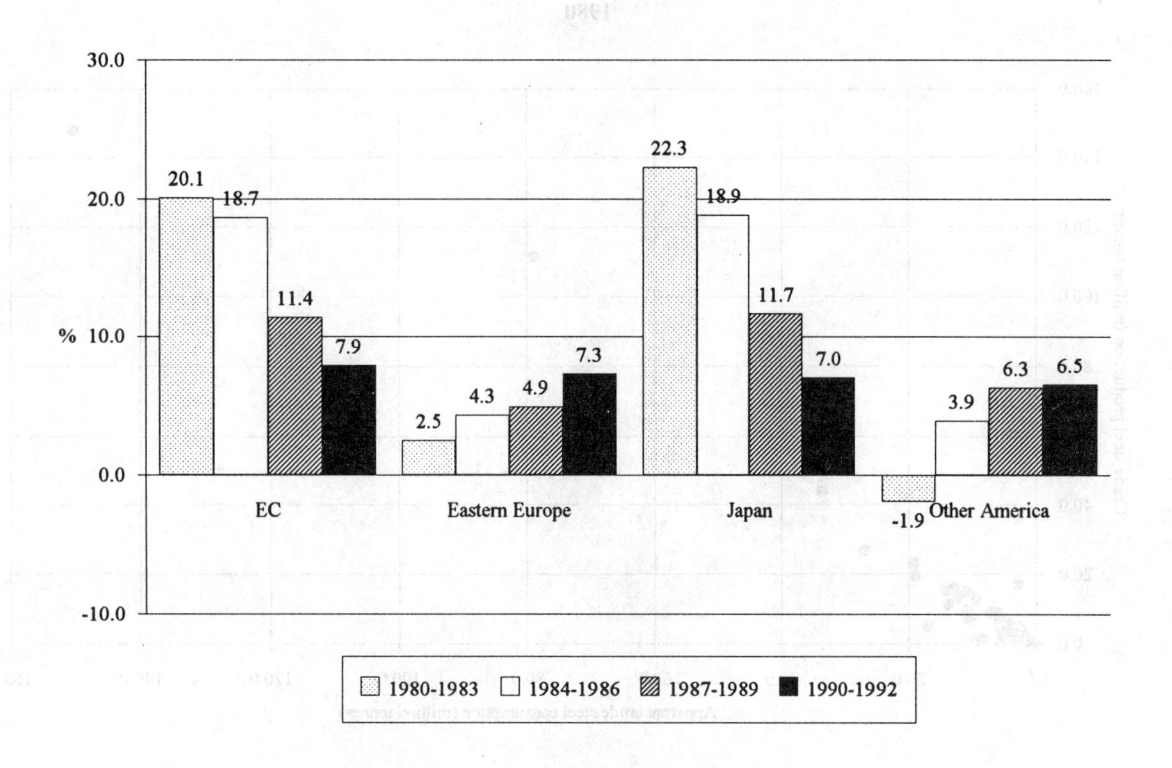

Net importers

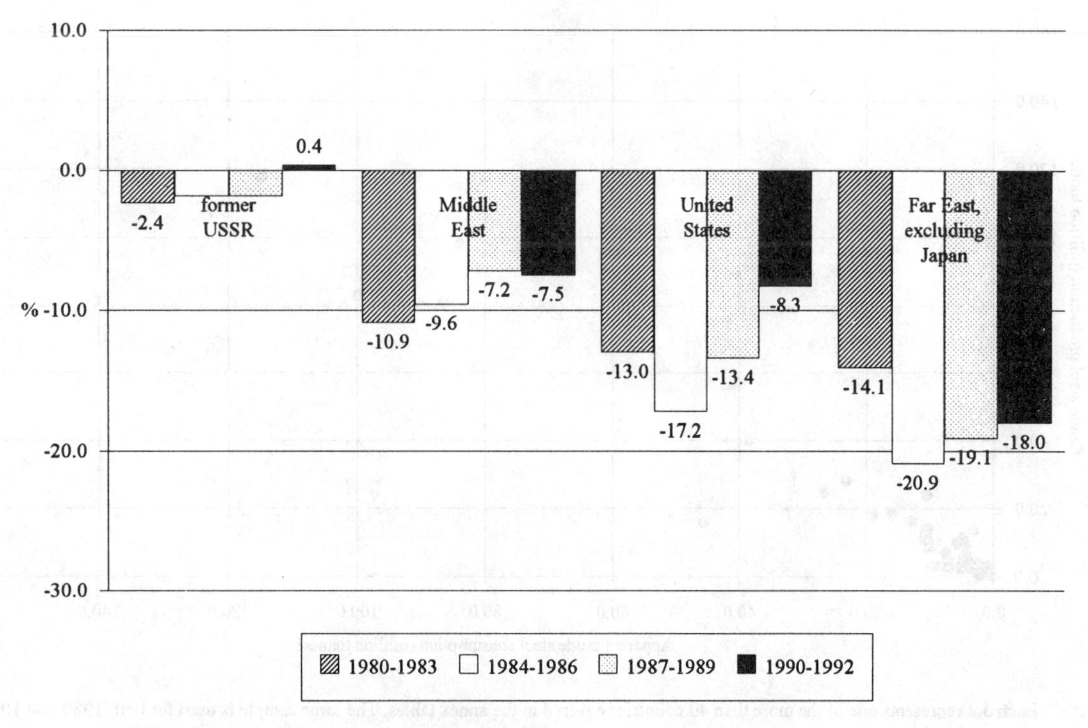

a/ Figures are annual average for period specified.

Figure 1-7-1 Relationship between consumption and production in steel a/

1980

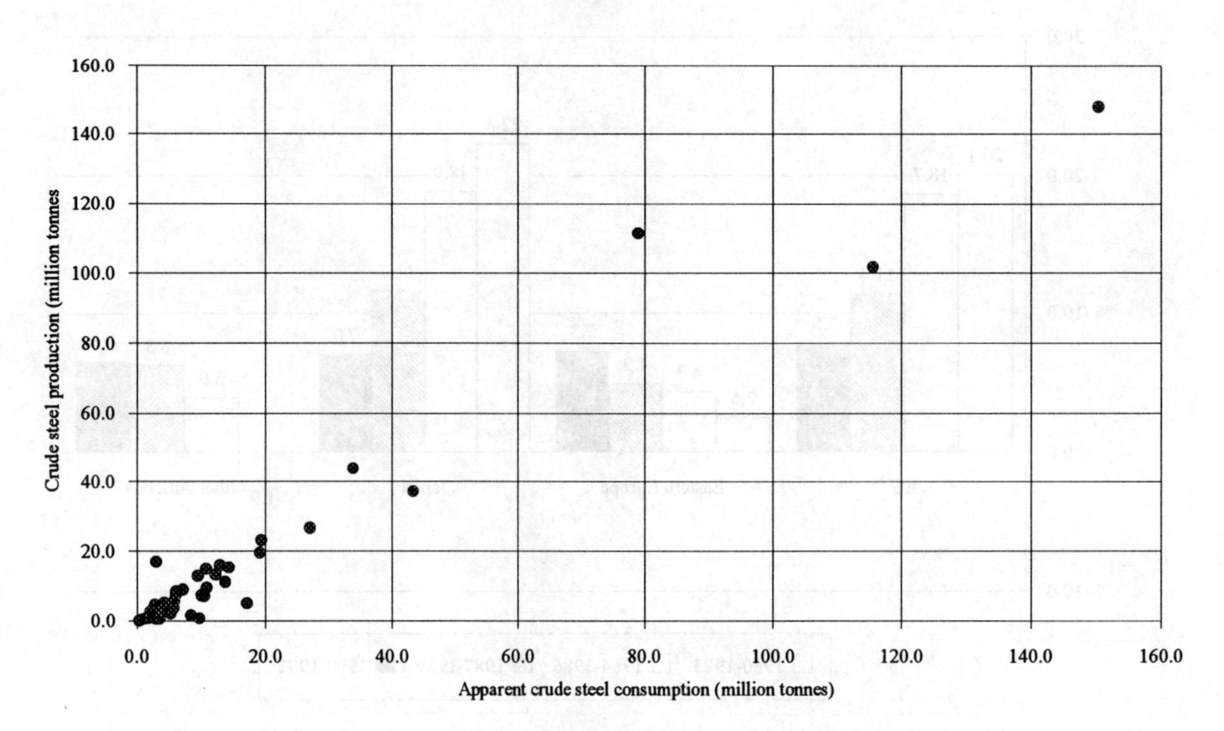

1992

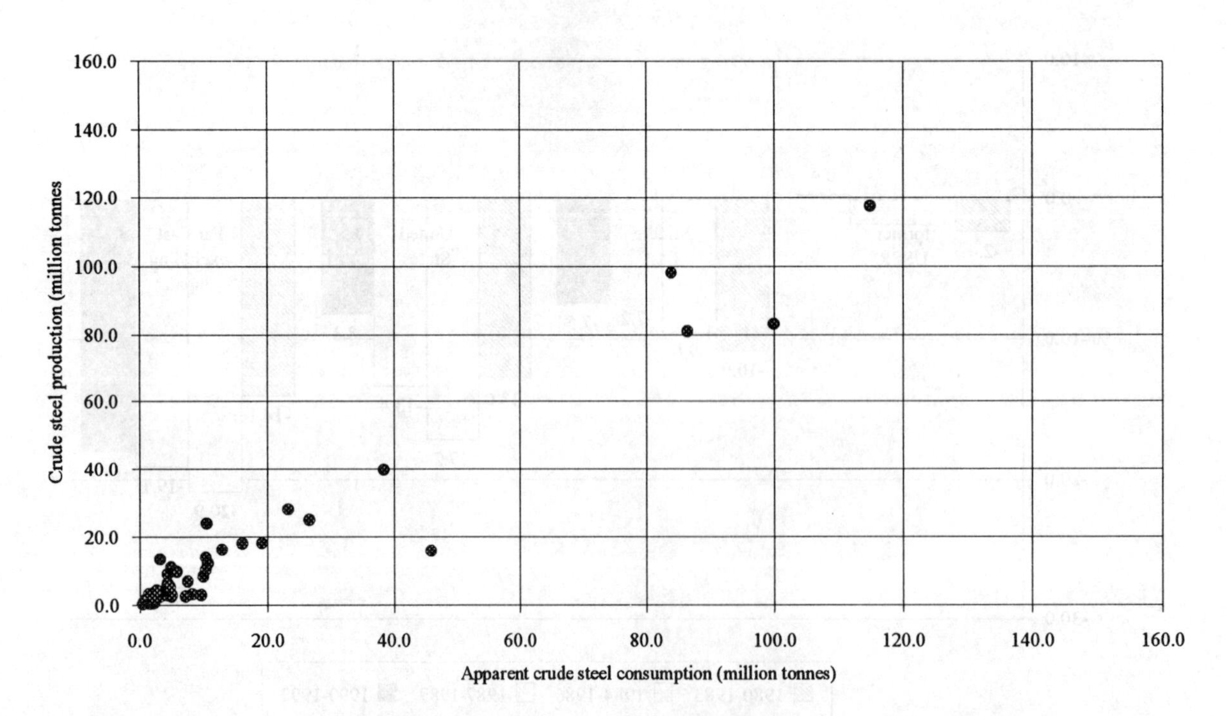

a/ Each dot represents one of the more than 40 countries covered in the annex tables. The same sample is used for both 1980 and 1992.

Figure 1-7-2 Relationship between consumption and production in steel a/

Excluding largest steel producers

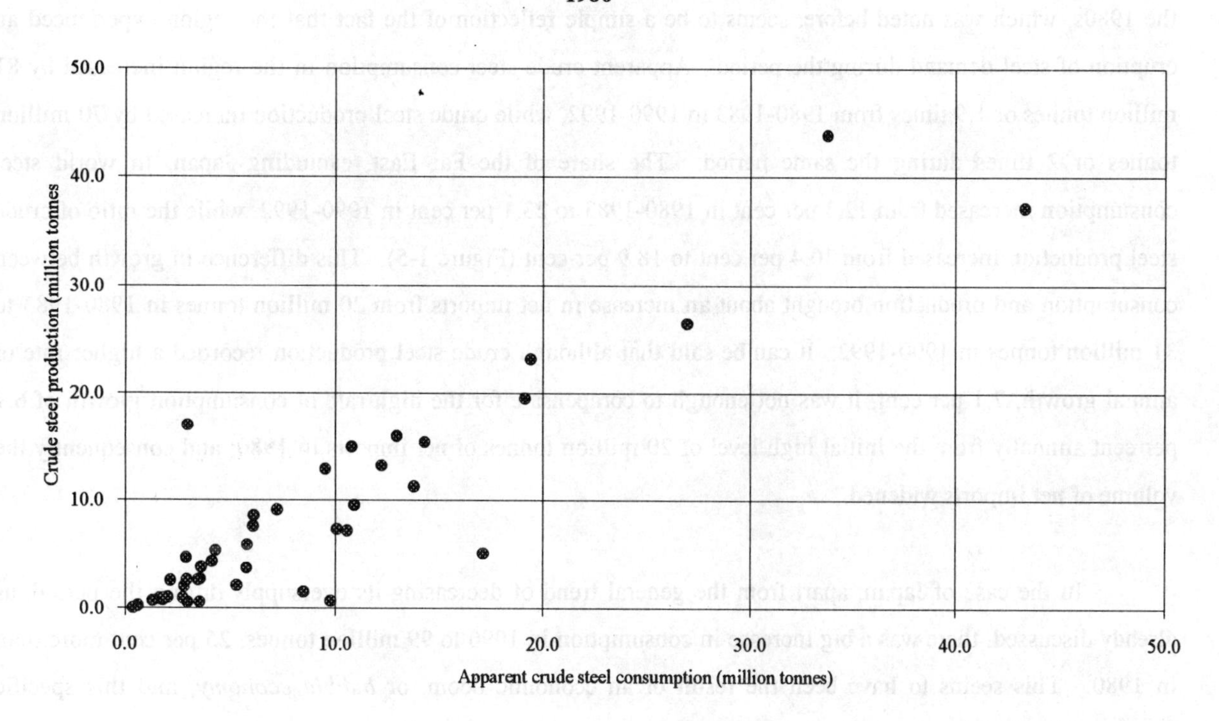

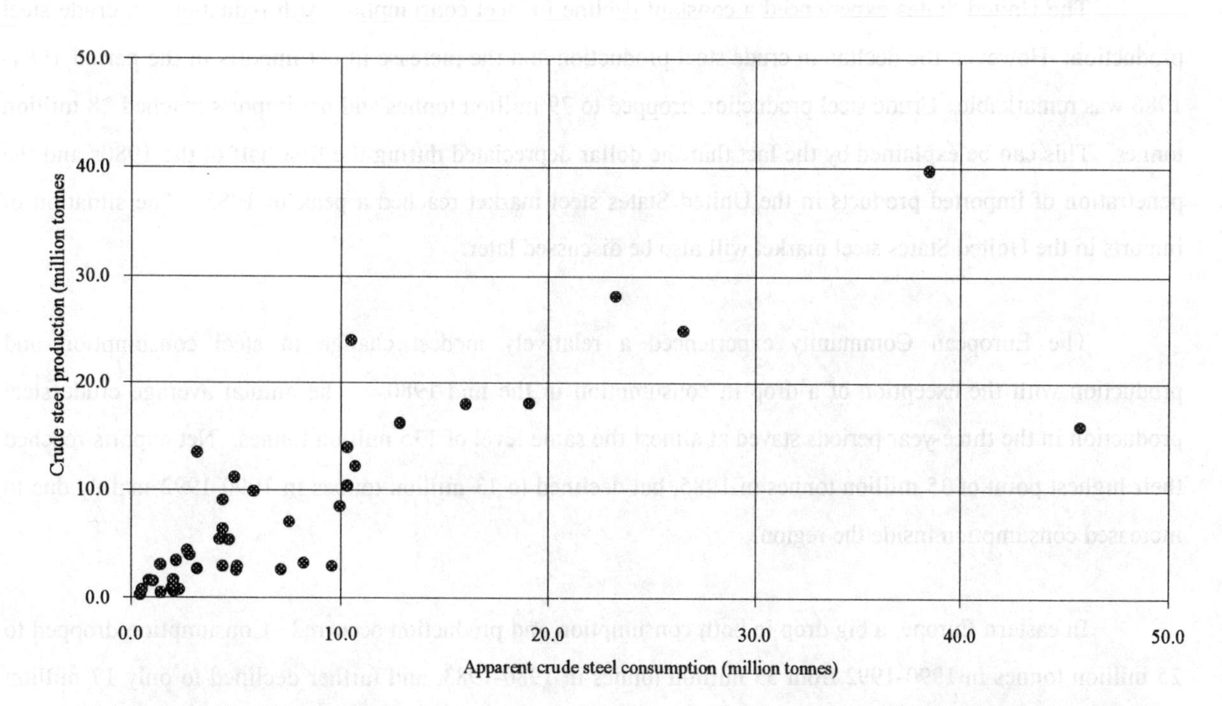

a/ Each dot represents one of the more than 40 countries covered in the annex tables, excluding the three largest steel producers, the former USSR, Japan and the United States in 1980, and the four largest steel producers, the former USSR, Japan, the United States and China in 1992.

1.2. Consumption, production and net exports by major regions and countries

Figure 1-8 summarizes the changing balance between consumption, production and net exports by major regions and countries. The rapid increase in crude steel production in the Far East, excluding Japan, in the 1980s, which was noted before, seems to be a simple reflection of the fact that the region experienced an eruption of steel demand during the period. Apparent crude steel consumption in the region increased by 81 million tonnes or 1.9 times from 1980-1983 to 1990-1992, while crude steel production increased by 70 million tonnes or 2 times during the same period. The share of the Far East, excluding Japan, in world steel consumption increased from 12.3 per cent in 1980-1983 to 23.1 per cent in 1990-1992, while the ratio of crude steel production increased from 10.4 per cent to 18.9 per cent (Figure 1-5). This difference in growth between consumption and production brought about an increase in net imports from 20 million tonnes in 1980-1983 to 31 million tonnes in 1990-1992. It can be said that although crude steel production recorded a higher rate of annual growth, 7.1 per cent, it was not enough to compensate for the high rate of consumption growth of 6.7 per cent annually from the initial high level of 20 million tonnes of net imports in 1980, and consequently the volume of net imports widened.

In the case of Japan, apart from the general trend of decreasing its oversupply during the period, as already discussed, there was a big increase in consumption in 1990 to 99 million tonnes, 25 per cent more than in 1980. This seems to have been the result of an economic boom, or *bubble economy*, and this specific situation will be discussed in detail in chapter 2.

The United States experienced a constant decline in steel consumption with reductions in crude steel production. However, the decline in crude steel production and the increase in net imports in the period 1984-1986 was remarkable. Crude steel production dropped to 79 million tonnes and net imports reached 28 million tonnes. This can be explained by the fact that the dollar depreciated during the first half of the 1980s and the penetration of imported products in the United States steel market reached a peak in 1985. The situation of imports in the United States steel market will also be discussed later.

The European Community experienced a relatively modest change in steel consumption and production with the exception of a drop in consumption in the mid-1980s. The annual average crude steel production in the three-year periods stayed at almost the same level of 135 million tonnes. Net exports reached their highest point of 35 million tonnes in 1985, but declined to 13 million tonnes in 1990-1992 mainly due to increased consumption inside the region.

In eastern Europe, a big drop in both consumption and production occurred. Consumption dropped to 25 million tonnes in 1990-1992 from 55 million tonnes in 1980-1983, and further declined to only 17 million tonnes in 1992. However, because the rate of reduction in crude steel production was not bigger than the reduction in consumption, net exports from the region increased sharply to 12 million tonnes in 1990-1992.

Figure 1-8 Consumption, production and net exports a/

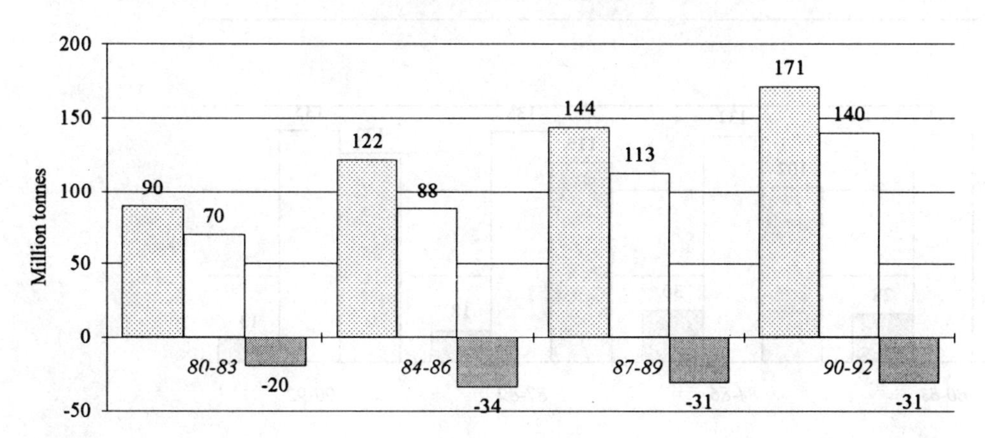

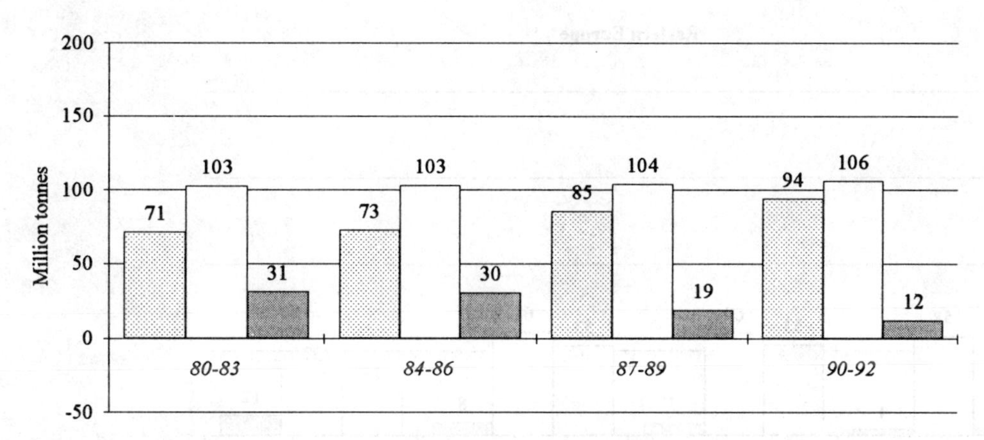

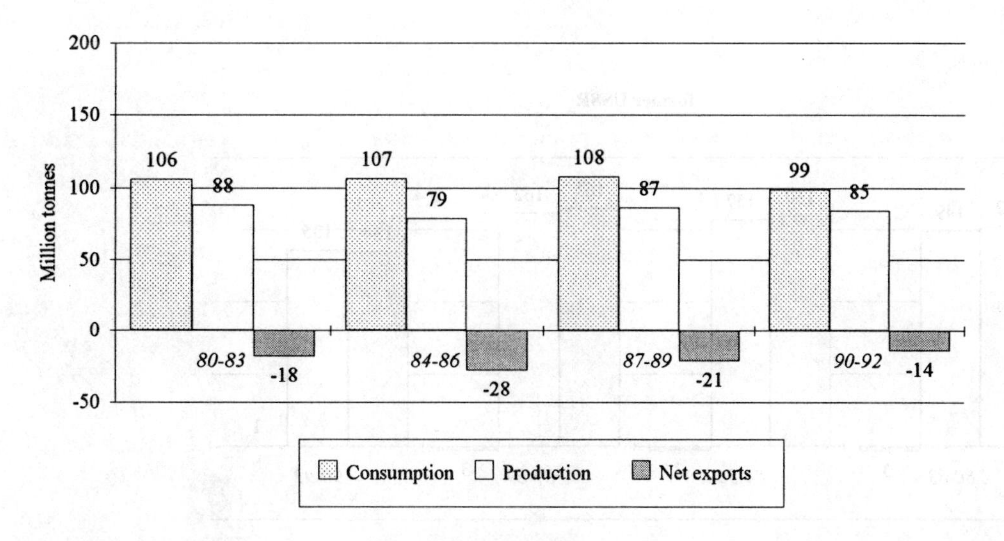

a/ Figures are annual average for period specified.

Figure 1-8 Consumption, production and net exports (continued)

EC

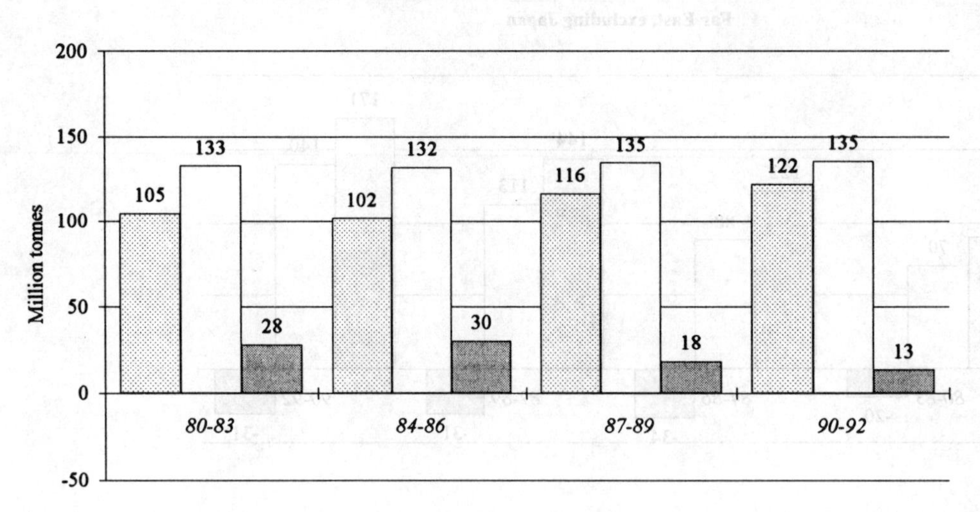

Eastern Europe

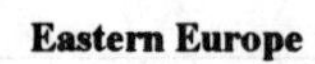

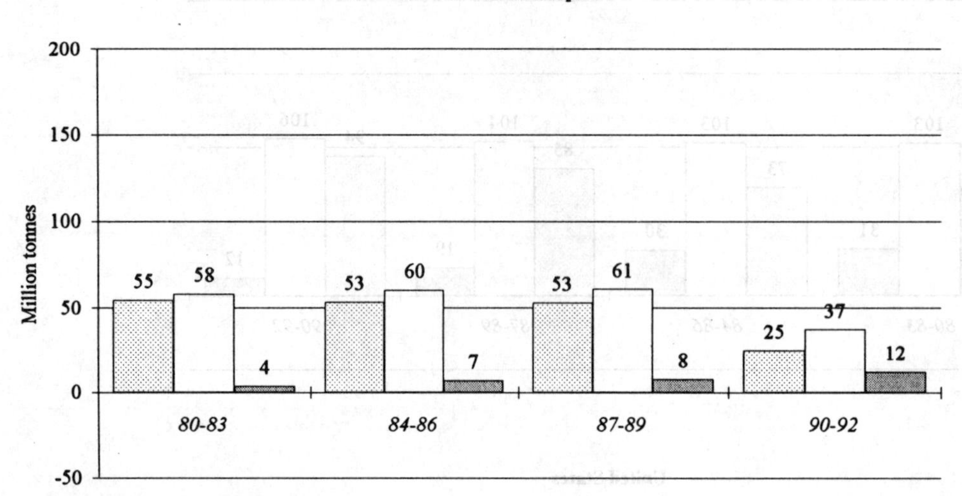

former USSR

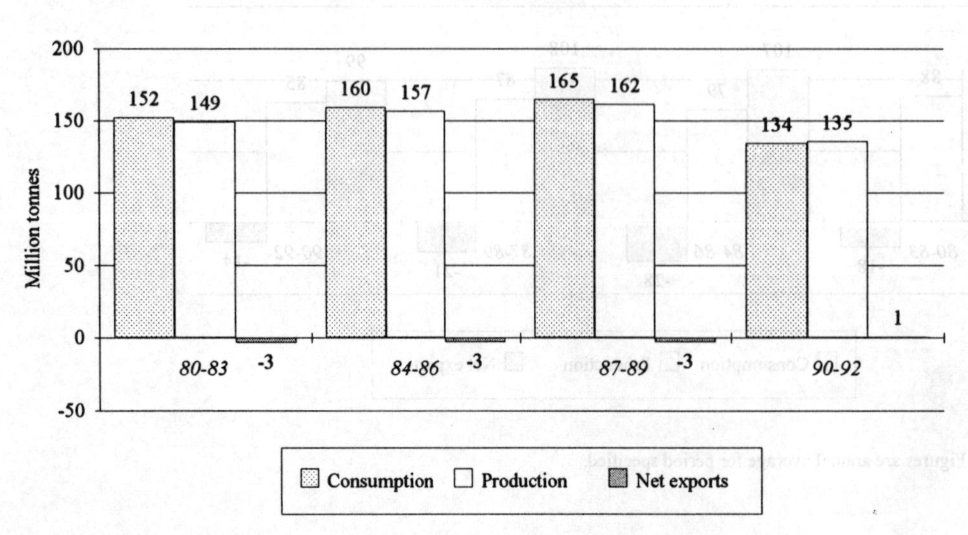

Figure 1-8 Consumption, production and net exports (continued)

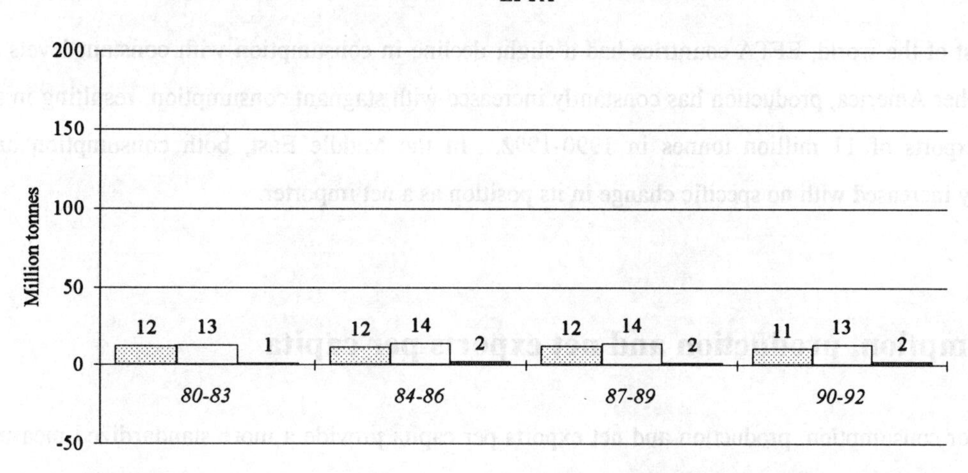

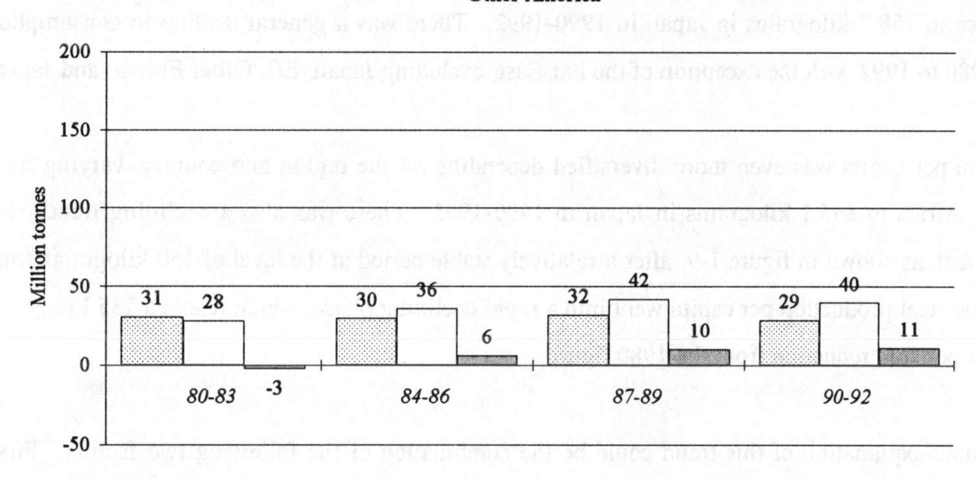

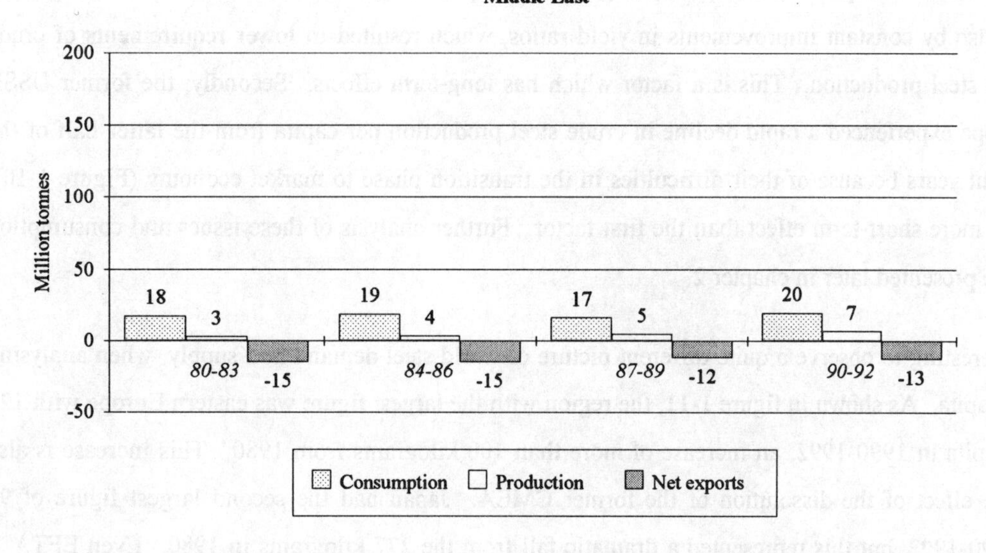

The former USSR maintained constant levels of consumption and production up to 1990. However, because of the difficult times following the dissolution of the country, consumption and production declined by 42 million tonnes and 36 million from 1990 to 1992, respectively.

In the rest of the world, EFTA countries had a slight decline in consumption with constant levels of production. In Other America, production has constantly increased with stagnant consumption, resulting in an increase of net exports of 11 million tonnes in 1990-1992. In the Middle East, both consumption and production steadily increased with no specific change in its position as a net importer.

1.3. Consumption, production and net exports per capita

Figures for consumption, production and net exports per capita provide a more standardized measure regardless of the size and population of the country. Table 1-3 summarizes consumption, production and net exports per capita in major regions and countries. Consumption per capita varied from the lowest of 20.6 kilograms in Africa to 758.7 kilograms in Japan in 1990-1992. There was a general decline in consumption per capita from 1980 to 1992 with the exception of the Far East, excluding Japan, EC, Other Europe and Japan.

Production per capita was even more diversified depending on the region and country, varying from 19.5 kilograms in Africa to 855.1 kilograms in Japan in 1990-1992. There was also a declining trend from 1980 to 1992. In fact, as shown in figure 1-9, after a relatively stable period at the level of 150 kilograms from 1984 to 1990, crude steel production per capita went into a rapid declining phase, which reached 135 kilograms per capita, an 18.2 per cent reduction from the 1980 figure.

The possible explanation of this trend could be the combination of the following two factors. First, there was a general decline in crude steel production per capita in most developed countries, caused not only by stagnant growth in consumption per capita brought about by the move towards more service-oriented economies, but also by constant improvements in yield ratios, which resulted in lower requirements of crude steel for finished steel production. This is a factor which has long-term effects. Secondly, the former USSR and eastern Europe experienced a rapid decline in crude steel production per capita from the latter half of the 1980s up to recent years because of their difficulties in the transition phase to market economy (Figure 1-10). This might have more short-term effect than the first factor. Further analysis of these issues and consumption per capita will be presented later in chapter 2.

It is interesting to observe a quite different picture of world steel demand and supply, when analysing net exports per capita. As shown in figure 1-11, the region with the largest figure was eastern Europe with 122 kilograms per capita in 1990-1992, an increase of more than 100 kilograms from 1980. This increase is also explained by the effect of the dissolution of the former CMEA. Japan had the second largest figure of 96 kilograms in 1990-1992, but this represented a dramatic fall from the 277 kilograms in 1980. Even EFTA, a relatively small steel producing region with only a 1.8 per cent share of crude steel production, had significant net exports per capita in 1992. However, it should be mentioned that because of the small size of its steel consumption and production sudden fluctuations of these series may occur, as in the case of Oceania in table 1-3. The United States remained one of the largest net importers of crude steel along with the Middle East during the period.

Table 1-3 Consumption, production and net exports per capita a/

(Crude steel equivalent)

(kilograms)

Regions and countries	1980-1983			1984-1986		
	Cons.	Prod.	N.exp.	Cons.	Prod.	N.exp.
AFRICA	31.6	21.9	-9.7	25.4	20.9	-4.5
MIDDLE EAST	127.4	17.1	-110.3	122.8	22.9	-99.8
FAR EAST	66.5	71.2	4.8	75.3	74.0	-1.2
Japan	604.7	868.6	263.9	601.4	854.1	252.7
NORTH AMERICA	457.2	395.5	-61.7	447.4	349.1	-98.3
United States	454.9	377.4	-77.4	439.9	325.9	-114.0
OTHER AMERICA	86.2	78.6	-7.6	76.1	92.3	16.2
OCEANIA	371.4	388.9	17.5	349.0	357.5	8.5
EUROPE	417.1	450.2	33.0	412.3	455.7	43.4
EC	327.6	415.6	88.0	316.0	409.5	93.6
EFTA	388.6	404.1	15.5	365.2	437.9	72.8
EASTERN EUROPE	497.2	528.7	31.5	477.9	539.7	61.8
former USSR	566.2	553.7	-12.5	574.8	564.0	-10.8
OTHER EUROPE	133.8	100.1	-33.7	141.2	128.4	-12.8
WORLD TOTAL	153.0	153.0	-0.1	151.7	150.5	-1.2

Regions and countries	1987-1989			1990-1992		
	Cons.	Prod.	N.exp.	Cons.	Prod.	N.exp.
AFRICA	22.6	20.5	-2.1	20.6	19.5	-1.1
MIDDLE EAST	97.0	30.0	-67.0	106.6	38.5	-68.1
FAR EAST	83.9	79.6	-4.3	92.4	85.9	-6.6
Japan	696.8	850.0	153.2	758.7	855.1	96.4
NORTH AMERICA	444.8	369.1	-75.6	388.7	346.5	-42.1
United States	434.2	347.8	-86.4	387.8	332.1	-55.7
OTHER AMERICA	77.1	101.0	23.9	65.0	90.3	25.3
OCEANIA	359.7	353.4	-6.2	269.0	352.4	83.4
EUROPE	435.7	461.9	26.2	359.1	396.2	37.1
EC	359.4	415.6	56.1	362.9	402.9	39.9
EFTA	385.4	433.4	48.0	334.2	403.9	69.7
EASTERN EUROPE	471.8	540.5	68.7	235.8	357.5	121.8
former USSR	580.5	569.7	-10.8	464.7	466.7	2.0
OTHER EUROPE	191.6	156.5	-35.1	125.9	157.9	32.0
WORLD TOTAL	156.3	153.5	-2.8	141.6	141.6	-0.1

a/ Figures are annual average for period specified. Consumption is apparent crude steel consumption. Production is crude steel production. Figures for net exports are calculated as production minus consumption. Regions and countries with a negative value of net exports are net importers. The WORLD TOTAL of net exports does not always equal zero because of small estimation errors for apparent crude steel consumption.

Figure 1-9 World crude steel production per capita

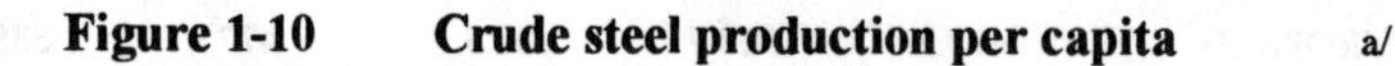

Figure 1-10 Crude steel production per capita a/

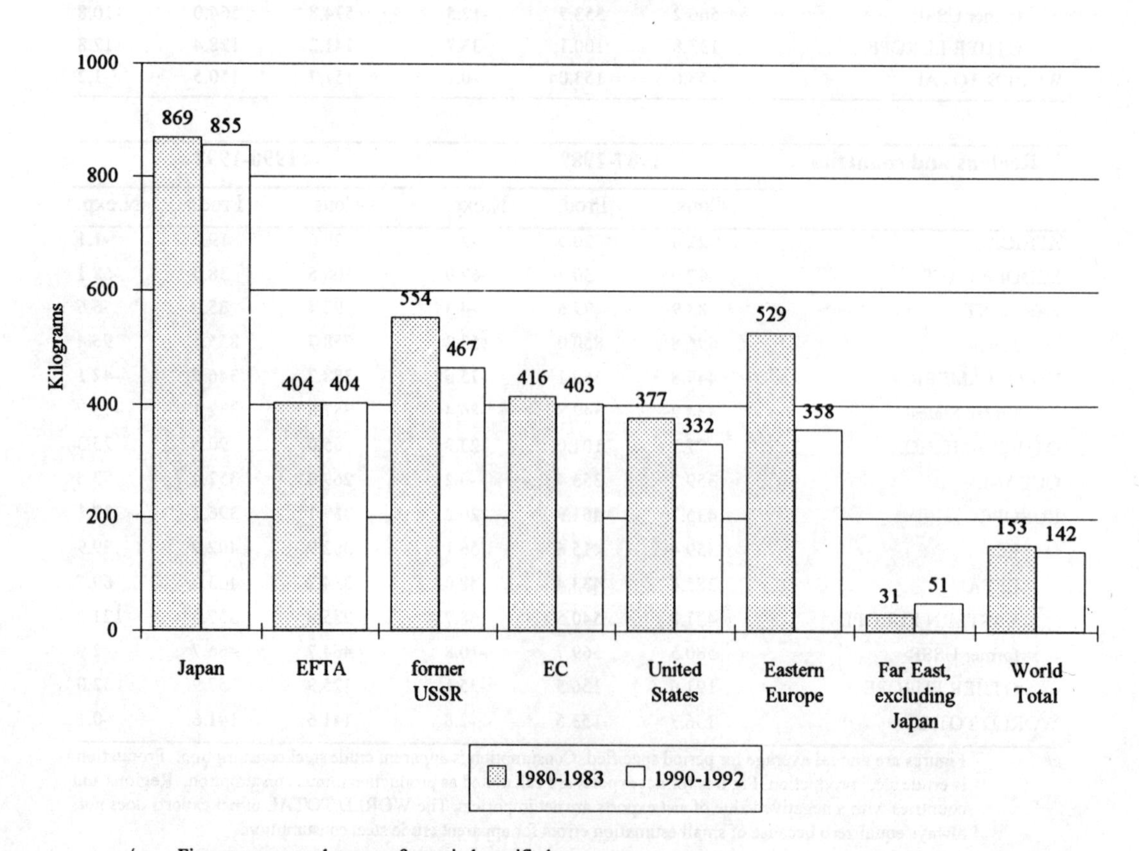

a/ Figures are annual average for period specified.

Figure 1-11 Net exports per capita a/

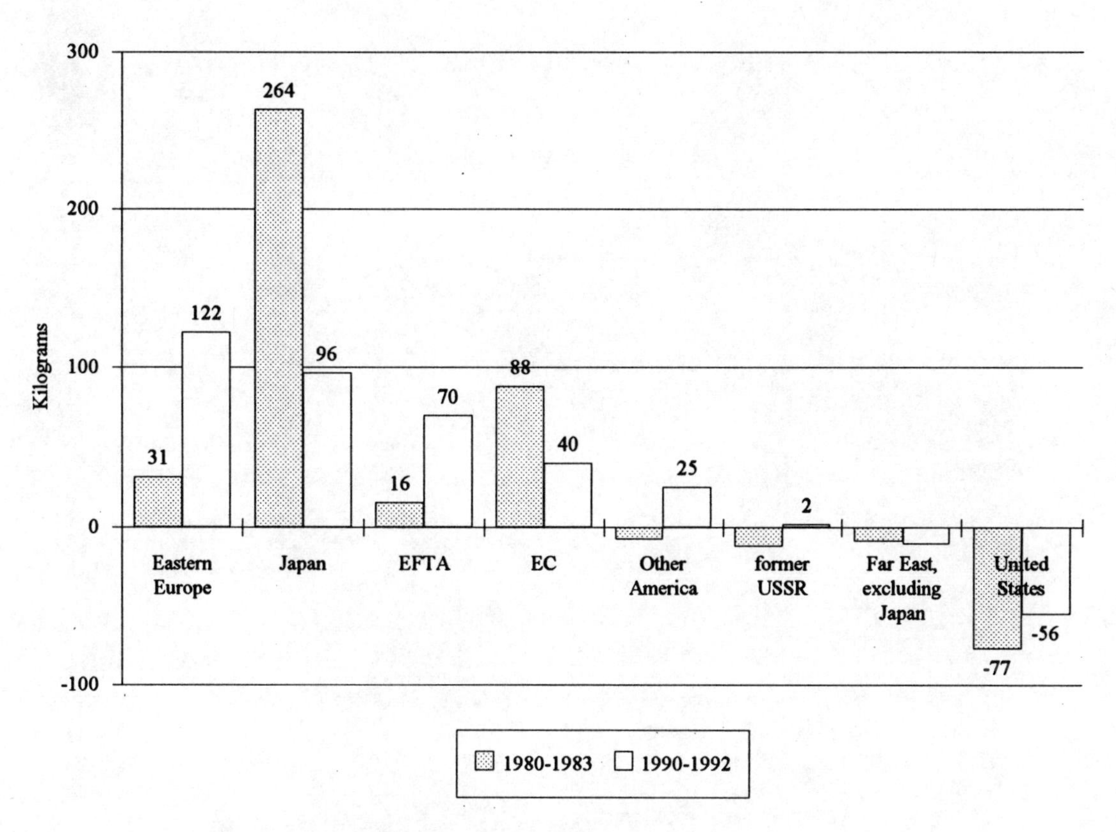

a/ Figures are annual average for period specified.

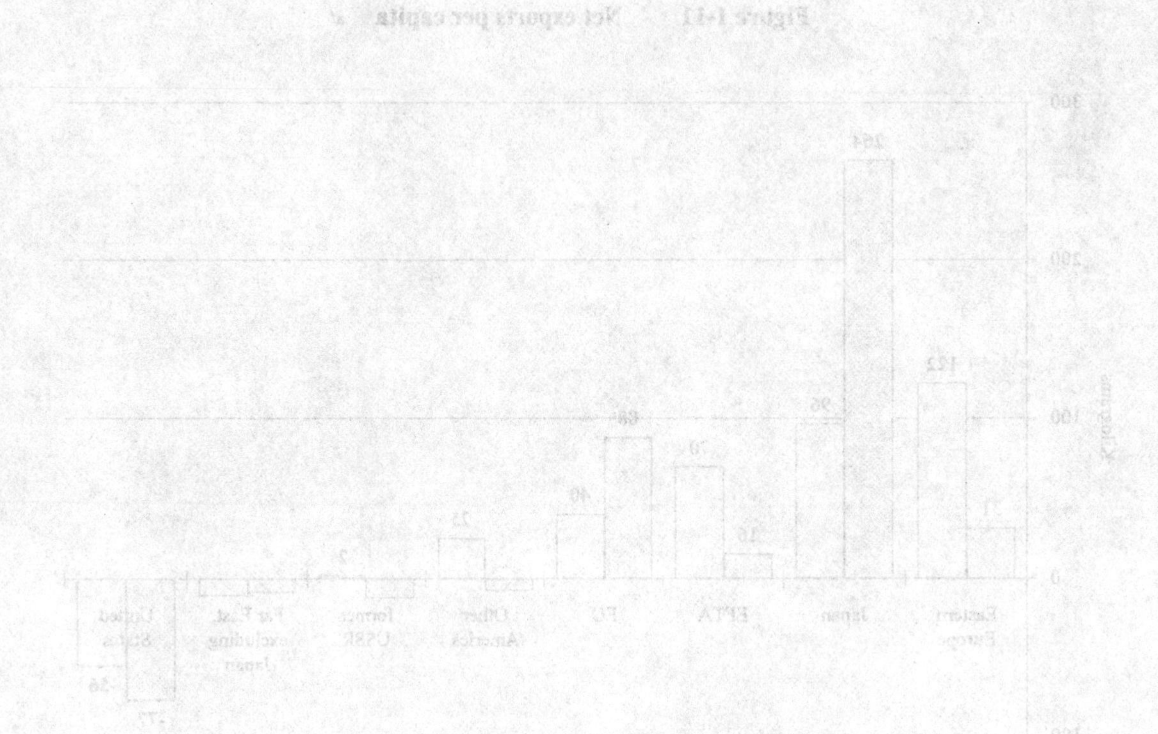

CHAPTER 2
STRUCTURAL CHANGES IN WORLD STEEL CONSUMPTION

2.1. General evolution of world finished steel consumption

2.1.1. Apparent steel consumption

World finished steel consumption in the 1980s started with a declining trend followed by seven years of consecutive growth until 1989, and another declining phase in the early 1990s (Figure 2-1). However, on the basis of a three-year average, it showed a modest, but constant increase (Table 2-1). As a result, world steel consumption started at 546 million tonnes in 1980-1983 and increased to 657 million tonnes in 1990-1992. This represents an increase of 111 million tonnes, or 20.3 per cent.

Among the major regions and countries, the Far East, excluding Japan, recorded a remarkable increase of 51.1 million tonnes during the period, or an annual growth rate of 7.2 per cent (Figure 2-2). The share of that region in the world total increased from 12.9 per cent in 1980-1983 to 18.5 per cent in 1990-1992 (Figure 2-3). It even increased to 25.1 per cent in 1992. Both Japan and the European Community had boom periods in consumption in the late 1980s and the early 1990s, with long-term trends showing a constant increase in consumption. Japan's share increased to 13.3 per cent in the early 1990s, while the EC's share remained unchanged at the level of 16 per cent. The United States experienced slight, but constant growth in steel consumption with some annual fluctuations. Against the background of a general increase in finished steel consumption on a worldwide basis, the share of the United States dropped from 15.0 per cent in the early 1980s to 13.7 per cent in the early 1990s. Steel consumption in eastern Europe remained unchanged until 1989, but has suffered a sharp drop since then. The region's share declined from 7.9 per cent in 1980 to only 2.2 per cent in 1992. The former USSR showed constant growth until 1989, but has been in a rapid declining phase since then. Its share has shrunk from 20.0 per cent in 1980 to 14.3 per cent in 1992.

The growth pattern of finished steel consumption ran parallel to that of apparent crude steel consumption, but the magnitude of growth in finished steel consumption was larger than in apparent crude steel consumption. The most important factor explaining this difference is the constant increase in the continuous casting ratio throughout the world. The continuously cast steel ratio increased from 28.1 per cent in 1980 to 65.0 per cent in 1992, resulting in a decline of the ratio of apparent crude steel consumption to apparent finished steel consumption from 1.261 in 1980 to 1.176 in 1992 (Table 2-2 and Figure 2-4). All regions succeeded in achieving sharp increases in their continuously cast steel ratios except eastern Europe and the former USSR where there was a slow pace in introducing this technology. Most advanced steel producing countries seem to be approaching the maximum point of utilization of continuous casting, with ratios of more

Figure 2-1 World finished steel consumption and the growth rate of the world economy a/

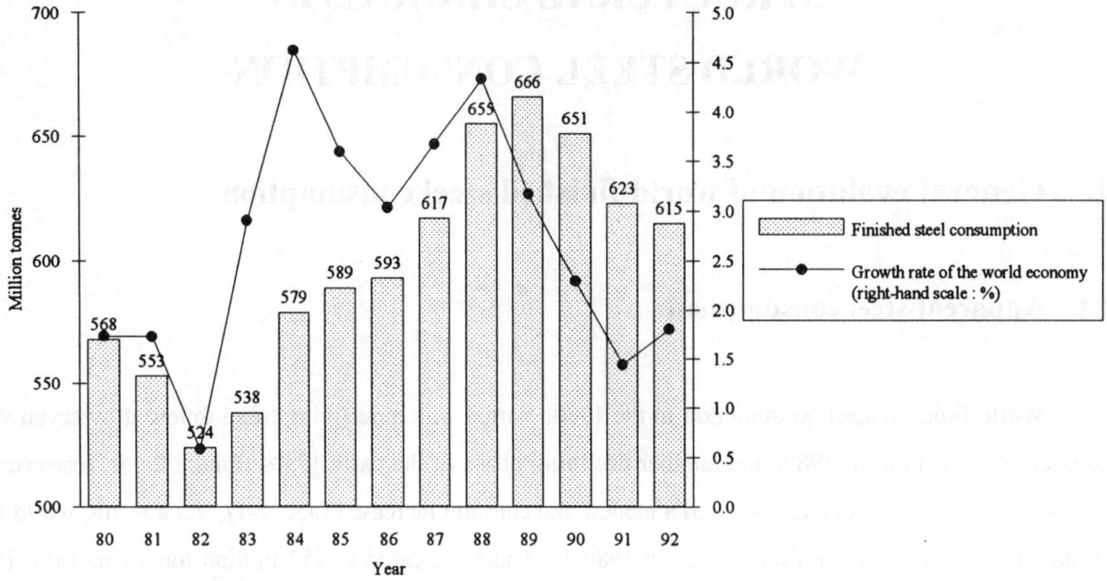

a/ The growth rate is the percentage change from the previous year.

Table 2-1 Finished steel consumption in major regions and countries a/

Regions and countries	Amount (Million tonnes)				Share (%)			
	80-83	84-86	87-89	90-92	80-83	84-86	87-89	90-92
AFRICA	11.0	10.2	9.9	10.1	2.0	1.7	1.5	1.5
MIDDLE EAST	15.1	16.6	14.5	15.4	2.8	2.8	2.2	2.3
FAR EAST	135.5	164.4	195.9	209.2	24.8	28.0	30.3	31.8
Japan	64.8	68.9	80.7	87.4	11.9	11.7	12.5	13.3
NORTH AMERICA	91.2	96.9	101.3	101.2	16.7	16.5	15.7	15.4
United States	81.8	86.3	89.3	89.9	15.0	14.7	13.8	13.7
OTHER AMERICA	25.3	24.6	26.9	25.2	4.6	4.2	4.2	3.8
OCEANIA	5.2	5.2	6.1	6.0	1.0	0.9	0.9	0.9
EUROPE	262.3	268.7	291.2	289.8	48.0	45.8	45.1	44.1
EC	88.0	87.4	100.2	106.4	16.1	14.9	15.5	16.2
EFTA	9.9	10.1	10.7	11.0	1.8	1.7	1.7	1.7
EASTERN EUROPE	41.5	41.2	41.8	37.2	7.6	7.0	6.5	5.7
former USSR	115.4	121.1	125.6	124.3	21.1	20.6	19.4	18.9
OTHER EUROPE	7.5	8.8	12.9	11.0	1.4	1.5	2.0	1.7
OTHERS	0.1	0.2	0.2	0.3	0.0	0.0	0.0	0.0
WORLD TOTAL	545.9	586.8	646.0	657.2	100.0	100.0	100.0	100.0

a/ Figures are annual average for period specified.

Figure 2-2 Finished steel consumption by major regions and countries a/

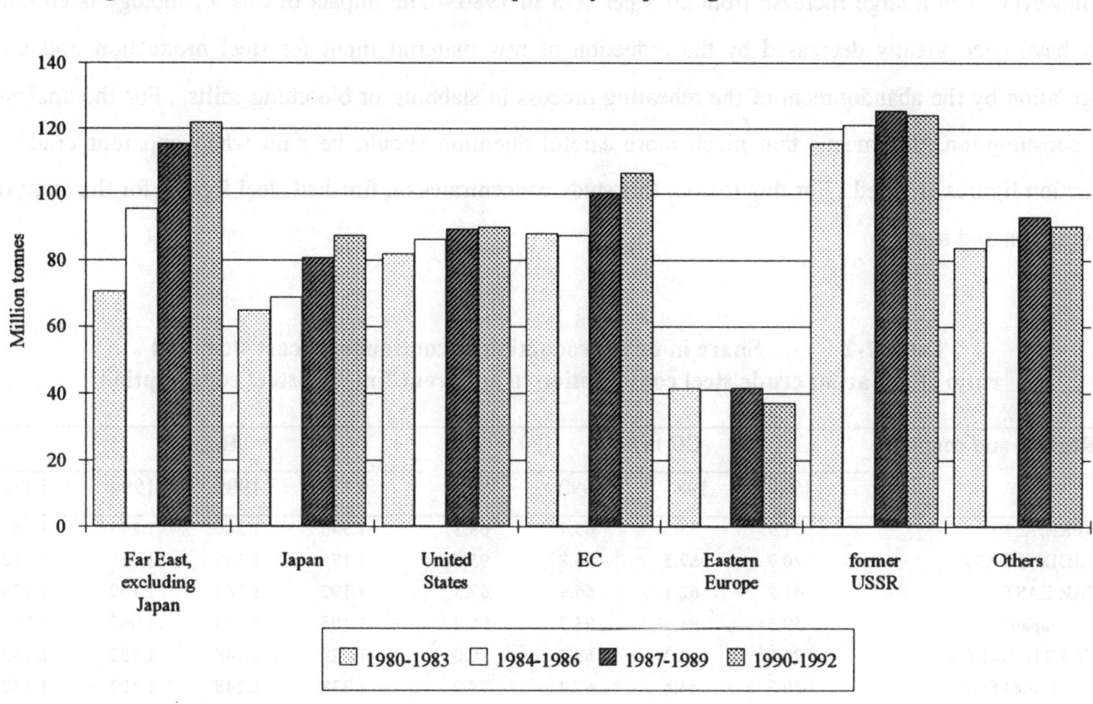

a/ Figures are annual average for period specified.

Figure 2-3 Share of finished steel consumption by major regions and countries a/

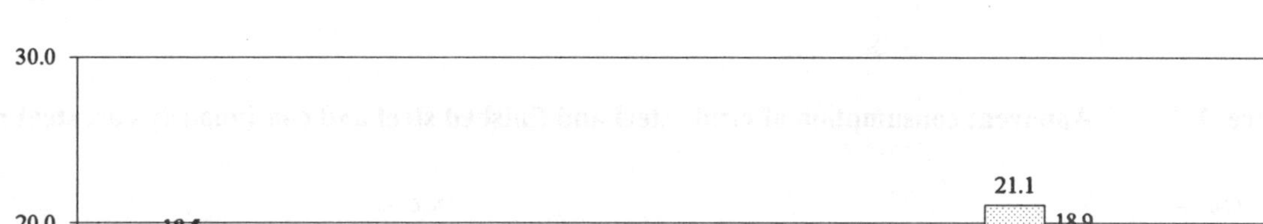
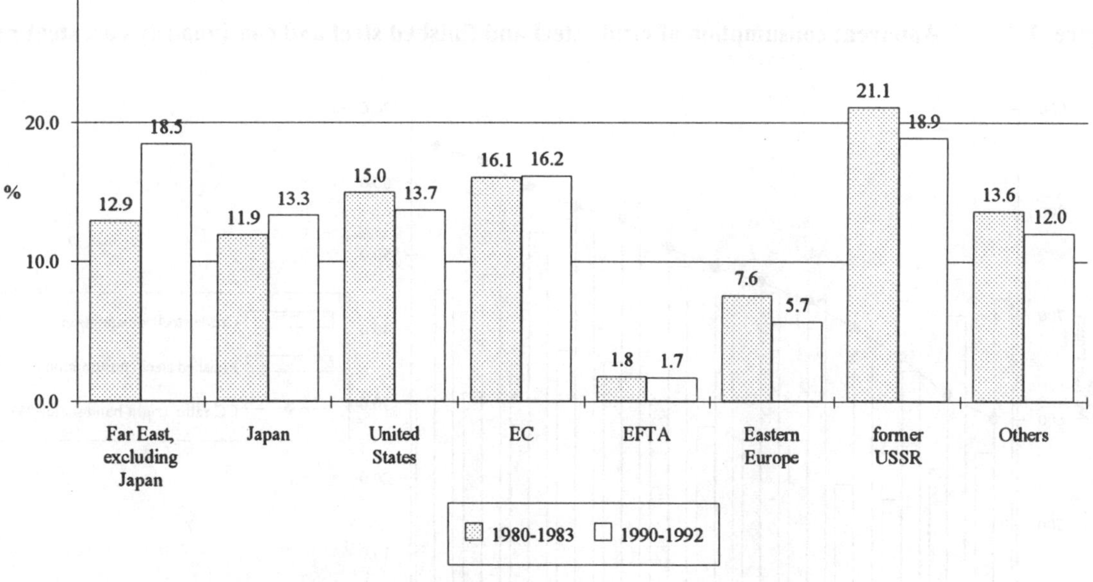

a/ Figures are annual average for period specified.

than 90 per cent. The United States still has room for improvement with a figure of 78.9 per cent in 1992. It has, however, seen a large increase from 20.3 per cent in 1980. The impact of this technology is enormous. Costs have been greatly decreased by the reduction of raw material input for steel production and energy conservation by the abandonment of the reheating process in slabbing or blooming mills. For the analysis of steel consumption, this means that much more careful attention should be paid when apparent crude steel production figures are used. For this reason, this study concentrates on finished steel figures for the analysis of consumption and trade.

Table 2-2 Share in total production of continuously cast steel and ratio of apparent crude steel consumption to apparent finished steel consumption

Regions and countries	CC ratio (%)				Ratio			
	1980	1985	1990	1992	1980	1985	1990	1992
AFRICA	51.9	53.5	65.7	68.3	1.346	1.280	1.243	1.240
MIDDLE EAST	20.7	67.5	96.8	95.6	1.175	1.145	1.111	1.112
FAR EAST	41.2	62.1	66.9	67.5	1.192	1.184	1.159	1.158
Japan	59.5	91.1	93.9	95.4	1.108	1.050	1.067	1.063
NORTH AMERICA	21.0	44.3	68.6	80.0	1.329	1.248	1.182	1.149
United States	20.3	44.4	67.4	78.9	1.338	1.248	1.193	1.158
OTHER AMERICA	32.4	48.5	61.3	63.2	1.240	1.202	1.179	1.170
OCEANIA	10.3	29.5	85.0	86.8	1.312	1.269	1.131	1.129
EUROPE	23.6	38.9	50.4	57.1	1.274	1.251	1.232	1.211
EC	40.0	69.8	89.7	92.0	1.195	1.161	1.150	1.144
EFTA	55.7	86.9	92.9	94.1	1.267	1.137	1.166	1.066
EASTERN EUROPE	9.8	19.5	22.2	23.7	1.321	1.291	1.275	1.275
former USSR	10.7	13.6	17.9	18.8	1.324	1.318	1.309	1.307
OTHER EUROPE	24.5	60.3	85.9	90.0	1.261	1.173	1.132	1.124
WORLD TOTAL	28.1	46.5	59.3	65.0	1.261	1.227	1.194	1.176

Figure 2-4 Apparent consumption of crude steel and finished steel and continuously cast steel ratio

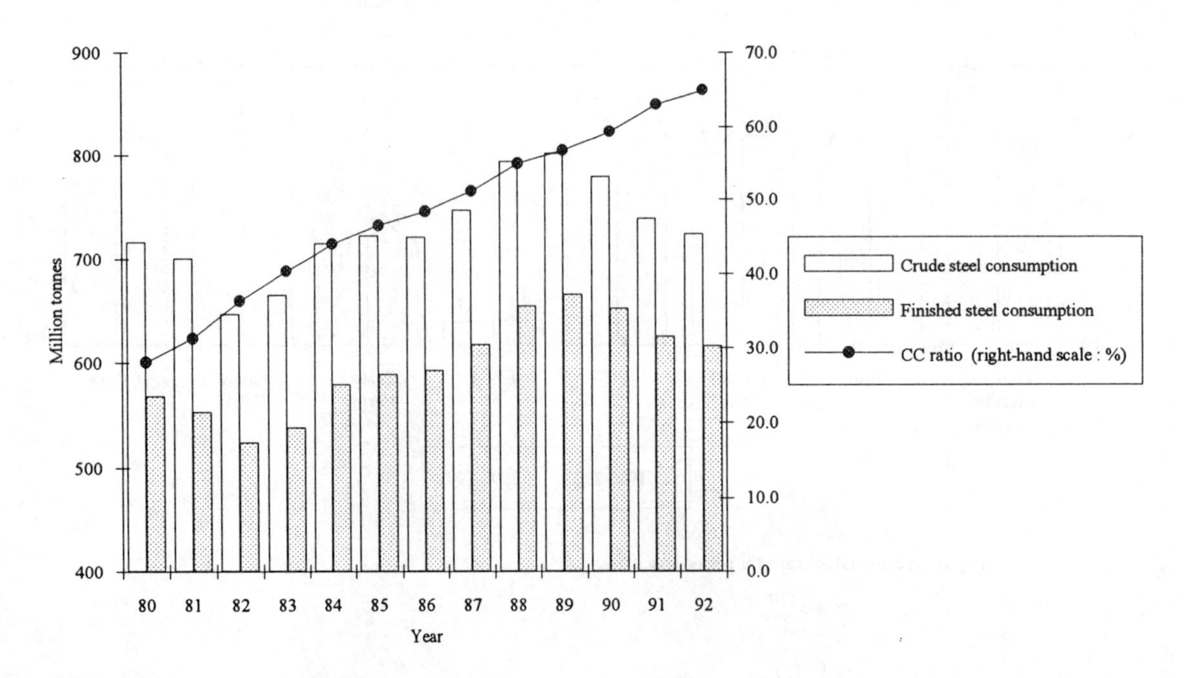

2.1.2. Steel consumption per capita

World finished steel consumption per capita dropped from 130.7 kilograms in 1980 to 115.2 kilograms in 1992. However, this downward trend was not a general one on a worldwide basis. In 1989, it was much higher with a figure of 131.1 kilograms (Table 2-3). The major factor for this drop was the rapid decline in eastern Europe and the former USSR after 1990 (Figure 2-5). As a matter of fact, many countries recorded increases in consumption per capita during the period. Those countries are not only the emerging Far Eastern economies such as China, Indonesia and the Republic of Korea, but also elsewhere in the world such as Egypt, Iran, Belgium-Luxembourg, Denmark, Greece, Italy, Portugal, Spain, Austria, Turkey and even Japan. This trend will be analysed in more detail in the section on steel intensity.

Among the major regions and countries, the Republic of Korea had constant growth of steel consumption per capita except for a drop from 1991 to 1992 (Figure 2-5). Another interesting observation in figure 2-5 is that Japan experienced a rapid increase in consumption from 1987 to 1991, while it had stable or rather stagnant consumption in the first half of the 1980s and faced a big drop from 1991 to 1992. Considering the degree of economic maturity of the nation, the boom in steel consumption in Japan from 1987 to 1991 could have been caused by some specific short-term factors. The European Community experienced sluggish steel consumption per capita up to 1987, but enjoyed modest growth between 1988 and 1990. North America, including the United States, experienced fluctuations with a declining trend on a long-term basis.

Table 2-3 Finished steel consumption and GDP per capita in major regions and countries

Regions and countries	Consumption (Kilograms)				GDP (US$)*		
	1980	1985	1989	1992	1980	1985	1989
AFRICA	26.5	21.0	17.4	16.1	891	793	763
MIDDLE EAST	100.1	114.8	84.2	101.4	3 142	2 702	2 321
FAR EAST	58.6	64.6	75.7	80.0	805	960	1 128
Japan	610.5	578.5	718.4	634.3	9 109	10 635	12 381
NORTH AMERICA	379.3	366.8	359.1	337.1	11 643	12 746	14 119
United States	374.1	360.2	350.2	333.0	11 710	12 829	13 204
OTHER AMERICA	85.0	62.2	63.9	58.4	2 144	2 011	2 022
OCEANIA	285.9	276.2	312.1	209.9	10 221	11 150	11 829
EUROPE	345.5	324.7	357.8	266.1	6 516	6 977	7 728
EC	304.6	267.8	335.8	311.9	9 827	10 421	11 738
EFTA	347.4	305.4	350.9	287.0	13 314	14 531	16 154
EASTERN EUROPE	407.8	367.8	356.3	128.1	4 455	4 849	5 149
former USSR	428.0	430.0	444.7	301.1	3 750	4 302	4 822
OTHER EUROPE	109.6	119.3	136.0	105.9	1 933	1 964	2 054
WORLD TOTAL	130.7	124.1	131.1	115.0	2 705	2 828	3 035

* 1980 constant US dollars.

Figure 2-5 Finished steel consumption per capita by major regions and countries

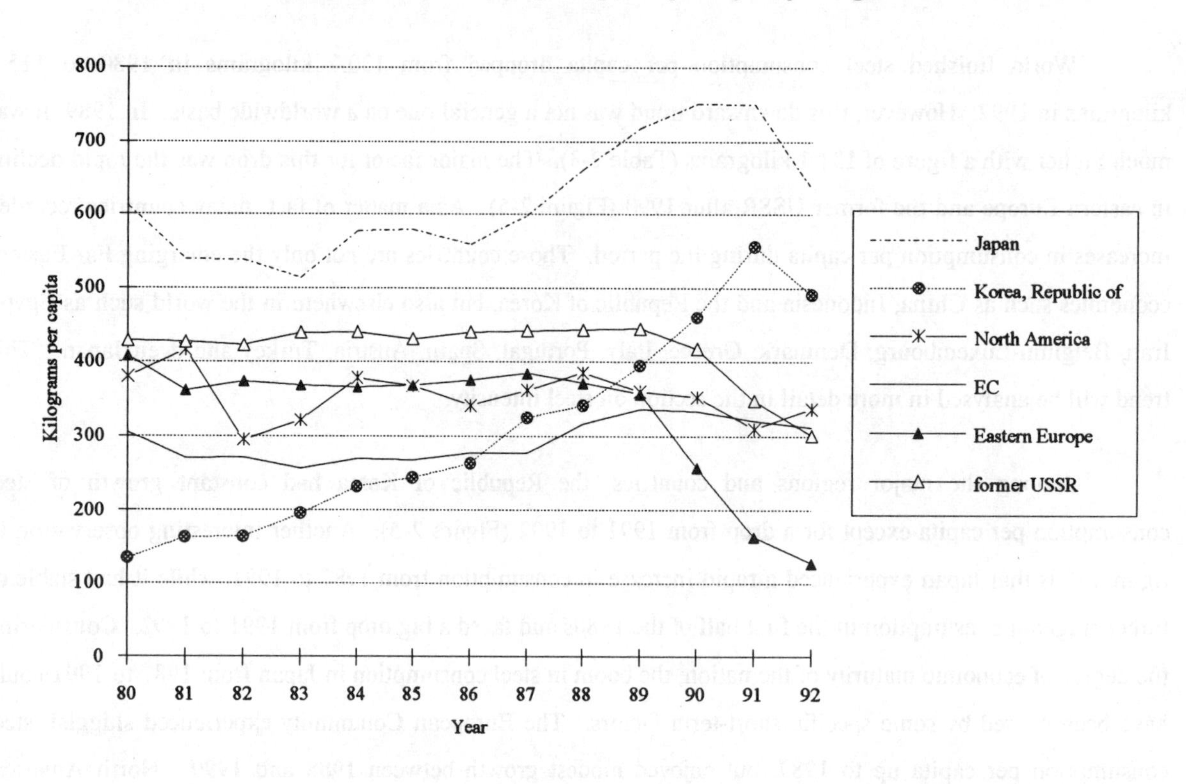

Figure 2-6-1 shows the relationship between steel consumption per capita and GDP per capita in 1980 and 1989. Because of the complexity of calculating GDP data at constant US dollar prices, 1989 is the most recent year when data for all the countries analysed are available. Each dot in figure 2-6-1 represents one country whose X-axis value is GDP per capita and Y-axis value is steel consumption per capita. Life cycle theory points to four sequential stages in these graphs, namely an accelerated growth period, a slow growth period, a mature declining period and a decreased, but stable period.[1/] Because the data in this figure are for different countries with different economic and social structures, the adoption of the general life cycle theory might not be really appropriate However, it is still interesting to see some unique patterns of growth in steel consumption per capita, because the countries covered are at different stages of economic growth.

For a more detailed analysis, figure 2-6-2 excludes eastern Europe and the former USSR which have different economic and steel consuming structures. In the figure for 1980, three groups of countries are easily identifiable. One is the group of countries with less than US$8000 GDP per capita and less than 250 kilograms of steel consumption per capita. These developing or developed countries show an increasing linear trend in steel consumption per capita accompanied by an increase in GDP per capita. The second group is a collection of highly developed countries with more than US$8000 GDP per capita and steel consumption between 200 and 500 kilograms per capita. There is no clear trend in this group probably because of the different steel

---

1/ *The Evolution of the Specific Consumption of Steel*, pp. 24, Economic Commission for Europe, 1984; and "New games, new rules" in *The McKinsey Quarterly*, 1993, Number 2, 1993.

Figure 2-6-1 Relationship between steel consumption per capita and GDP per capita in 1980 and 1989 a/

1980

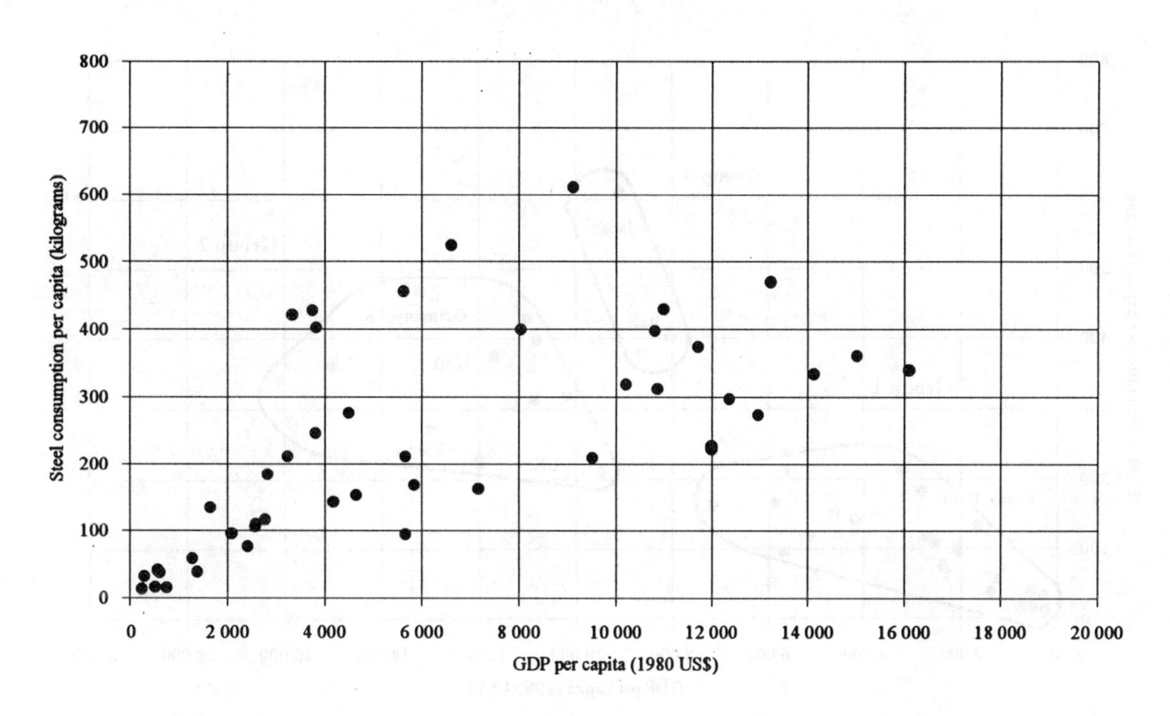

1989

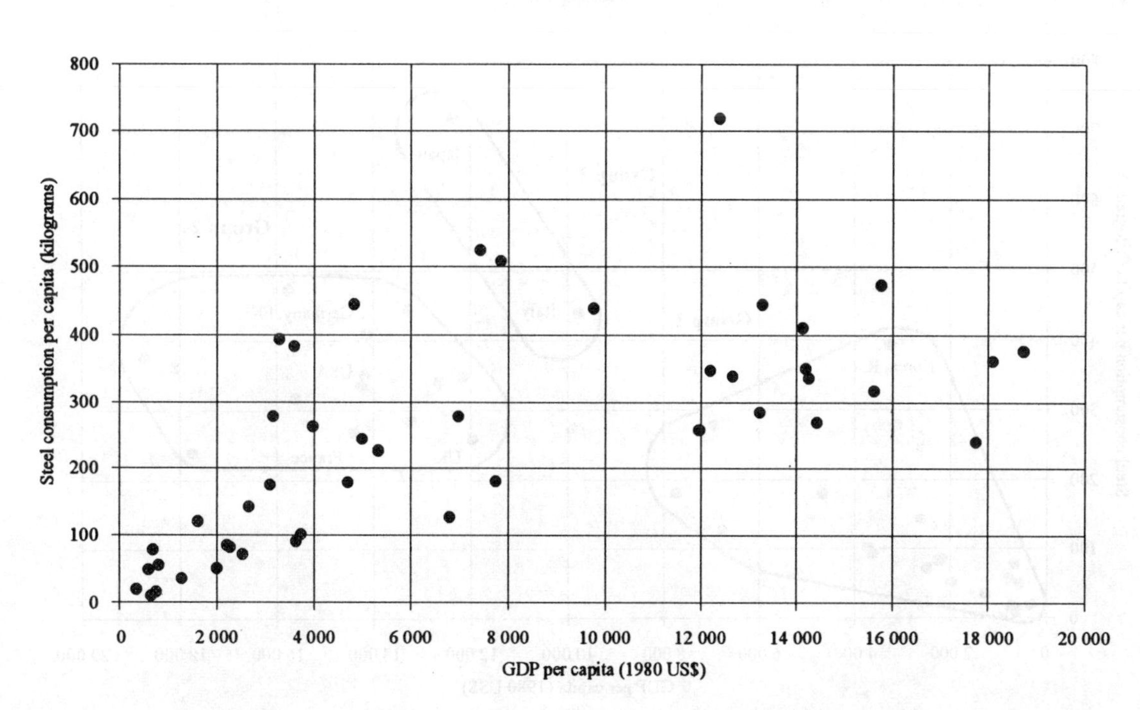

a/ Each dot represents one of the more than 40 countries in the annex tables. The same sample is used for both 1980 and 1989.

Figure 2-6-2 Relationship between steel consumption per capita and GDP per capita in 1980 and 1989 ***excluding Eastern Europe and former USSR*** a/

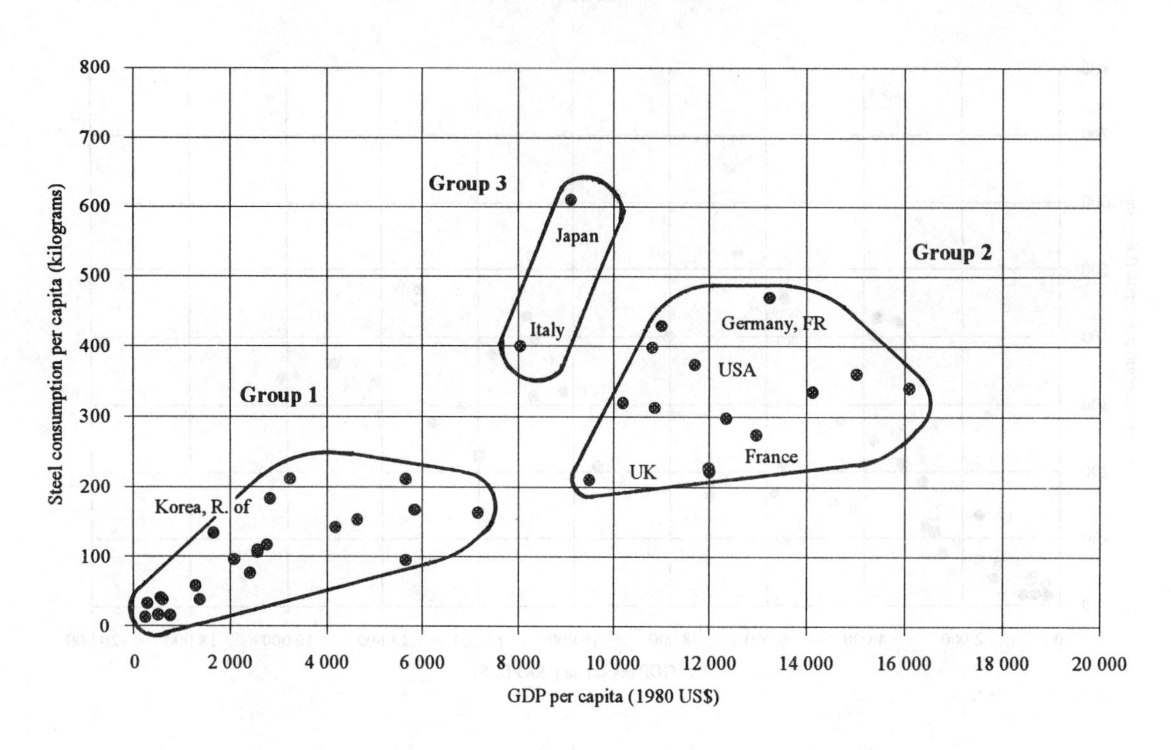

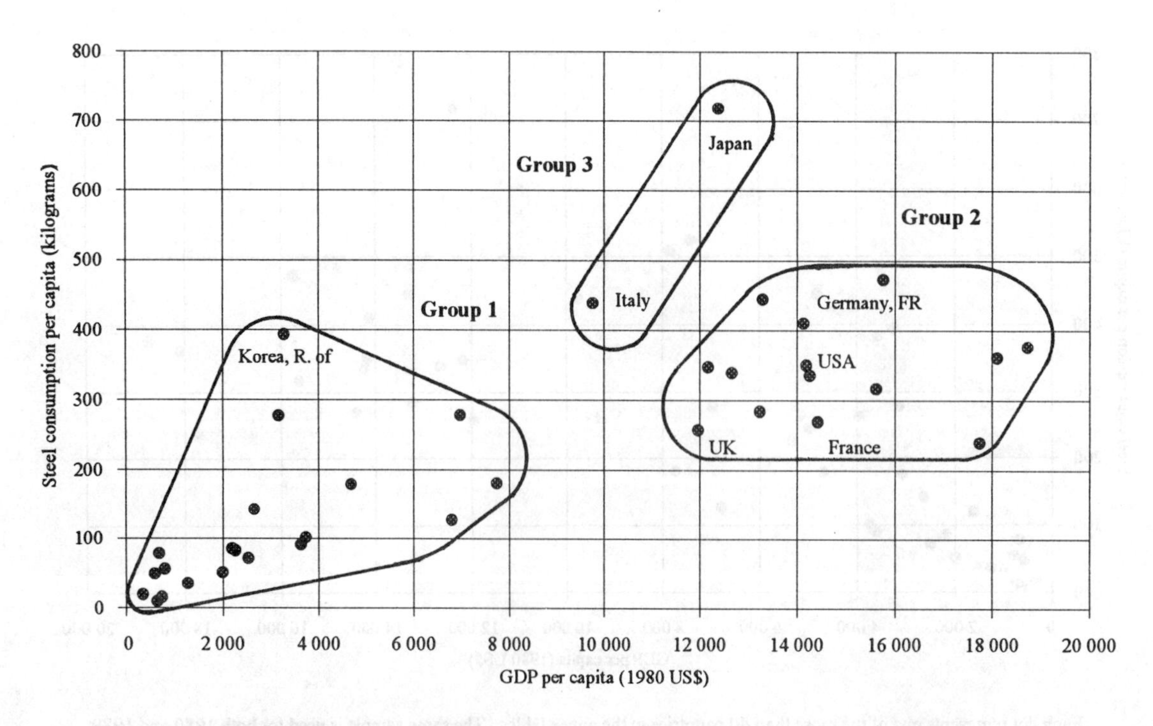

a/ Each dot represents one of the more than 40 countries covered in the annex tables, excluding countries in Eastern Europe and the former USSR. The same sample is used for both 1980 and 1989.

Figure 2-6-3 Relationship between steel consumption per capita and GDP per capita in 1980 and 1989

Eastern Europe and former USSR

1980

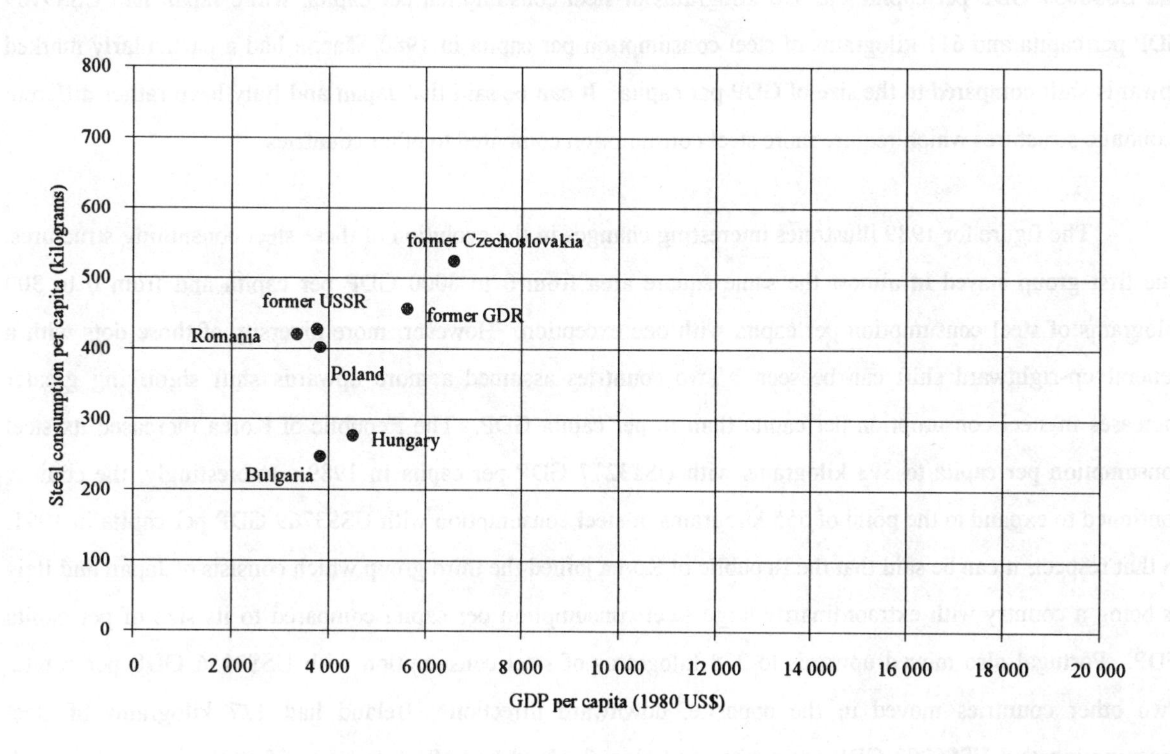

1989 a/

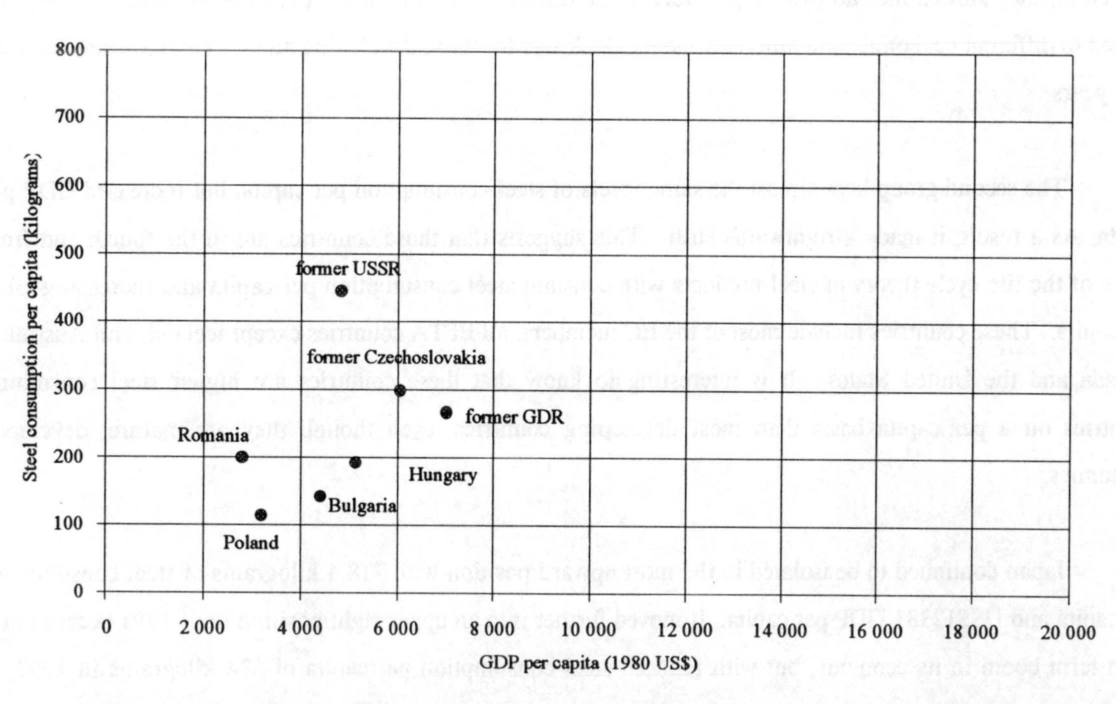

a/ Data for the former USSR is for 1989, data for Bulgaria, the former GDR and Hungary are for 1990, and data for the former Czechoslovakia, Poland and Romania are for 1991.

consuming structure in each. However, it is certain that these developed countries have some range of steel consumption per capita regardless of the size of per capita GDP. The third group consists of only Japan and Italy which had relatively high steel consumption per capita compared with the size of GDP per capita. Italy had US$8037 GDP per capita and 399 kilograms of steel consumption per capita, while Japan had US$9109 GDP per capita and 611 kilograms of steel consumption per capita in 1980. Japan had a particularly marked upwards shift compared to the size of GDP per capita. It can be said that Japan and Italy have rather different economic structures which require more steel consumption compared to other countries.

The figure for 1989 illustrates interesting changes in the evolution of these steel consuming structures. The first group stayed in almost the same square area from 0 to 8000 GDP per capita and from 0 to 300 kilograms of steel consumption per capita with one exception. However, more diversity of those dots with a general up-rightward shift can be seen. Two countries assumed a more upwards shift signifying greater increases in steel consumption per capita than in per capita GDP. The Republic of Korea increased its steel consumption per capita to 393 kilograms with US$3277 GDP per capita in 1989. Interestingly, the country continued to expand to the point of 555 kilograms of steel consumption with US$3769 GDP per capita in 1991. In that respect, it can be said that the Republic of Korea joined the third group which consists of Japan and Italy as being a country with extraordinarily large steel consumption per capita compared to its size of per capita GDP. Portugal also moved upwards to 278 kilograms of steel consumption with US$3156 GDP per capita. Two other countries moved in the opposite, downward direction. Ireland had 127 kilograms of steel consumption and US$6793 GDP per capita, and New Zealand had 181 kilograms of steel consumption with US$7737 GDP per capita. Steel consumption per capita in these two countries continued to fall until 1992. Therefore, they shifted into downward positions. The different patterns in the evolution of steel consumption related to different economic structures became much clearer for those developing and developed countries after nine years.

The second group kept almost the same levels of steel consumption per capita, but increased GDP per capita. As a result, it made a rightwards shift. This suggests that those countries are in the fourth and final phase of the life cycle theory of steel products with constant steel consumption per capita and increasing GDP per capita. These countries include most of the EC members, all EFTA countries except Iceland, and Australia, Canada and the United States. It is interesting to know that these countries are higher steel consuming countries on a per capita basis than most developing countries, even though they are mature, developed economies.

Japan continued to be isolated in the most upward position with 718.4 kilograms of steel consumption per capita and US$12381 GDP per capita. It moved further into an upper right position until 1991 because of a short-term boom in its economy, but with reduced steel consumption per capita of 634 kilograms in 1992, it should come down close to the second group. However, it will continue to be a little isolated above the second group because of its high steel intensity in construction which will be discussed later. Italy which had US$9769 GDP per capita and 438.4 kilograms of steel consumption per capita also stayed a little away from the second group. However, given the higher growth rate in GDP per capita of 21.6 per cent than the 9.8 per cent for steel consumption per capita for these nine years, the proximity to the second group seems to be closer in 1992.

Figure 2-6-3 illustrates the relationship between steel consumption per capita and GDP per capita in eastern Europe and the former USSR. These countries occupied upward positions in the first group in figure 2-6-2, which means that they have relatively high steel consumption per capita compared with the size of their per capita GDP. However, there is also a big diversity among the former CMEA countries. Bulgaria and Hungary occupy positions nearest to the developing and developed market economy group with values of steel consumption per capita of less than 300 kilograms. The former USSR, Poland and Romania have the most upward positions with more than 400 kilograms and relatively small GDP per capita. The former GDR and the former Czechoslovakia have the highest figures for both steel consumption per capita and GDP per capita.

This can be explained as a combination of the following two factors. One is the past economic history in the region. Both the former GDR and the former Czechoslovakia were the central region for heavy industry from the beginning of the modern, industrialized society. Another is the result of centralized planning under the former CMEA system, which made the former USSR, Poland and Romania the most intensive steel consumers and the former GDR and the former Czechoslovakia the second most intensive consumers of steel products, as the most mature industrialized countries of the region.

However, according to the most recent data available for those countries from 1989 to 1991, the situation has changed dramatically. Except for the former USSR for which 1989 data are the latest available, steel consumption per capita decreased in all those countries to between 100 and 300 kilograms and almost joined the trend line for the first group of developing market economies in figure 2-6-2. Even the former USSR reduced its steel consumption per capita to 301 kilograms in 1992, although its GDP data are not available and cannot be plotted on the figure. From the point of view of the relationship between steel consumption per capita and GDP per capita, the difficult transition phase to market economy in those countries also brought about a transition of the steel consuming structure. This was a time when their dots moved to the ordinary trend line for developing market economy countries. In that respect, even if the transition phase of those countries progresses successfully, there is no reason to assume that they will return to their earlier higher steel consuming structures.

2.1.3. Ratio of production and imports to consumption

Table 2-4 shows the ratios of production and imports to steel consumption. As described already in chapter 1 and figure 2-7, the general trend in the ratio of production to consumption is a shift towards equilibrium. Africa, the Middle East and the Far East, excluding Japan, increased their ratios to become more self-dependent in terms of the volume of steel production, while the former large producers such as Japan and the EC reduced their ratios. One exception to this is eastern Europe which increased its ratio to 157.9 in 1990-1992 caused by a sharp drop in steel consumption during the preceding period. While more detailed analysis will be provided later, it is interesting to note that the import ratio has also been increasing on a worldwide basis. Given the general situation that many regions and countries are becoming more self-dependent in steel, this suggests that the exchange of different products is increasing.

Table 2-4 Ratio of production and imports to steel consumption a/

Regions and countries	Production ratio b/				Import ratio c/			
	80-83	84-86	87-89	90-92	80-83	84-86	87-89	90-92
AFRICA	69.7	82.7	90.6	95.0	49.4	45.6	40.5	40.8
MIDDLE EAST	13.5	18.8	30.9	36.0	91.0	86.8	76.5	70.6
FAR EAST	107.2	98.4	94.9	92.9	18.8	23.0	19.9	20.8
Japan	143.8	142.0	122.5	112.9	2.9	4.9	7.8	8.4
NORTH AMERICA	86.1	78.1	83.0	89.2	18.5	23.8	20.4	18.8
United States	82.5	74.2	80.1	85.6	19.0	24.5	19.7	18.0
OTHER AMERICA	94.2	121.4	131.3	139.0	24.6	17.5	16.7	24.6
OCEANIA	104.8	102.5	98.4	132.0	23.4	23.0	21.1	19.8
EUROPE	107.9	110.5	106.0	110.5	25.7	26.8	29.5	32.3
EC	127.0	129.7	116.1	111.0	44.2	47.5	51.5	57.9
EFTA	104.5	120.1	112.8	121.6	64.6	67.9	71.9	75.8
EASTERN EUROPE	106.4	112.9	114.6	157.9	24.2	23.2	23.3	15.6
former USSR	97.8	98.1	98.1	100.7	8.1	8.4	8.1	4.2
OTHER EUROPE	75.1	91.0	83.6	125.6	36.2	44.8	50.4	33.1
WORLD TOTAL	99.9	99.2	98.2	100.0	25.0	26.9	25.8	26.9

a/ Figures are annual average for period specified.
b/ Ratio of crude steel production to apparent crude steel consumption.
c/ Ratio of steel imports to finished steel consumption. Figures for the EC and EFTA include intra-regional trade (imports).

Figure 2-7 Ratio of steel production to apparent crude steel consumption a/

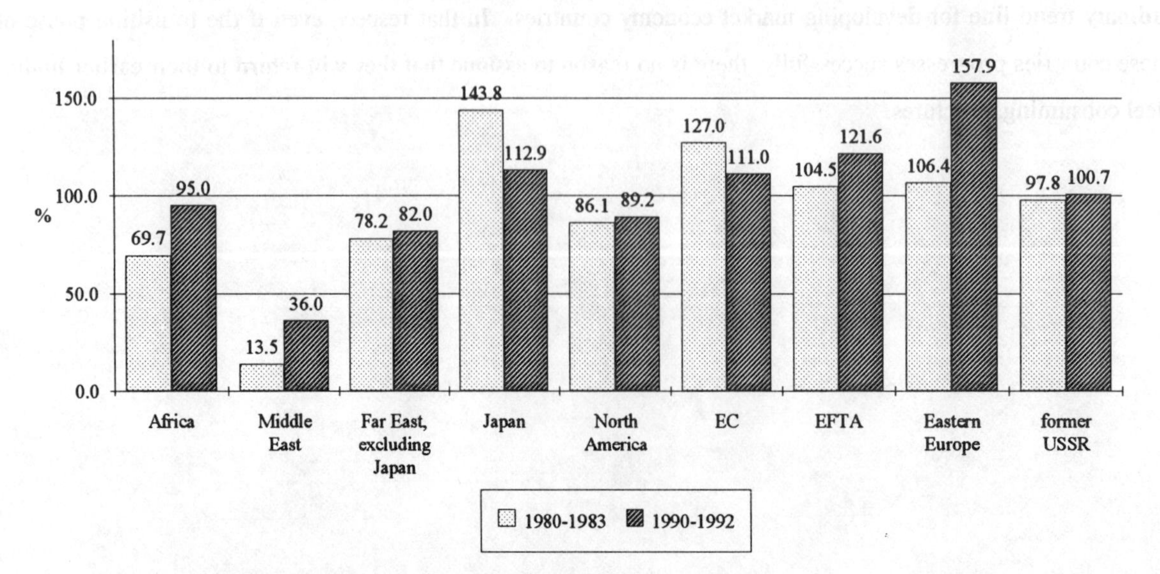

a/ Figures are annual average for period specified.

2.1.4. Steel intensity of GDP

The steel intensity of GDP provides a direct measure of the importance of steel in the total economy of regions or countries. Table 2-5 shows the steel intensity of GDP in major regions and countries. As shown in figure 2-8, the world steel intensity of GDP declined from 48.3 grams per 1980 United States dollar in 1980 to 43.2 grams in 1992. However, the evolution can be divided into two parts. One is a rapid declining phase from 1980 to 1982. Another is a rather stable period from 1982 to 1989. One possible explanation for the rapid decline at the beginning of the 1980s is that all countries made serious efforts to achieve more efficient use of steel products in their economies in the atmosphere of energy saving following the second oil crisis. The structure of world steel consumption in the total economy changed structurally at the beginning of the 1980s, with steel becoming less important in the total economy. Interestingly, the importance of steel did not change significantly throughout the rest of the 1980s. As a study by the International Iron and Steel Institute notes, overall decline in steel intensity is thought to be caused by a decrease in the importance of the main steel-using industries and sectors in the generation of national wealth, i.e. to a structural changes in the economy rather than the decline of steel as an industrial material.[2/]

Figure 2-9 shows the different patterns of development in the steel intensity of GDP by regions. Developed market economies show more or less the same trend as the world total. Both the EC and Japan experienced modest increases in the late 1980s. However, the nominal value for Japan is almost double the figure for the EC and North America. The Far East, excluding Japan, maintained high figures of steel intensity of around 70 to 80 grams per 1980 United States dollar during the period, while Other America and Africa faced a declining trend and the Middle East experienced a cyclical upward and downward trend. Eastern Europe and the former USSR had extraordinary steel intensity of more than 80 grams per 1980 United States dollar at the beginning of the 1980s. However, as a result of constant decline and a particularly sharp drop after 1989 for eastern Europe, those levels seem to have come down to close to the world average level in recent years. This is a reflection of the fact that these countries are making more efficient use of steel products in their economies in an effort to catch up with the market economy countries as well as of the general decline in steel-using heavy industrial sectors such as the defence industry.

Figure 2-8 World steel intensity of GDP

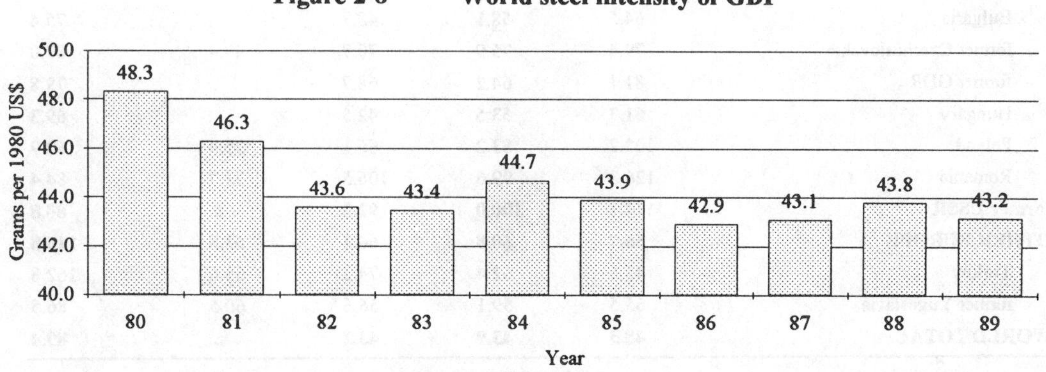

[2/] *Intermaterial Competition: An Economic Analysis of General Trends*, pp. 2-9, International Iron and Steel Institute, 1989.

Table 2-5 Steel intensity of GDP in major regions and countries

Regions and countries	Steel intensity a/				Index of 89
	1980	1985	1989	1991	(1980=100)
AFRICA	**29.7**	**26.5**	**22.8**	**21.8**	**76.7**
MIDDLE EAST	**31.8**	**42.5**	**36.3**	**...**	**114.0**
Egypt	72.1	86.9	113.5	116.0	157.4
Iran	31.7	36.9	36.4	52.5	114.6
FAR EAST	**72.8**	**67.3**	**67.1**	**67.3**	**92.2**
China	108.8	114.5	81.7	73.5	75.1
Japan	67.0	54.4	58.0	55.4	86.6
Korea, Republic of	81.1	105.1	120.1	147.4	148.0
NORTH AMERICA	**32.6**	**28.8**	**25.4**	**23.1**	**78.1**
Canada	39.1	36.0	33.4	27.2	85.6
United States	31.9	28.1	24.7	22.7	77.2
OTHER AMERICA	**39.6**	**30.9**	**31.6**	**27.4**	**79.8**
Argentina	41.6	29.1	25.3	28.3	60.8
Brazil	46.1	35.7	38.5	26.7	83.5
Mexico	42.2	30.1	28.1	32.8	66.6
Venezuela	33.0	23.0	27.1	29.5	82.3
OCEANIA	**28.0**	**24.8**	**26.4**	**20.6**	**94.3**
Australia	28.7	24.9	26.8	20.8	93.2
New Zealand	22.7	24.0	23.3	18.4	102.7
EC	**31.0**	**25.7**	**28.6**	**26.2**	**92.3**
Belgium-Luxembourg	18.5	23.9	23.5	22.8	126.9
Denmark	21.1	21.5	20.3	19.7	96.1
France	24.1	17.1	18.7	17.8	77.7
former FR Germany	35.5	31.3	30.0	28.3	84.5
Greece	34.2	31.8	38.1	52.0	111.5
Ireland	16.7	16.5	18.6	14.7	111.6
Italy	49.7	38.9	44.9	42.2	90.3
Netherlands	19.0	19.1	21.5	17.4	113.3
Portugal	42.6	34.6	87.9	48.0	206.5
Spain	37.3	24.3	39.9	35.0	107.0
United Kingdom	22.1	19.8	21.5	17.9	97.6
EFTA	**26.1**	**21.0**	**21.7**	**17.1**	**83.3**
Austria	31.3	24.4	28.5	26.8	91.2
Finland	36.8	26.5	29.0	19.7	79.0
Iceland	15.9	10.7	9.3	9.4	58.3
Norway	23.7	17.5	13.5	6.9	57.0
Sweden	23.9	20.6	20.0	16.4	83.4
Switzerland	21.1	18.5	20.1	15.6	95.2
EASTERN EUROPE	**91.2**	**75.5**	**68.8**	**48.5**	**75.5**
Bulgaria	64.5	58.1	48.7	...	75.4
former Czechoslovakia	79.4	75.9	70.7	49.4	89.0
former GDR	81.1	64.2	64.7	...	79.8
Hungary	61.3	53.5	42.5	...	69.3
Poland	105.2	83.2	66.1	35.4	62.9
Romania	126.3	99.6	106.5	71.7	84.4
former USSR	**114.1**	**100.0**	**92.2**	**...**	**80.8**
OTHER EUROPE	**56.7**	**60.8**	**66.2**	**63.8**	**116.8**
Turkey	45.6	62.4	74.2	65.6	162.8
former Yugoslavia	65.5	59.1	56.6	60.5	86.5
WORLD TOTAL	**48.3**	**43.9**	**43.2**	**...**	**89.4**

a/ Finished steel consumption in grams per US$ at 1980 prices.

Figure 2-9 Steel intensity of GDP by regions

Developed market economies

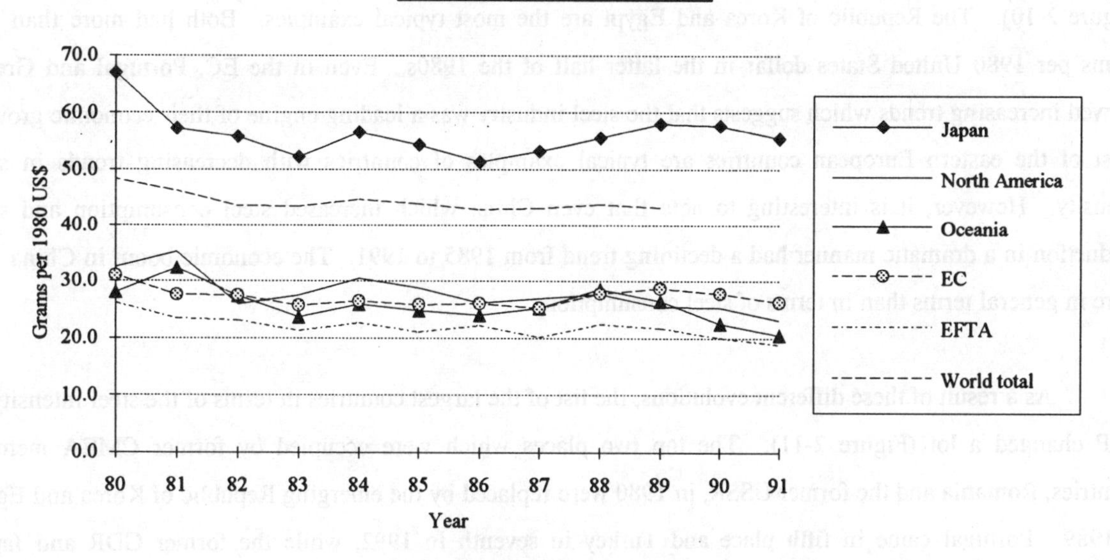

Developing market economies

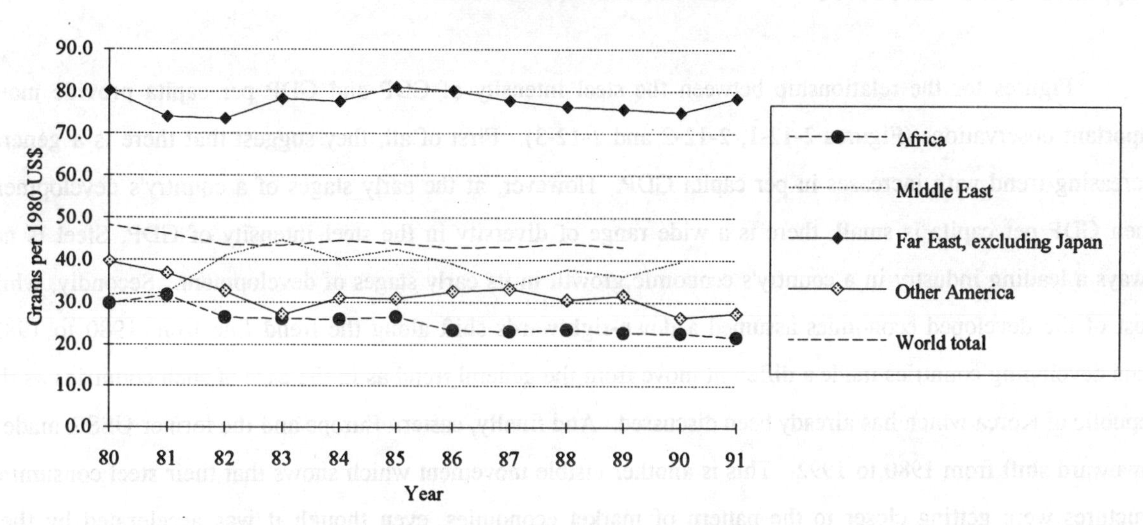

Eastern Europe and former USSR

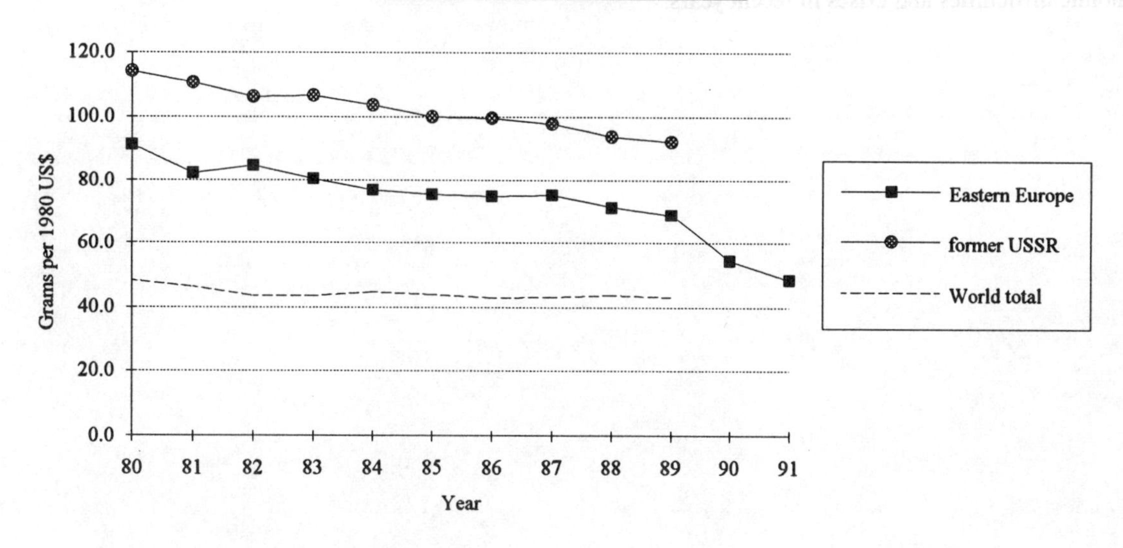

There are not many countries which show increasing trends in steel intensity except developing countries in the initial stages of economic development, but they do exist even in nearly developed economies (Figure 2-10). The Republic of Korea and Egypt are the most typical examples. Both had more than 100 grams per 1980 United States dollar in the latter half of the 1980s. Even in the EC, Portugal and Greece showed increasing trends which suggests that the steel industry was a leading engine of their economic growth. Most of the eastern European countries are typical examples of countries with decreasing trends in steel intensity. However, it is interesting to note that even China which increased steel consumption and steel production in a dramatic manner had a declining trend from 1985 to 1991. The economic boom in China was more in general terms than in terms of steel consumption.

As a result of these different evolutions, the list of the largest countries in terms of the steel intensity of GDP changed a lot (Figure 2-11). The top two places which were occupied by former CMEA member countries, Romania and the former USSR, in 1980 were replaced by the emerging Republic of Korea and Egypt in 1989. Portugal came in fifth place and Turkey in seventh in 1992, while the former GDR and Japan disappeared from the list in 1992.

Figures for the relationship between the steel intensity of GDP and GDP per capita provide many important observations (Figures 2-12-1, 2-12-2 and 2-12-3). First of all, they suggest that there is a general decreasing trend with increases in per capita GDP. However, at the early stages of a country's development when GDP per capita is small, there is a wide range of diversity in the steel intensity of GDP. Steel is not always a leading industry in a country's economic growth in its early stages of development. Secondly, while most of the developed economies assumed a down-rightwards shift along the trend line from 1980 to 1989, some developing countries made a different move from the general trend as in the case of such countries as the Republic of Korea which has already been discussed. And finally, eastern Europe and the former USSR made a downward shift from 1980 to 1992. This is another visible movement which shows that their steel consuming structures were getting closer to the pattern of market economies, even though it was accelerated by their economic difficulties and crises in recent years.

Figure 2-10 Steel intensity of GDP of selected countries

Countries showing increasing trend

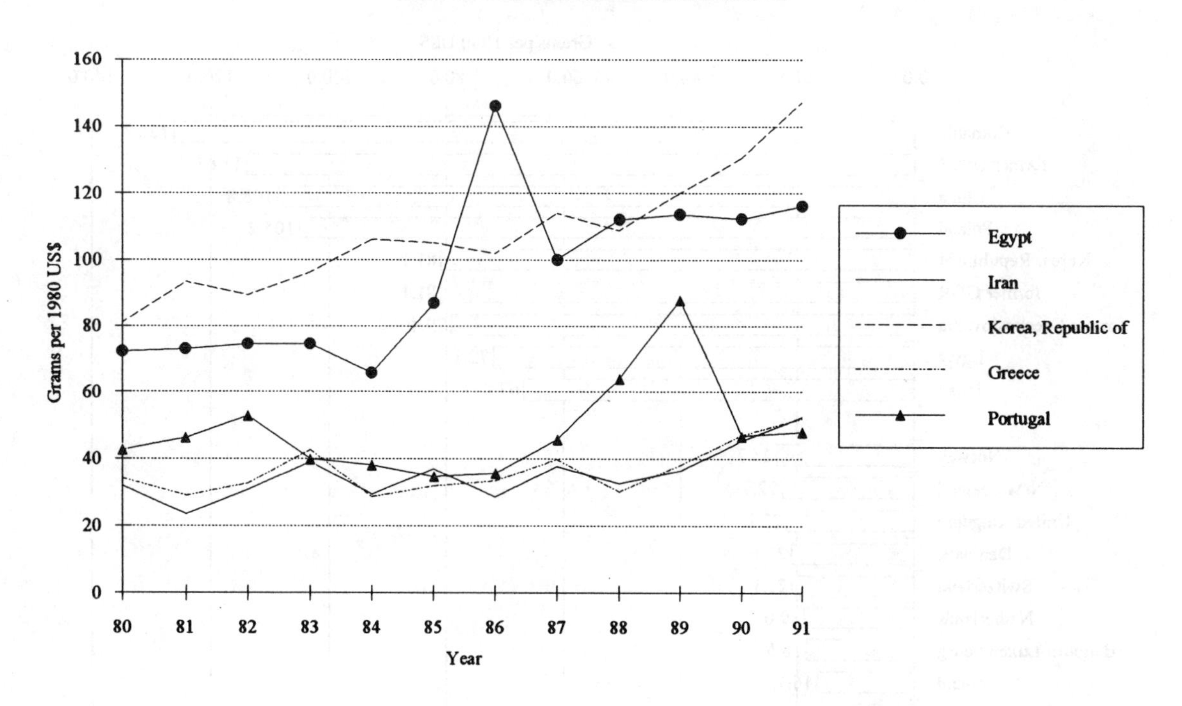

Countries showing decreasing trend

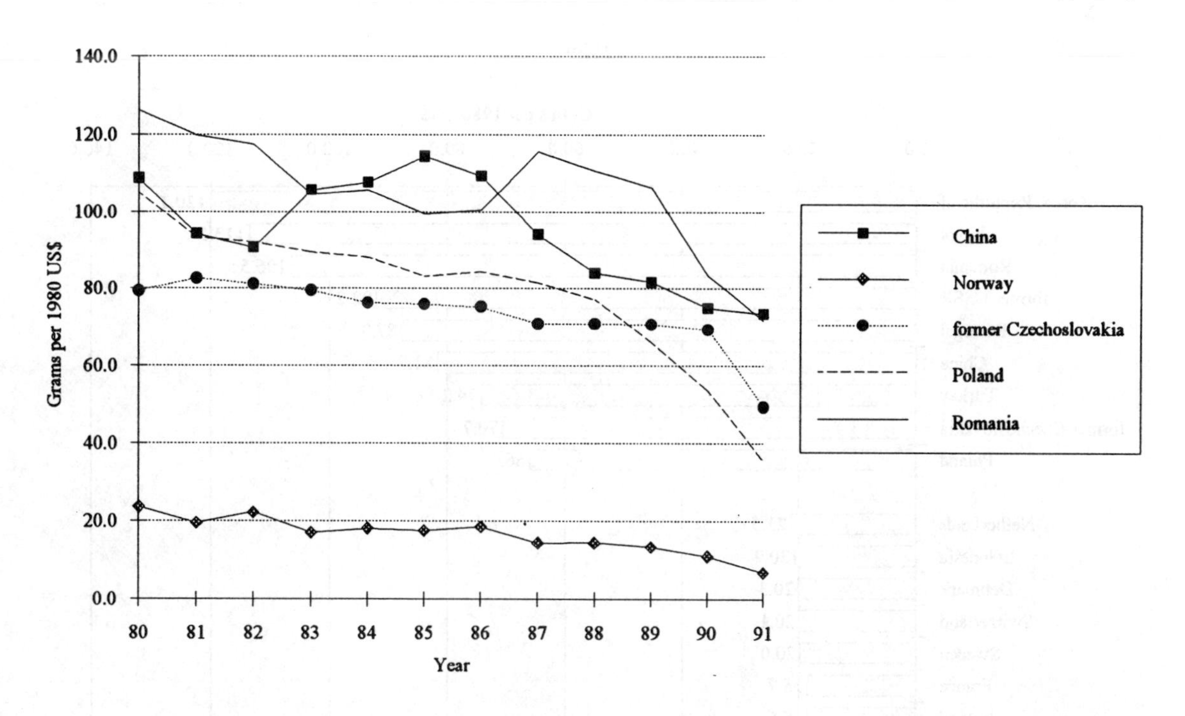

Figure 2-11 Largest and smallest countries in terms of steel intensity of GDP in 1980 and 1989

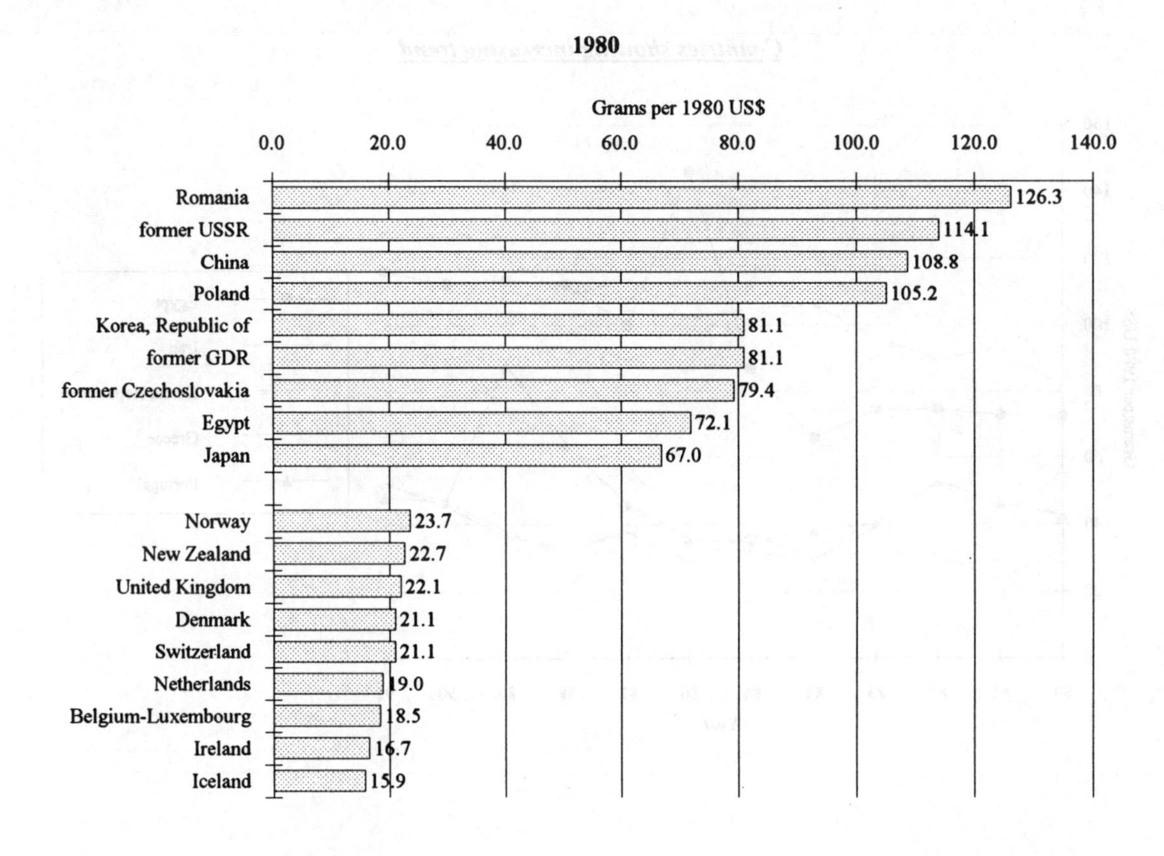

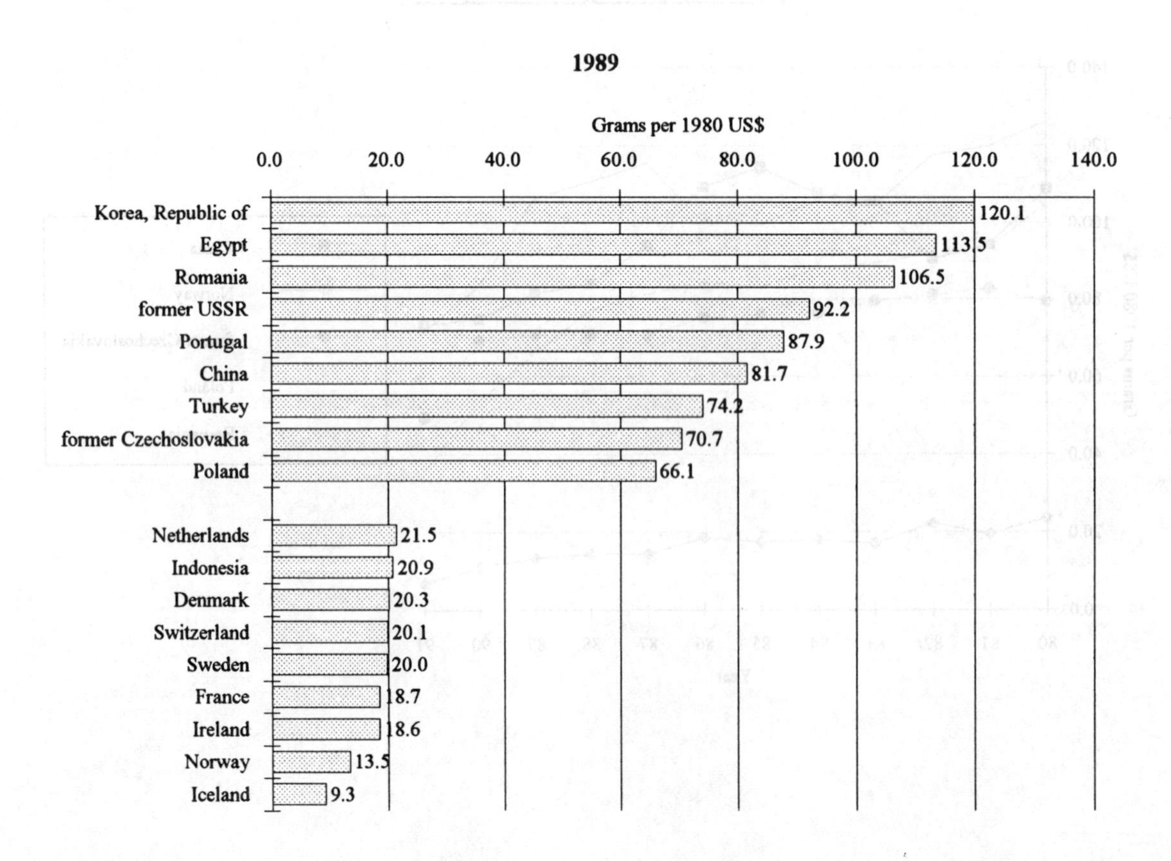

Figure 2-12-1 Relationship between steel intensity of GDP and GDP per capita in 1980 and 1989 a/

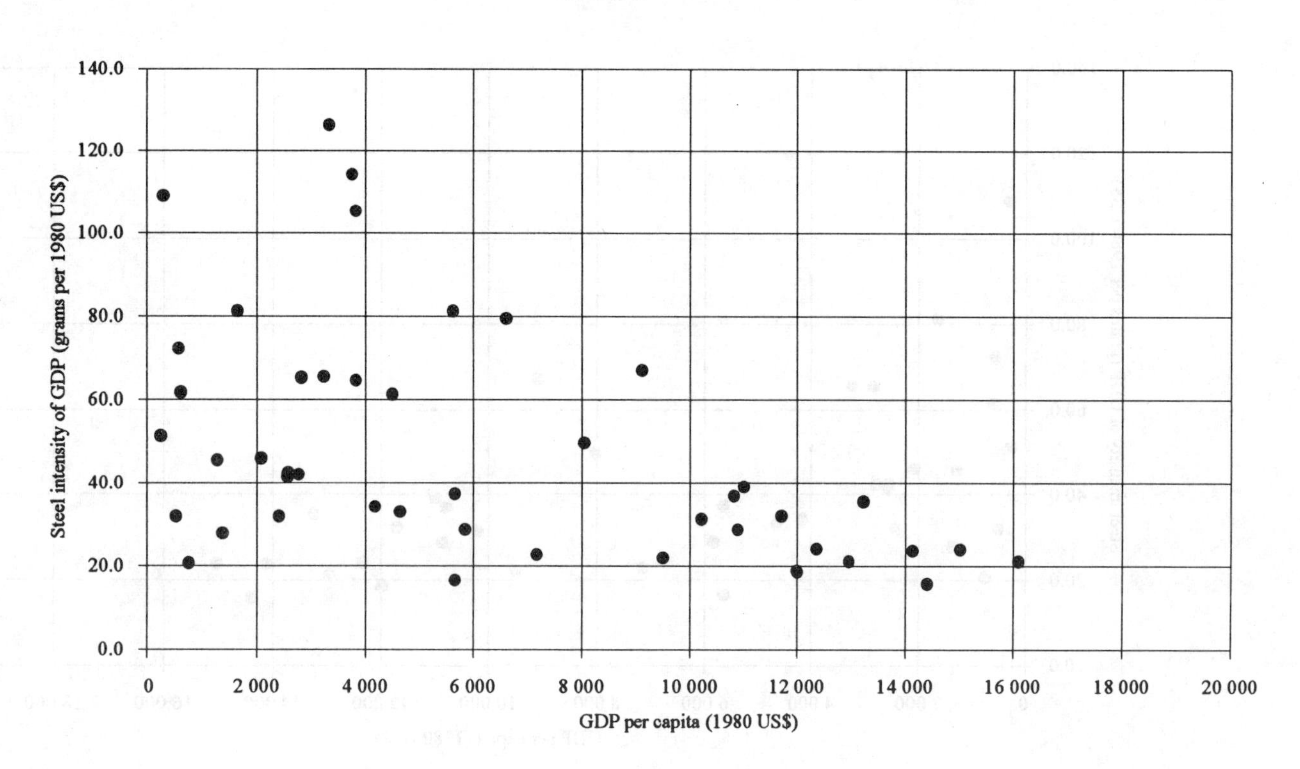

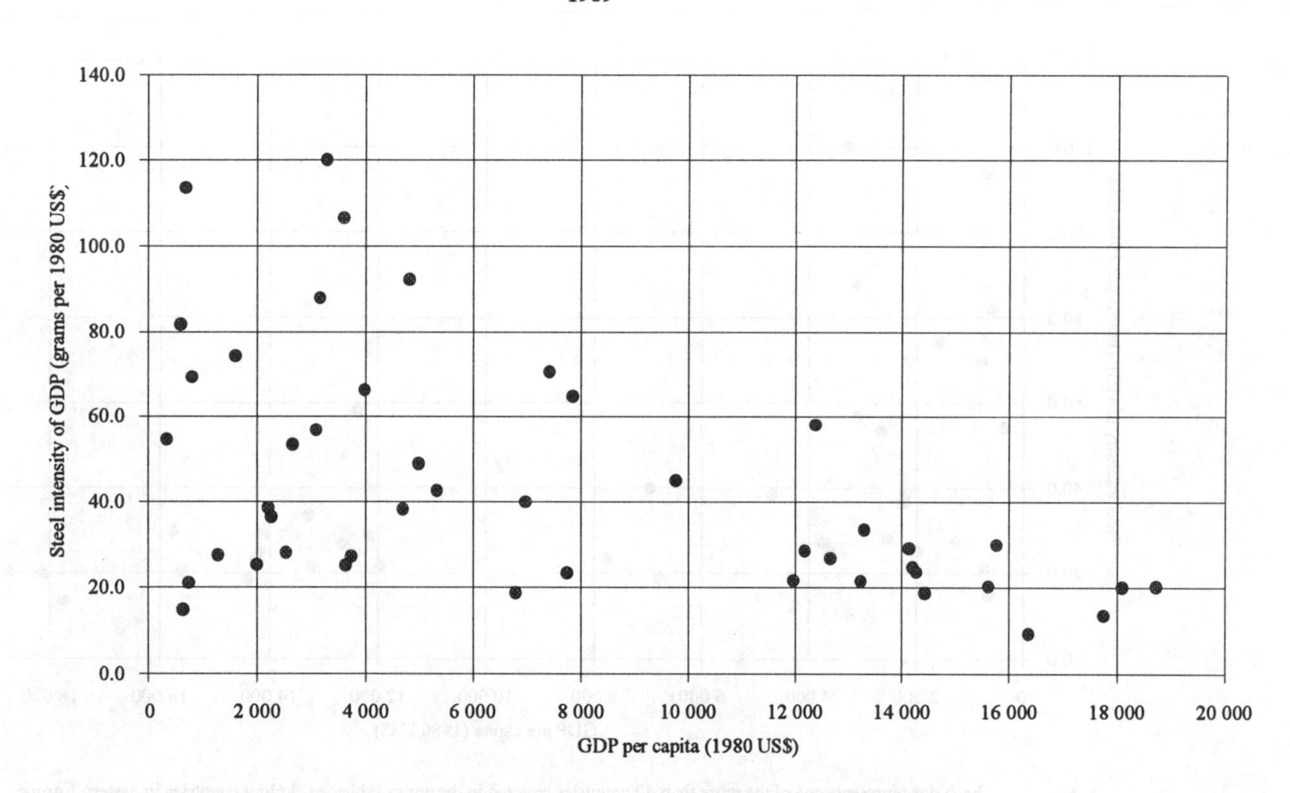

a/ Each dot represents one of the more than 40 countries covered in the annex tables. The same sample is used for both 1980 and 1989.

Figure 2-12-2 Relationship between steel intensity of GDP and GDP per capita in 1980 and 1989: *excluding Eastern Europe and former USSR* a/

1980

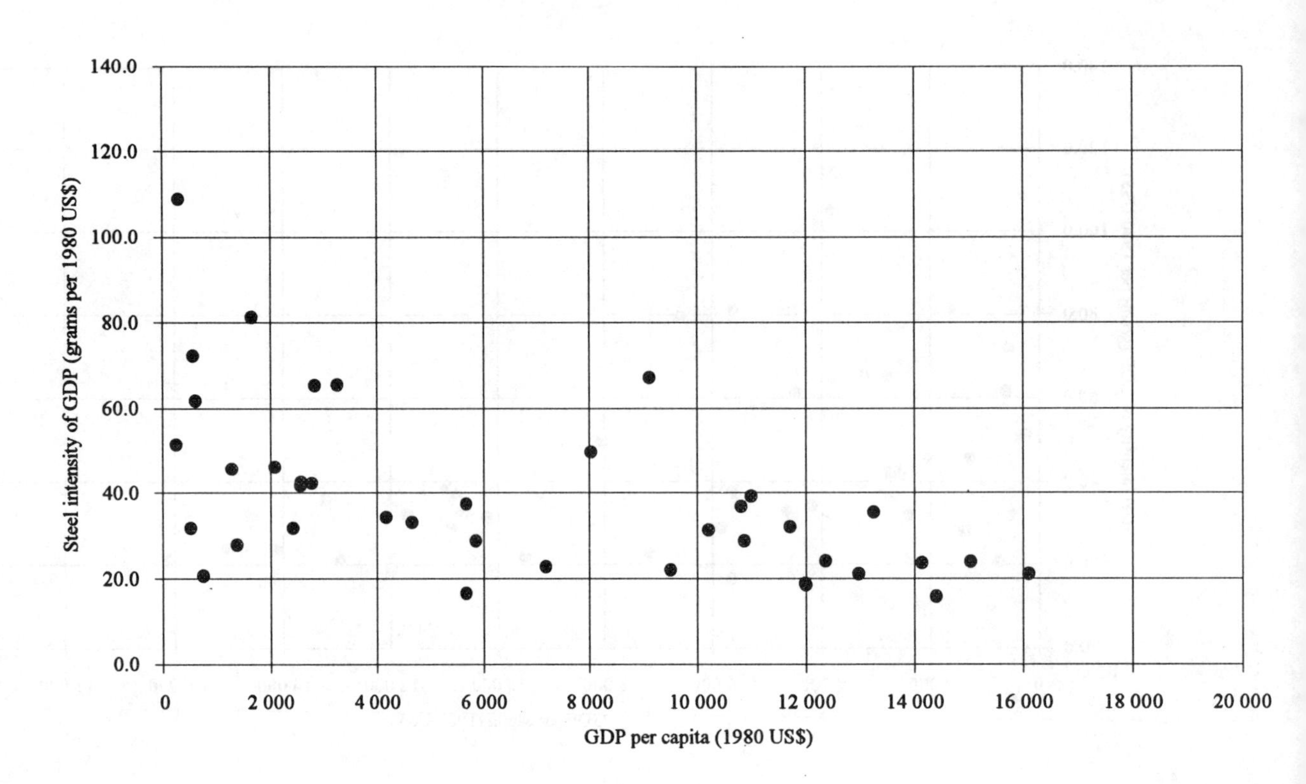

1989

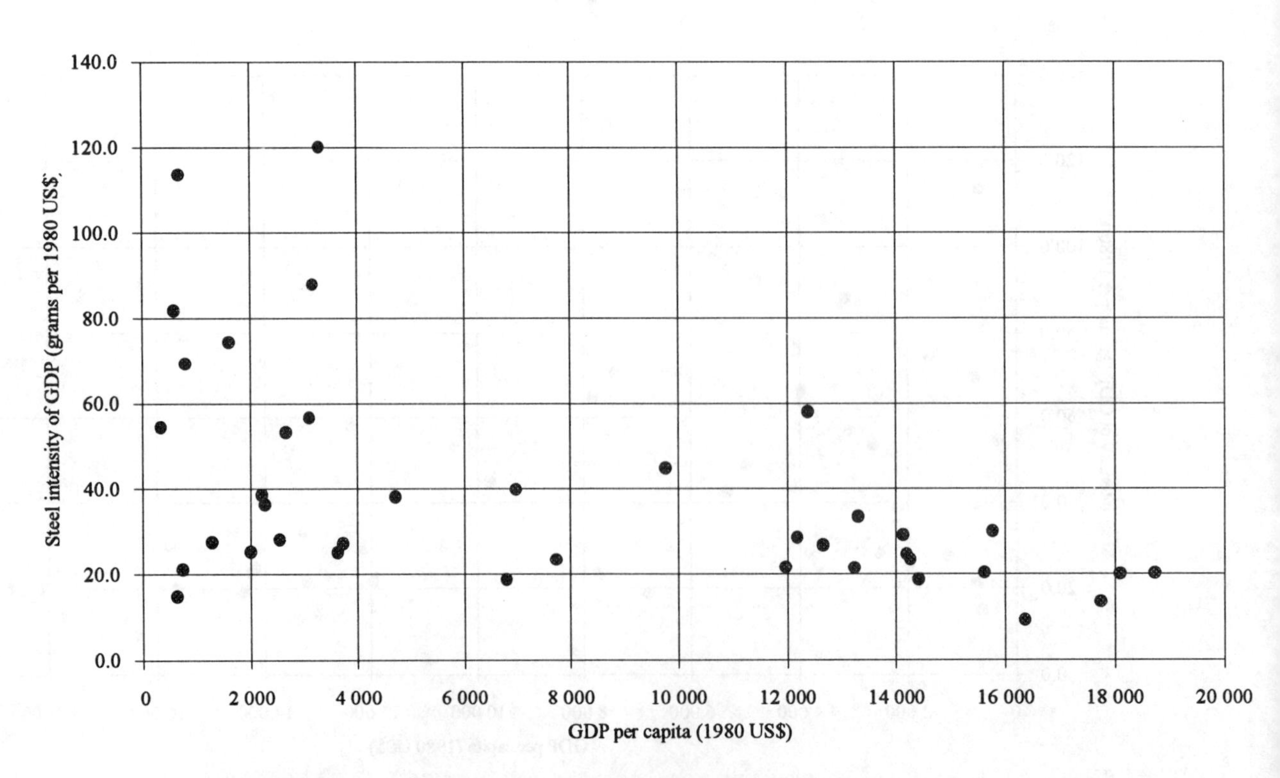

a/ Each dot represents one of the more than 40 countries covered in the annex tables, excluding countries in eastern Europe and the former USSR. The same sample is used for both 1980 and 1989.

Figure 2-12-3 Relationship between steel intensity of GDP and GDP per capita in 1980 and 1989: *Eastern Europe and former USSR* a/

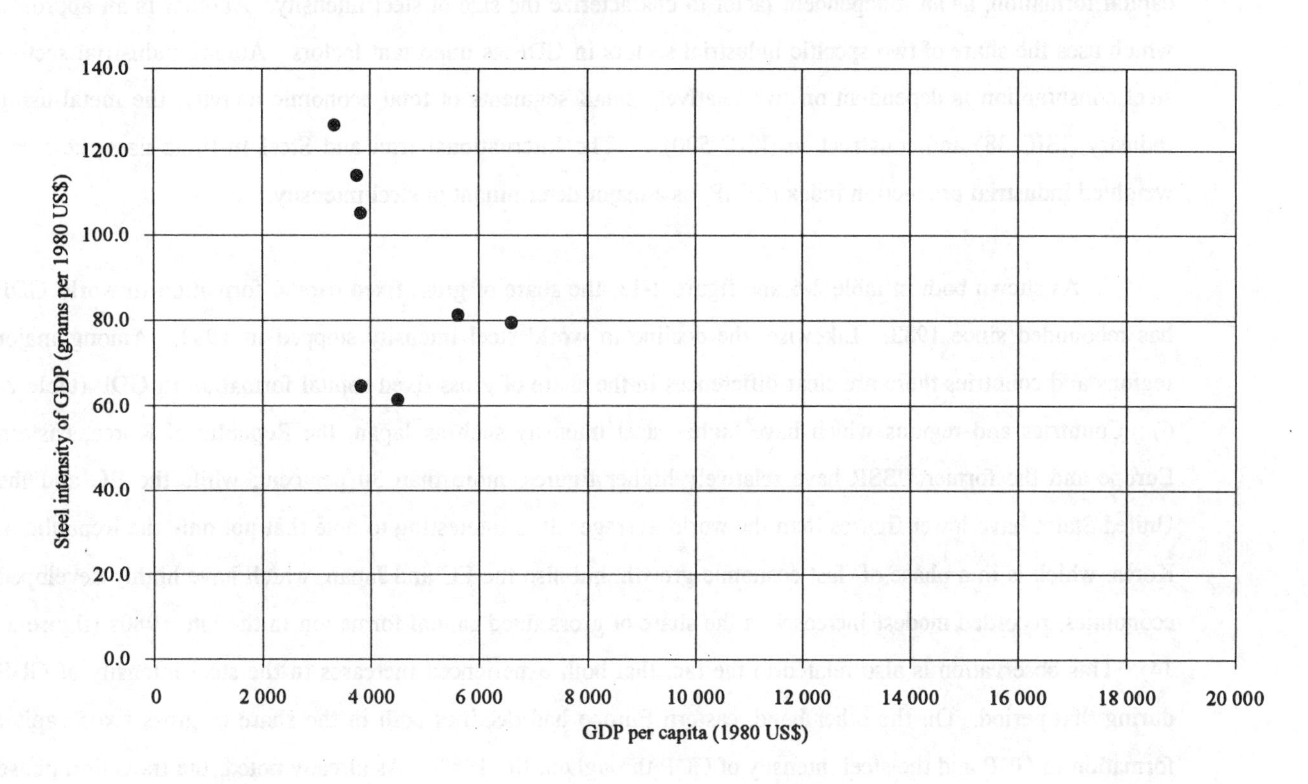

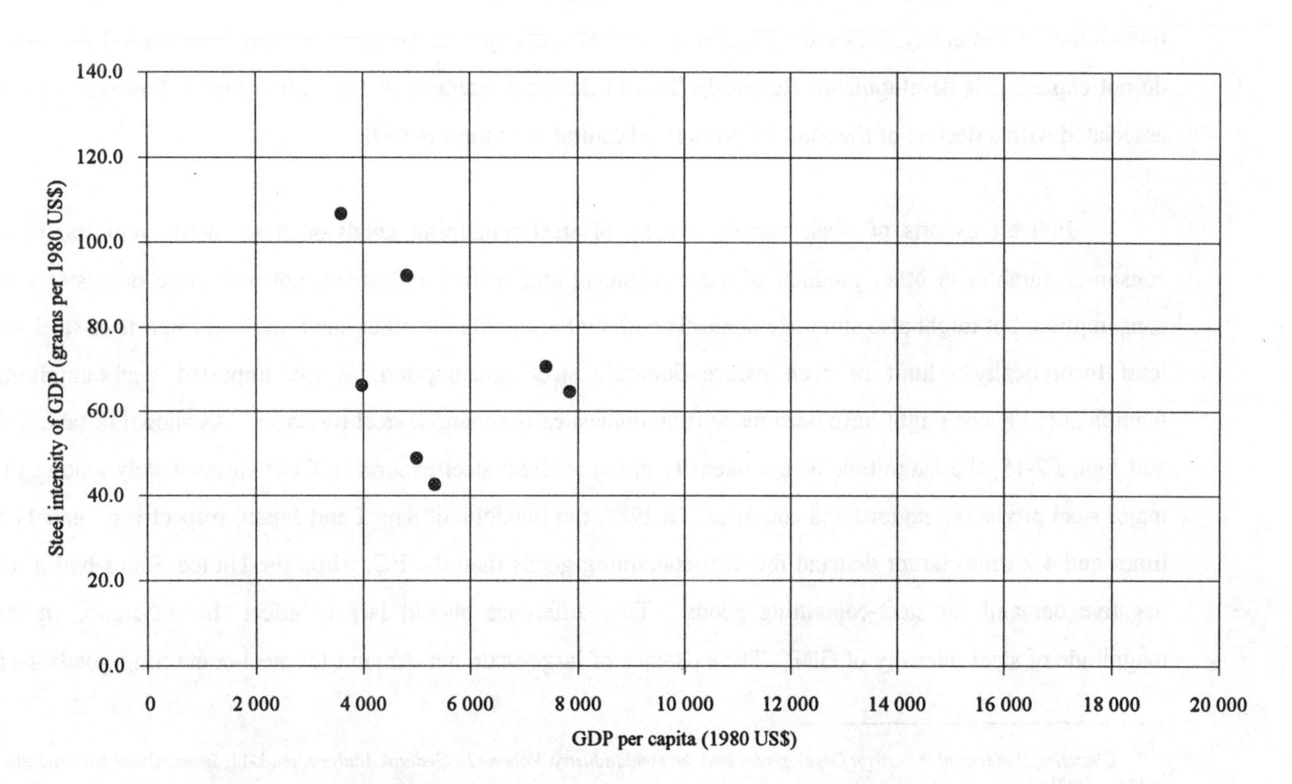

a/ Each dot represents one of the following seven countries: Bulgaria, the former Czechoslovakia, the former GDR, Hungary, Poland, Romania and the former USSR. The same sample is used for both 1980 and 1989.

The differences in the magnitude and development of steel intensity among countries are characterized by the different structure of their economies and growth patterns. There are several ways of assessing the differences in steel intensity among nations. One is to use the share of one component of GDP, gross fixed capital formation, as an independent factor to characterize the size of steel intensity. Another is an approach which uses the share of two specific industrial sectors in GDP as important factors. Among industrial sectors steel consumption is dependent on two relatively small segments of total economic activity, the metal-using industry (ISIC 38) and construction (ISIC 500).[3/] The International Iron and Steel Institute uses the steel-weighted industrial production index (SWIP) as a major determinant of steel intensity.[4/]

As shown both in table 2-6 and figure 2-13, the share of gross fixed capital formation in world GDP has rebounded since 1983. Likewise, the decline in world steel intensity stopped in 1983. Among major regions and countries there are clear differences in the share of gross fixed capital formation in GDP (table 2-6). Countries and regions which have higher steel intensity such as Japan, the Republic of Korea, eastern Europe and the former USSR have relatively higher figures, more than 30 per cent, while the EC and the United States have lower figures than the world average. It is interesting to note that not only the Republic of Korea, which is in a phase of fast economic growth, but also the EC and Japan, which have highly developed economies, recorded modest increases in the share of gross fixed capital formation in the late 1980s (figure 2-14). This observation is also related to the fact that both experienced increases in the steel intensity of GDP during that period. On the other hand, eastern Europe had declines both in the share of gross fixed capital formation in GDP and the steel intensity of GDP throughout the 1980s. As already noted, the transition phase to market economy in the region seemed to create an adjustment phase characterized by over-investment to the level of world standards, and consequently reduced the steel intensity. In the case of the former USSR, even though the steel intensity constantly declined in the 1980s, changes in the share of gross fixed capital formation do not explain this development. Recent declines in the steel intensity in the United States, however, can be associated with a decline of the share of gross fixed capital formation in GDP.

Indirect exports of steel, namely exports of steel-containing goods such as motor vehicles, other consumer durables or other products of the mechanical engineering industries, not only raise domestic steel consumption, but might also stimulate domestic output of steel. On the other hand, indirect imports of steel - at least theoretically - limit or even reduce domestic steel consumption, as the imported steel-containing manufactured goods might have been made from domestically produced steel instead.[5/] As shown in table 2-7 and figure 2-15, the magnitude of the intensity of net indirect steel exports in GDP varies widely among the major steel producing regions and countries. In 1989, the Republic of Korea and Japan, respectively, had 11.9 times and 4.2 times larger demand for steel-containing goods than the EC, while the United States had a net negative demand for steel-containing goods. This difference should largely affect the difference in the magnitude of steel intensity of GDP. The existence of large-scale net demand for steel-containing goods such

---

[3/] *Changing Patterns of Industrial Development and the Steel Industry, Volume I - General Analysis*, pp. 1-11, International Iron and Steel Institute, 1990.

[4/] Ibid.

[5/] Ibid, pp. 2-26.

as automobiles and electronic appliances makes both the Republic of Korea and Japan countries with high steel intensity of GDP.

Table 2-6 Share of gross fixed capital formation in GDP (%)

Year	World Total	EC	United States	Japan	Korea, Republic of	Eastern Europe	former USSR
1980	24.8	23.0	18.9	32.2	31.6	30.7	30.9
1981	24.4	20.5	19.8	31.8	30.1	28.2	30.2
1982	23.9	20.4	20.9	30.7	30.3	26.9	32.2
1983	22.9	20.0	18.0	29.2	30.6	25.9	32.9
1984	23.7	20.3	21.6	29.6	32.2	25.4	31.5
1985	23.9	20.1	21.0	30.0	31.1	25.1	32.1
1986	23.9	20.6	20.5	30.5	31.0	26.0	32.3
1987	24.3	21.0	20.8	31.8	31.9	25.3	32.2
1988	25.0	22.1	20.7	34.1	33.0	25.1	32.9
1989	25.6	23.0	20.5	35.6	37.7	25.6	34.0
1990	25.6	23.0	19.9	37.1	41.0	26.3	...
1991	24.7	22.5	18.0	36.7	44.0	...	...

Figure 2-13 Share of gross fixed capital formation in world GDP and world steel intensity

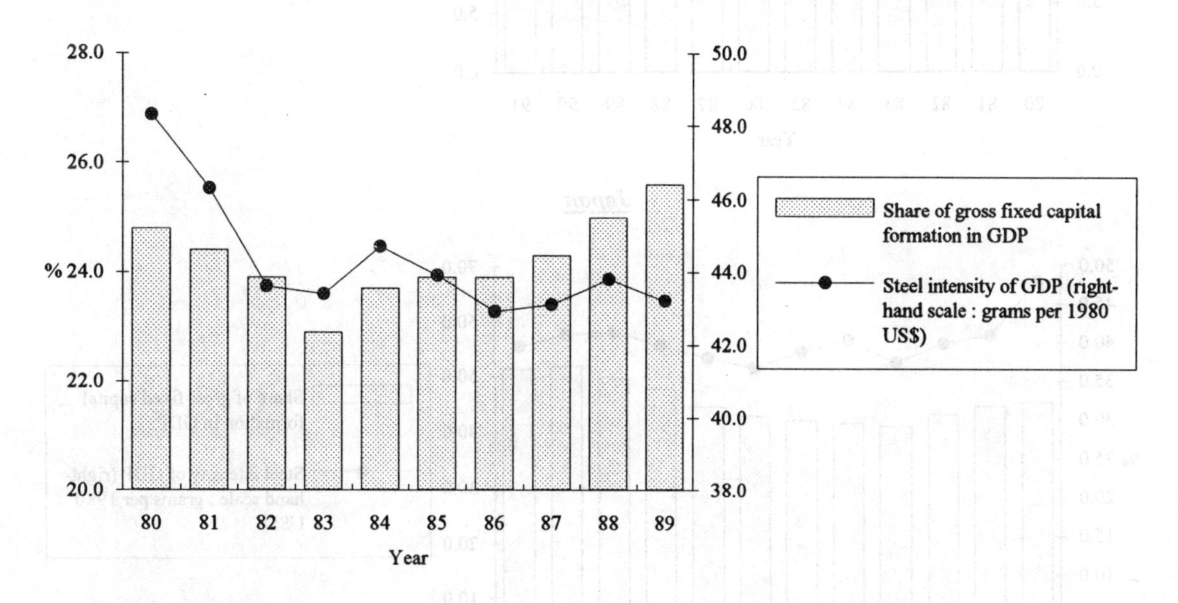

Figure 2-14 Share of gross fixed capital formation in GDP and steel intensity of GDP

EC

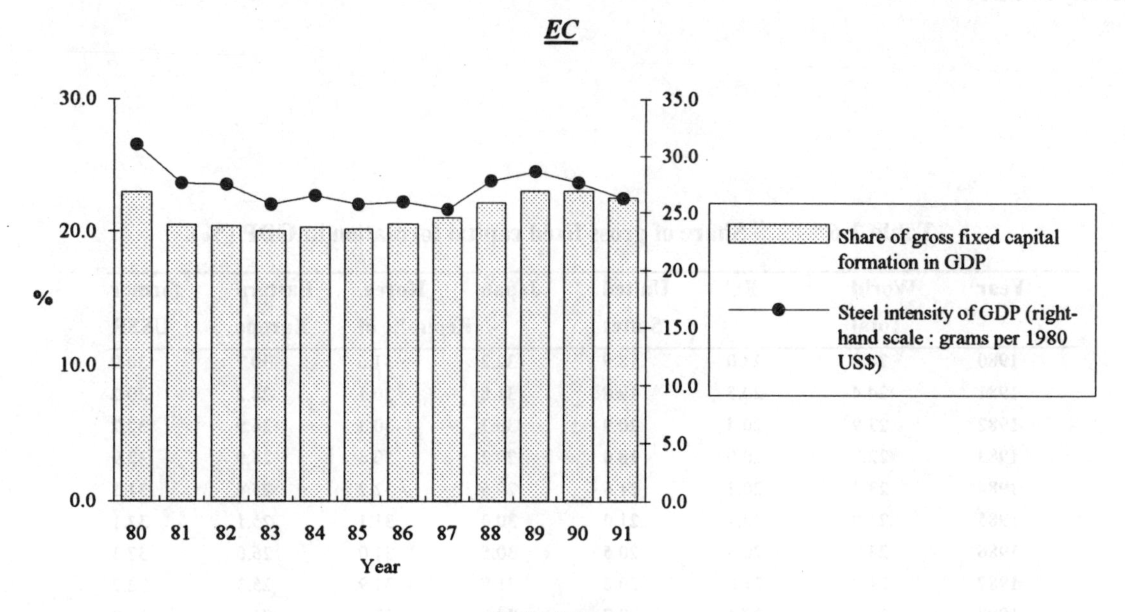

United States

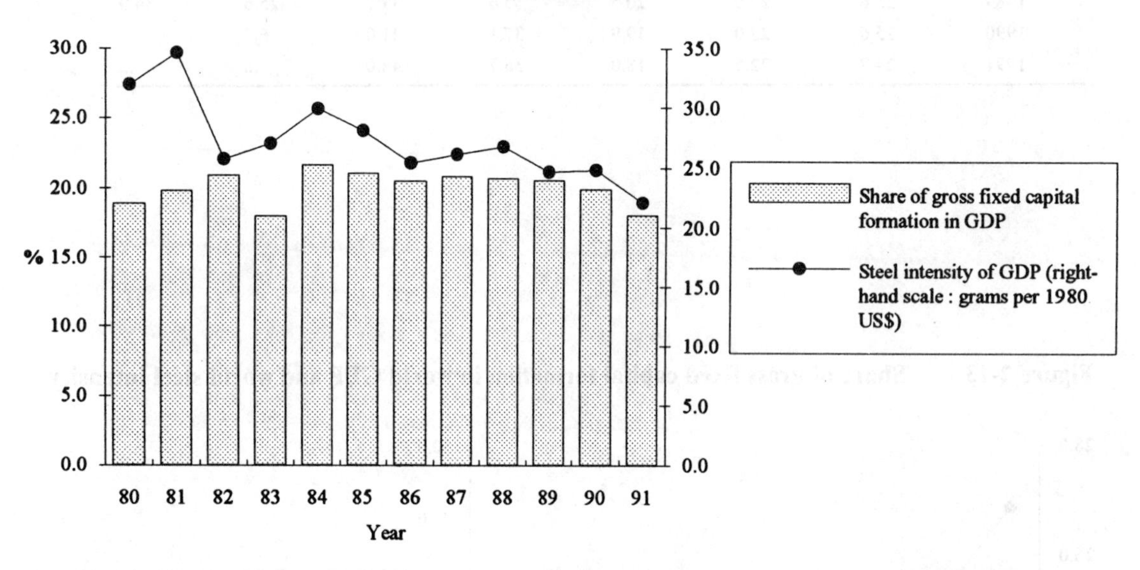

Japan

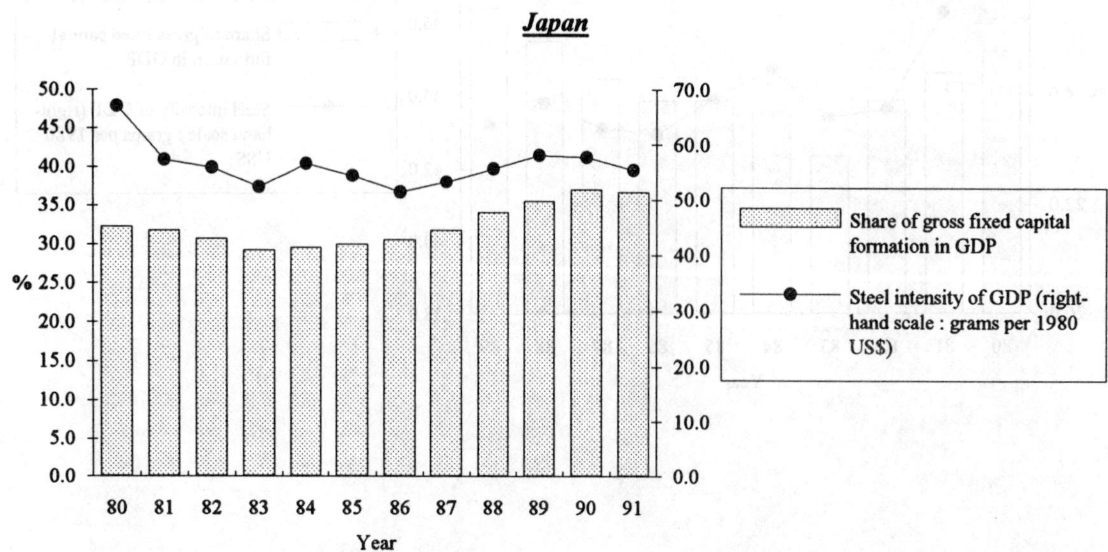

Figure 2-14 Share of gross fixed capital formation in GDP and steel intensity of GDP (continued)

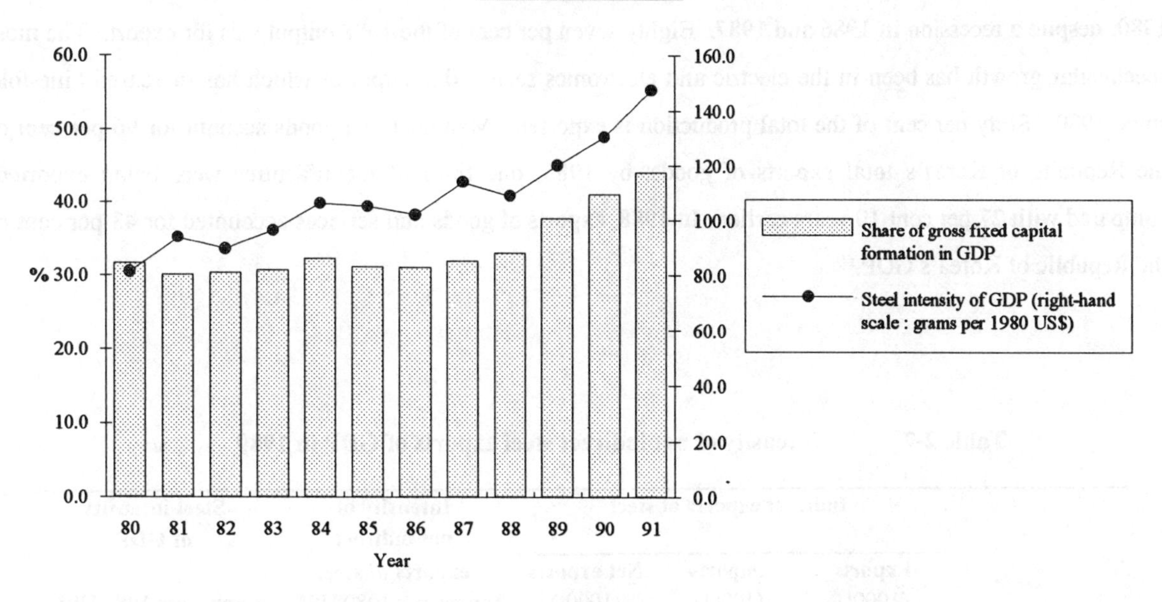

Eastern Europe

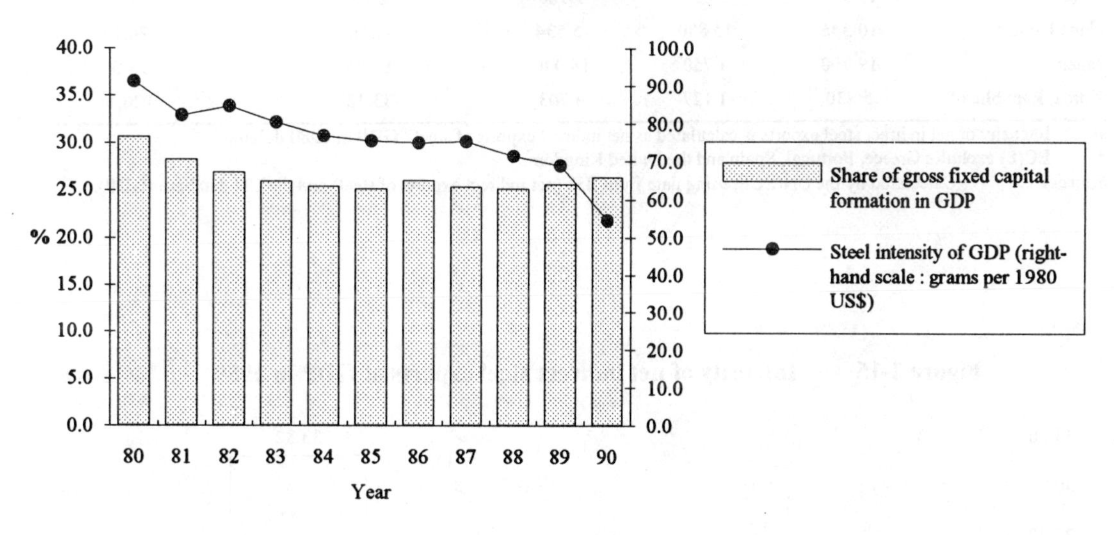

former USSR

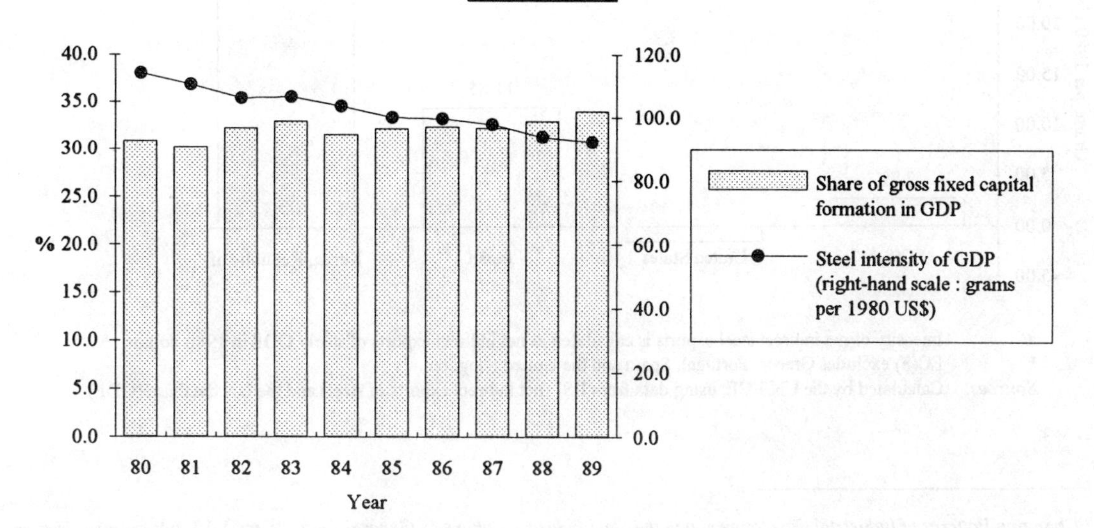

For example, automobile production in the Republic of Korea in 1988 was over 1 million units, 522 thousand for domestic use and 575 thousand for exports, while the country produced only 123 thousand units in 1980 with a quarter for exports. The production of the Korean shipbuilding industry has risen five-fold since 1980, despite a recession in 1986 and 1987. Eighty-seven per cent of the 1988 output was for export. The most spectacular growth has been in the electric and electronics sector, the output of which has increased nine-fold since 1980. Sixty per cent of the total production is exported. Manufactured goods account for 96 per cent of the Republic of Korea's total exports of goods; by 1985, one third of manufactures were being exported, compared with 27 per cent 10 years earlier. In 1988, exports of goods and services accounted for 43 per cent of the Republic of Korea's GDP.[6/]

Table 2-7 Intensity of net indirect steel exports of GDP in 1989 a/

	Indirect exports of steel			Intensity of net indirect exports of steel	Steel intensity of GDP
	Exports (1000t)	Imports (1000t)	Net exports (1000t)	(grams per 1980 US$)	(grams per 1980 US$)
EC(8)*	13 357	5 637	7 720	2.79	27.5
United States	10 346	15 880	-5 534	-1.56	24.7
Japan	19 790	1 750	18 040	11.85	58.0
Korea, Republic of	5 830	1 127	4 703	33.32	120.1

a/ Intensity of net indirect steel exports is calculated as net indirect exports of steel / GDP in 1980 dollars.
* EC(8) excludes Greece, Portugal, Spain and the United Kingdom.
Sources: Calculated by the UN/ECE, using data from IISI (net indirect exports of steel) and the UN database (GDP).

Figure 2-15 Intensity of net indirect steel exports of GDP in 1989 a/

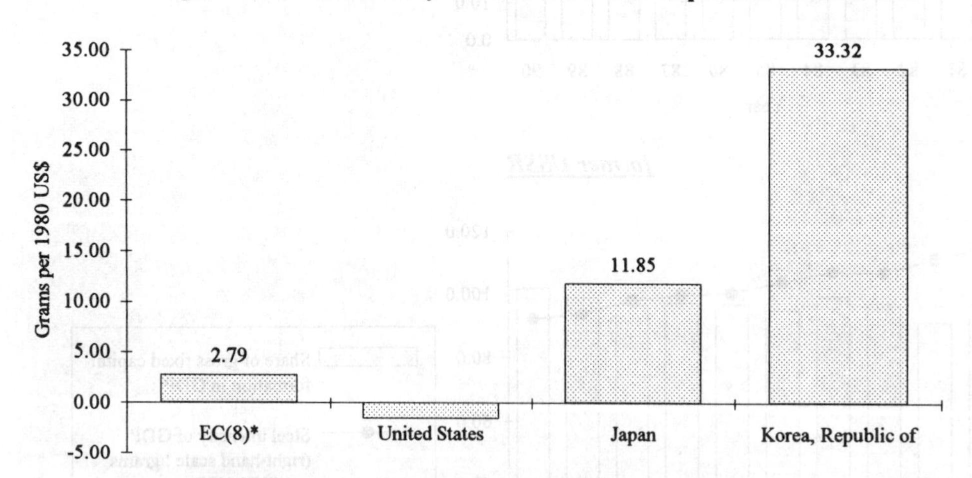

a/ Intensity of net indirect steel exports is calculated as net indirect exports of steel / GDP in 1980 dollars.
* EC(8) excludes Greece, Portugal, Spain and the United Kingdom.
Sources: Calculated by the UN/ECE, using data from IISI (net indirect exports of steel) and the UN database (GDP).

6/ *Changing Patterns of Industrial Development and the Steel Industry, Volume I - General Analysis*, pp. 1-17, International Iron and Steel Institute, 1990.

It is a common view that manufacturing industries and construction have lost ground to services as contributors to GDP in both the industrialized and the newly industrializing countries. As shown in figure 2-16, the share of construction in GDP declined in the 1980s in both developed market economies and developing market economies. However, it should be noted that the rate revived slightly from 1985 to 1989 in developed market economies, and in 1989 and 1990 in developing market economies. This trend was also clear for the EC, Japan and the Republic of Korea as is shown in figure 2-17-2 and was associated with an increased world share of gross fixed capital formation in GDP in the latter half of the 1980s. Even though there is a shortage of data available for the former CMEA countries, figures for Hungary show a constant decline in the share of construction. The relatively large share of construction (7-9 %) in both the Republic of Korea and Japan is one of the factors behind the higher steel intensity of those countries.

Contrary to the general perception, the share of manufacturing in GDP increased in the 1980s. While the growth rate was higher in the developing countries, even the developed countries experienced a slight increase or a stable share (figure 2-16 and figure 2-17-1). Of course, in the case of developed countries, the increase in the share of manufacturing does not necessarily mean a revival in the importance of steel-using manufacturing sectors. More advanced high-tech industries are emerging and changing the structure of the manufacturing industries. For example, in Japan, the production of television and communications equipment peaked in 1984 and since then manufacturers have been shifting production of some categories of product to countries outside Japan. The total 1988 value of high-tech products, which are of course very low in steel intensity, exceeded the value of road vehicles, an important steel-using sector. Even though the share of manufacturing in GDP is increasing, the manufacturing industry's product mix has changed to become much less material-intensive, and less steel-intensive, particularly in most developed countries. However, it should also be noted that some industrialized countries such as the United States and Spain experienced an increasing share of transport equipment in total metal-using industries. This could be explained by new plants constructed by foreign manufacturers seeking to globalize their supply bases. Increasing foreign direct investment by globalized or multinational companies has created country-specific factors for developments in steel intensity under the changing competitive advantage of nations.[7/]

In the manufacturing sector, the effect of material substitution should also be noted. In the transport equipment sector, the need to reduce the weight-power ratio to improve operating costs while not impairing the vehicle's safety and impact resistance is an important task for designers. Corrosion resistance, noise dampening and thermal insulation are also important requirements. Even though the main material used for car bodies is low carbon, cold-rolled sheet, major competitors for steel are aluminium and plastic materials and potentially, for engine parts, ceramics. On the other hand, the properties required in the building and construction sector are amply met by steel for load-bearing structures. Strength and rigidity are its foremost qualities and steel is, therefore, widely used in high-rise buildings, factories and bridges.[8/]

[7/] Ibid, pp. 1-13.

[8/] *Intermaterial Competition: An Economic Analysis of General Trends*, pp. 1-9, International Iron and Steel Institute, 1989.

Figure 2-16 Share of manufacturing and construction in GDP (%)

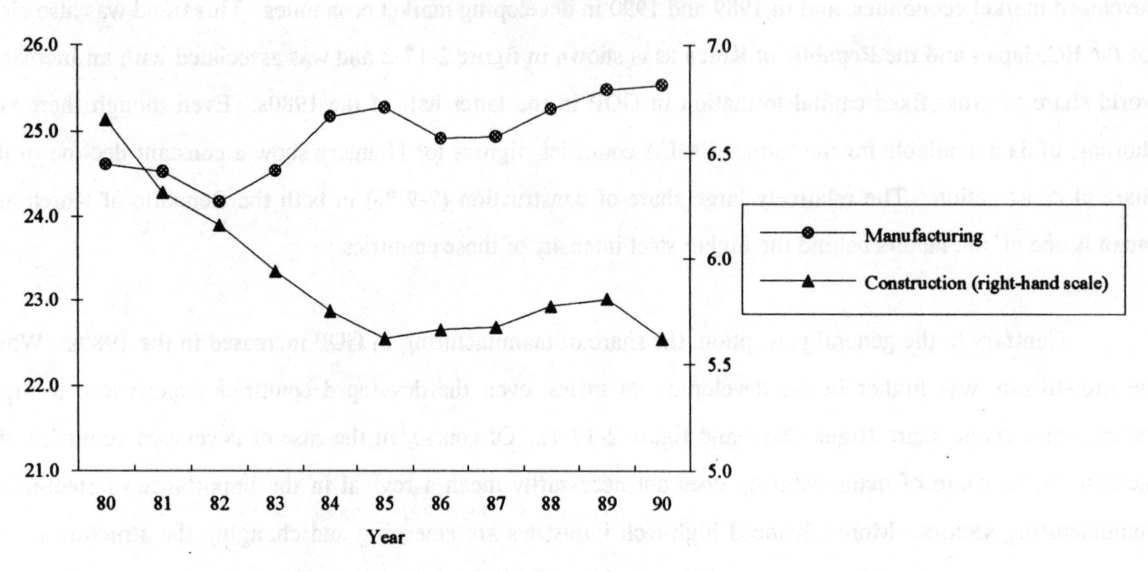

Source: *MSPA Data Bank of World Development Statistics*, United Nations.

Developing market economies

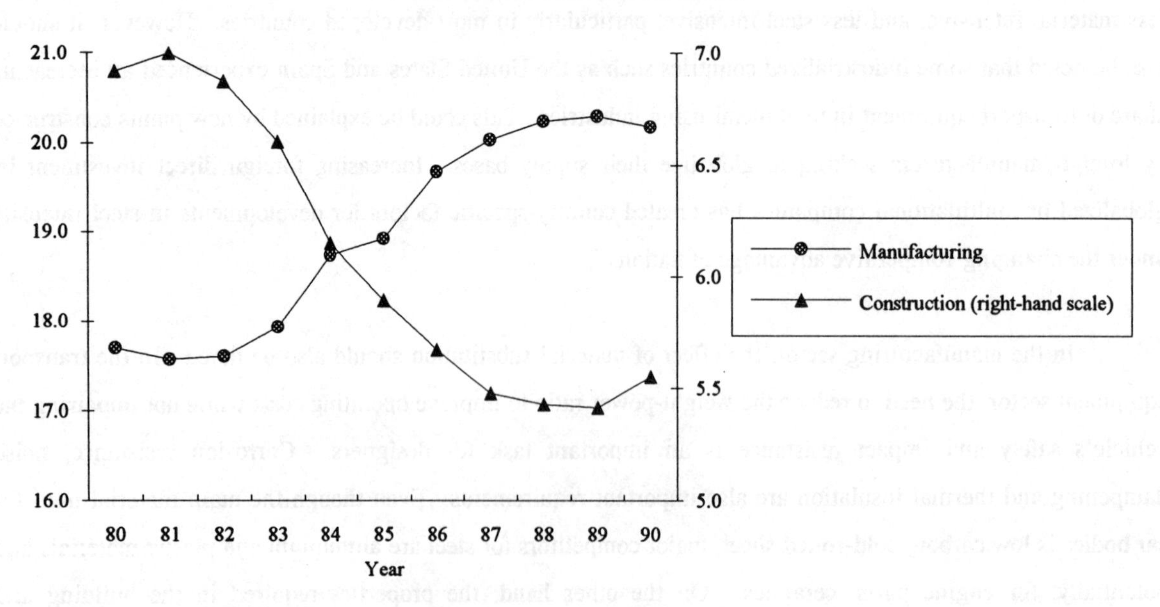

Source: *MSPA Data Bank of World Development Statistics*, United Nations.

Figure 2-17-1 Share of manufacturing in GDP by major regions and countries

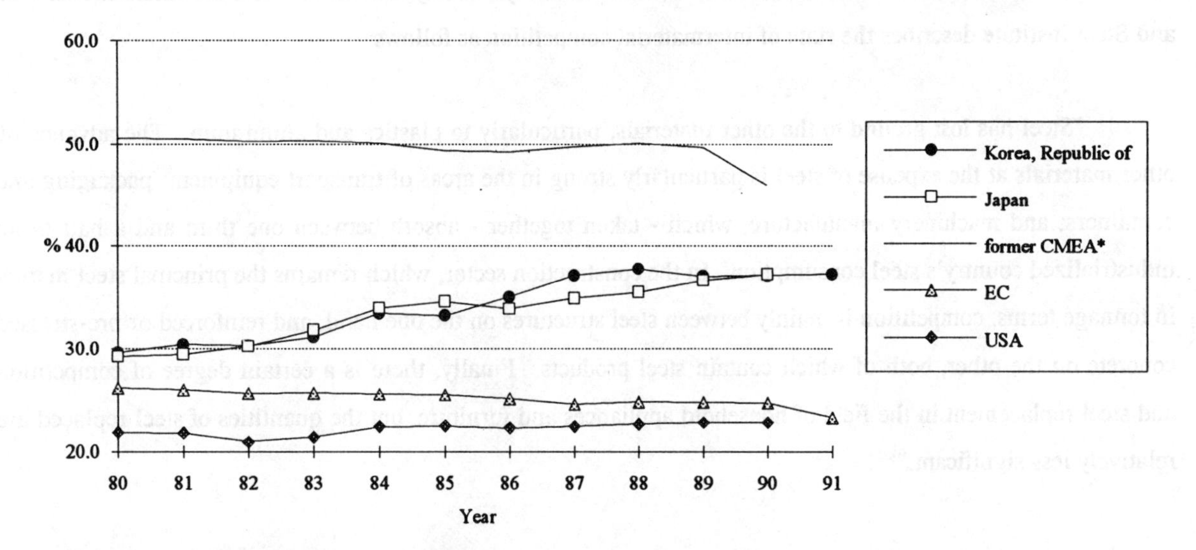

* Eastern Europe plus former USSR.

Source: *MSPA Data Bank of World Development Statistics*, United Nations.

Figure 2-17-2 Share of construction in GDP by major regions and countries

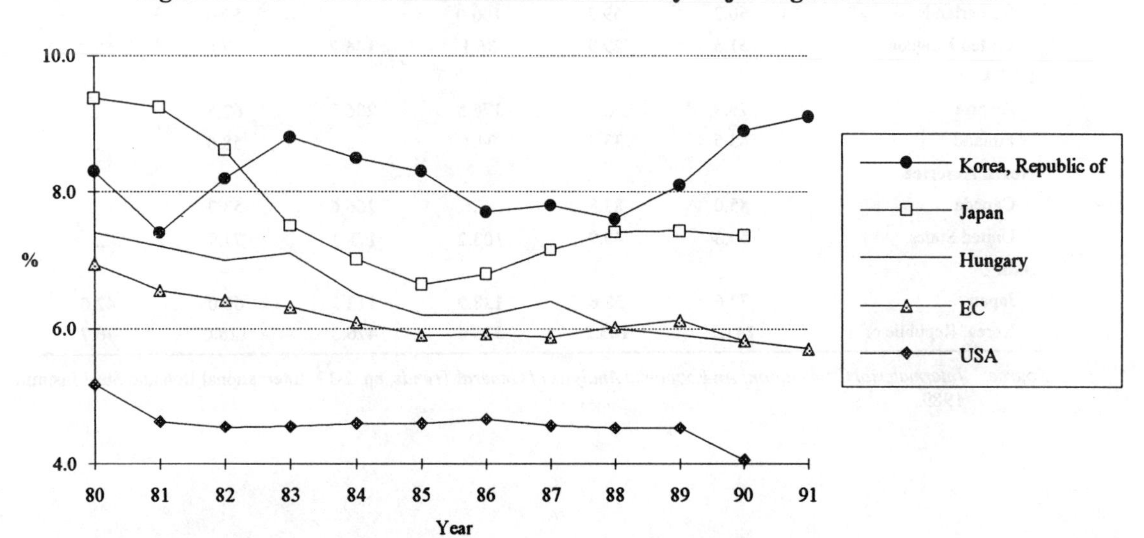

Source: *MSPA Data Bank of World Development Statistics*, United Nations.

As shown in table 2-8, aluminium and plastics increased their importance in GDP from 1968 to 1984, while steel and cast iron declined in terms of material intensity. Using data up to 1984, the International Iron and Steel Institute describes the state of intermaterial competition as follows:

"Steel has lost ground to the other materials, particularly to plastics and aluminium. The advance of other materials at the expense of steel is particularly strong in the areas of transport equipment; packaging and containers; and machinery manufacture; which - taken together - absorb between one third and a half of an industrialized country's steel consumption. In the construction sector, which remains the principal steel market in tonnage terms, competition is mainly between steel structures on the one hand, and reinforced or pre-stressed concrete on the other, both of which contain steel products. Finally, there is a certain degree of competition and steel replacement in the field of household appliances and furniture, but the quantities of steel replaced are relatively less significant."[9]

Table 2-8 Materials intensity of GDP, 1968 and 1984
(Indices, 1968=100)

Countries	Finished steel	Cast iron	Aluminium	Plastics	Cement	Industrial woods
EC						
Belgium-Luxembourg	70.1	...	107.7	...	57.0	...
France	55.0	46.2	125.2	217.9	49.1	...
former FR Germany	64.9	47.3	124.8	153.4	56.1	79.4
Italy	94.7	76.7	170.2	...	84.8	...
Netherlands	60.2	59.7	100.0	...	59.0	...
United Kingdom	51.5	29.9	86.1	146.7	57.8	58.9
EFTA						
Austria	79.9	...	176.5	206.3	62.8	...
Finland	83.5	33.4	94.1	...	58.6	...
North America						
Canada	85.0	83.8	...	266.6	53.7	...
United States	59.3	45.9	103.2	178.1	71.9	...
Asia						
Japan	72.6	38.6	138.9	112.2	66.0	42.0
Korea, Republic of	241.2	183.2	323.6	476.3	128.0	46.7

Source: *Intermaterial Competition: An Economic Analysis of General Trends*, pp. 2-19, International Iron and Steel Institute, 1989.

[9] *Intermaterial Competition: An Economic Analysis of General Trends*, pp. 3-5, International Iron and Steel Institute, 1989.

2.2. Changes in steel consumption of major steel-consuming sectors in selected countries

2.2.1. Measurements and methods of analysis

Steel intensity or specific steel consumption is a relation of the weight of steel consumed in specific industrial sectors to any kind of production measure of those sectors. If the measure of activity of steel-consuming sectors is also weight, the resulting relation is, in a narrow sense, the real specific steel consumption. If the available measure of a steel consuming activity is different, such as a value or an index of production, then it is more usual to call the resulting relation steel intensity.

Steel statistics, which are normally collected at the international level, do not allow for the straightforward calculation of specific steel consumption. Therefore, the data are collected by specifically designed questionnaires and the subject is researched in special studies. There are not many studies undertaken in this way on an international basis. One was published by the United Nations Economic Commission for Europe in 1984, *The Evolution of the Specific Consumption of Steel.* That study laid down the methodological framework for the analysis of steel intensity or specific steel consumption, defined the main and the most intensive steel-consuming sectors and studied them in the time period from 1960 to 1980.

The present section is a follow-up to that former study. It uses a similar methodology and studies the evolution of specific steel consumption in a similar sample of countries in the most intensive steel-consuming sectors for the period from 1980 to 1992. The most intensive steel-consuming sectors, as defined by ISIC (The International Standard Industrial Classification of all Economic Activities) are ISIC 380 (Manufacture of fabricated metal products, machinery and equipment) and ISIC 500 (Construction). The data for the present study, obtained from a questionnaire, also cover the following major groups of ISIC 380:

ISIC 381: Manufacture of fabricated metal products, except machinery and equipment;
ISIC 382: Manufacture of machinery except electrical;
ISIC 383: Manufacture of electrical machinery; and
ISIC 384: Manufacture of transport equipment.

Even though the questionnaire asked for an estimation of final steel consumption in each sector in real terms, the credibility of data may sometimes differ among countries because of the complexity of estimation. However, these data are still valuable and interesting for a comparison among countries. In addition to steel consumption data for the above sectors, data for the production of related steel-consuming sectors were also asked for in the questionnaire. Because not many countries could supply these data and constant United States dollar data are preferable as a standardized measure, United Nations statistics on national accounts which contain the origin of GDP in 1980 United States dollars are used for analysis when available.

The use of these data required some approximation, because the break-down of GDP by origin was not sufficiently detailed in the manufacturing sector. Only the value for total manufacturing (ISIC Division 3) was available as a component of GDP. Therefore, this value is used for the steel consumption of fabricated metal products, machinery and equipment (ISIC 380). From now on, the term manufacturing is used as a synonym of ISIC 380. This is rational, because ISIC 380 accounts for most of the steel consumption among manufacturing sectors.

Further approximation was necessary in the case of the former CMEA countries where only the value for total industry (the sum of ISIC 2, 3, 4 and 5, i.e. mining, manufacturing, utilities and construction) was available as a component of GDP. Therefore, these data are used for calculating the steel intensity of the sum of manufacturing and construction for the former CMEA countries. This is a reasonable approximation, because mining and utilities (ISIC Divisions 2 and 4) are negligible steel consumers in comparison with manufacturing and construction (ISIC Divisions 3 and 5).

Such regions or countries as the EC, France, the Netherlands, the United Kingdom, Finland, Sweden, the former Yugoslavia, Croatia, Slovenia, Hungary, the Russian Federation, Japan, the Republic of Korea and Brazil supplied particularly detailed answers to the questionnaire, and are therefore analysed in detail.

A complete set of basic and derived data on steel consumption and its shares, the production of steel-consuming sectors and steel intensities, is presented for all reporting countries in the annex tables. These tables sometimes contain data from the former UN/ECE study which illustrate developments before 1980.

2.2.2. Distribution of steel consumption among major steel-consuming sectors

Seventy to ninety per cent of all steel consumed is used in the manufacturing and construction sectors combined (Figure 2-18-1). In the United Kingdom and Japan the shares of manufacturing and construction are 80 to 90 per cent, while the share is lower at 75 to 80 per cent for the EC as a whole and is higher at a little more than 90 per cent in France. In other countries it is also in a similar range and only statistical fluctuations in some countries show this figure lower than 70 per cent or higher than 90 per cent. It is clear that manufacturing and construction are the predominant steel users at any stage of economic development in any country.

The general perception is that steel consumption in construction is more important in the initial stages of economic development, while manufacturing becomes the major steel consumer in more advanced stages of economic development. However, this proved to be erroneous. Japan, already a mature, highly developed country, and the Republic of Korea, a newly developed country, have an equal distribution of steel consumption between manufacturing and construction at around 40-50% (Figures 2-18-2 and 2-18-3). This is, of course, a

Figure 2-18-1 Share of steel consumption by industrial sectors in selected countries: *Manufacturing plus construction*

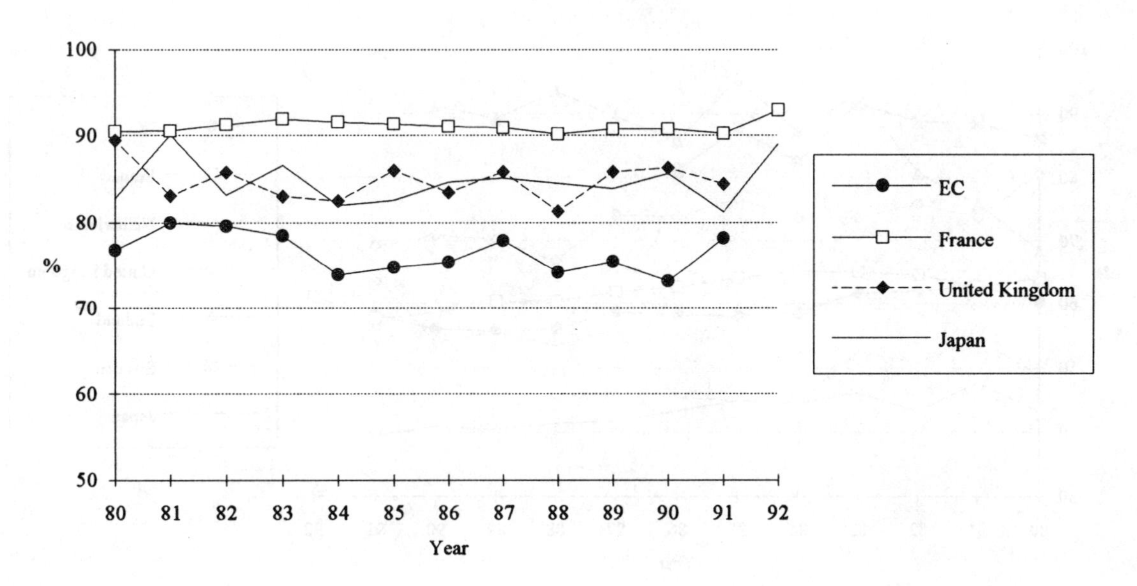

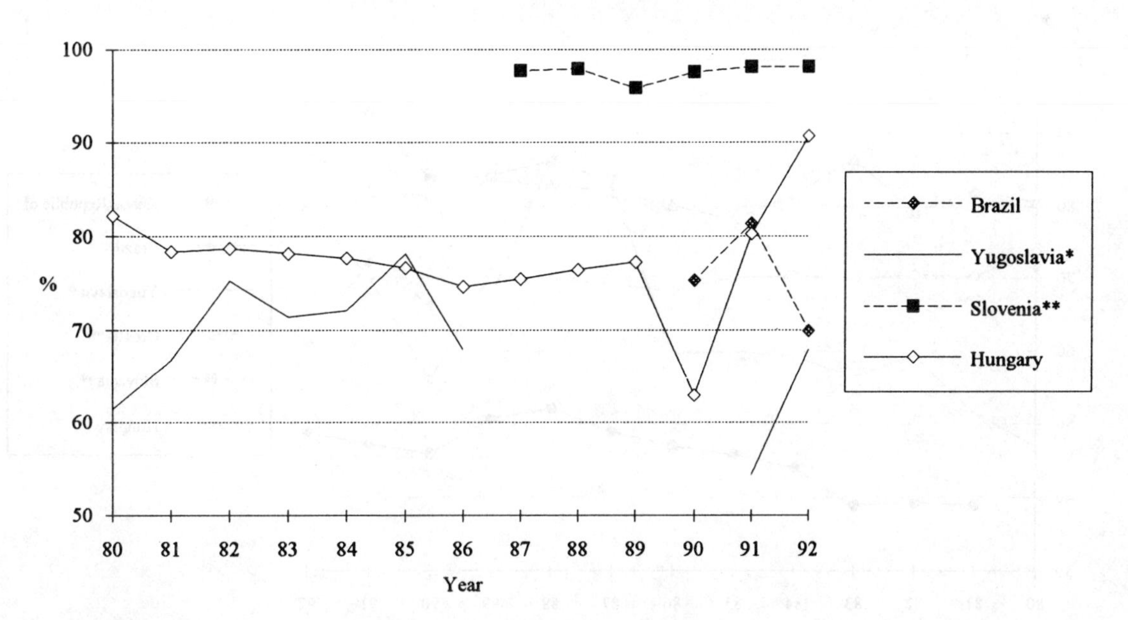

* Data for 1991 and 1992 are for the FR of Yugoslavia (Serbia and Montenegro). Data up to 1986 are for the former Yugoslavia.

** Slovenia has also reported data for years before it became an independent State.

Figure 2-18-2 Share of steel consumption by industrial sectors in selected countries: *Manufacturing*

Developed economies

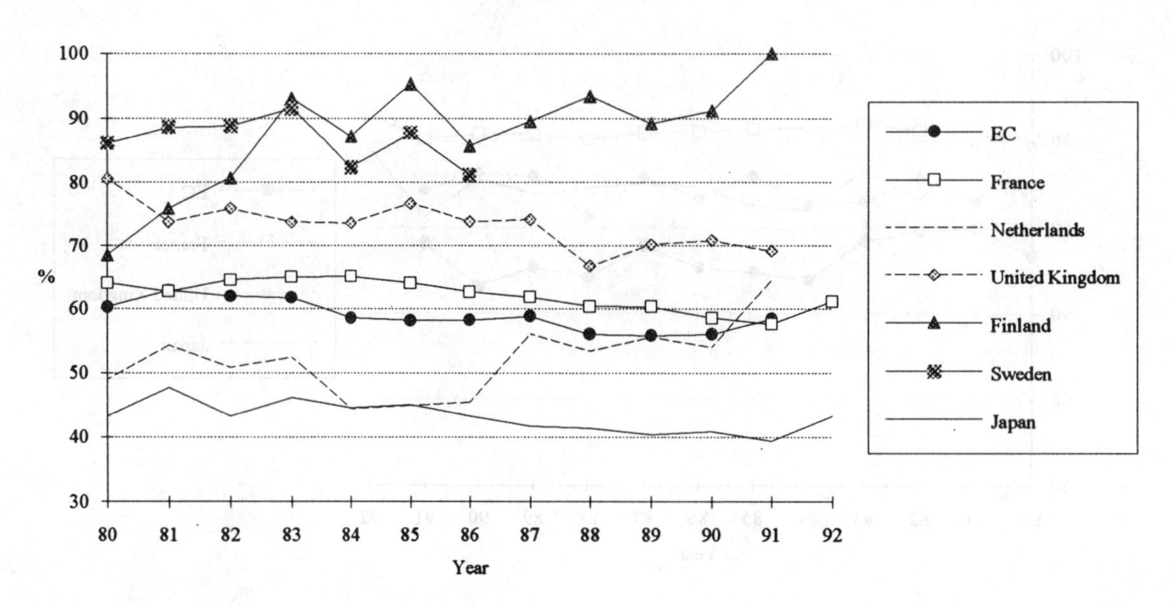

Developing economies and the former CMEA

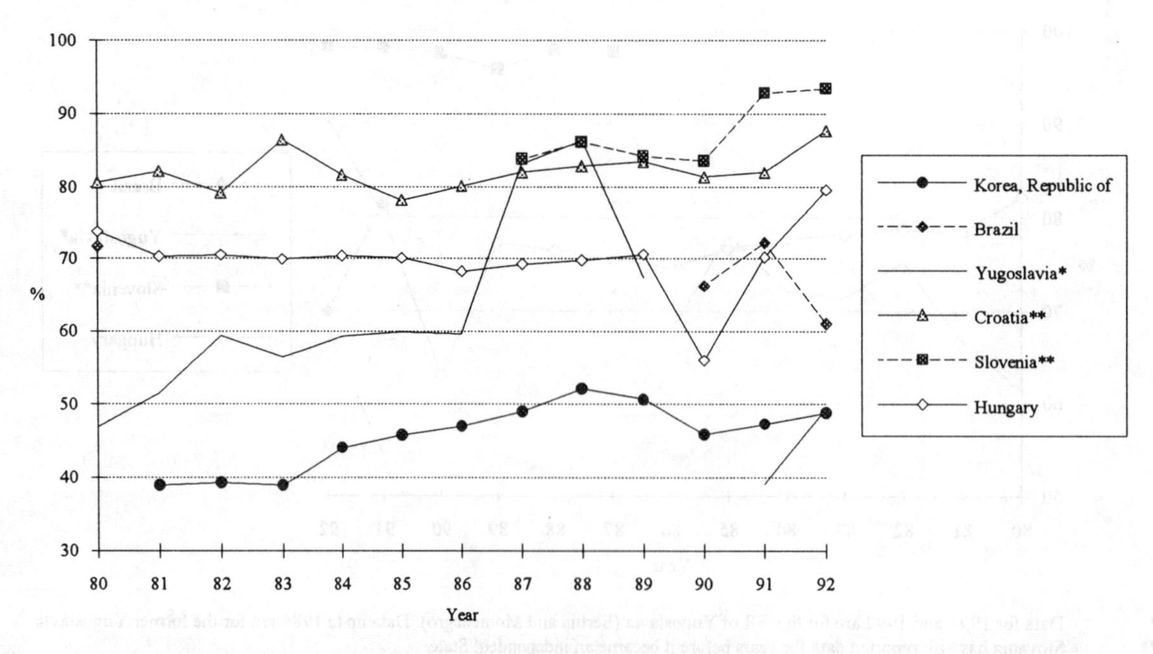

* Data for 1991 and 1992 are for the FR of Yugoslavia (Serbia and Montenegro). Data up to 1989 are for the former Yugoslavia.

** Croatia and Slovenia have reported data for years before they became independent States.

Figure 2-18-3 Share of steel consumption by industrial sectors in selected countries: *Construction*

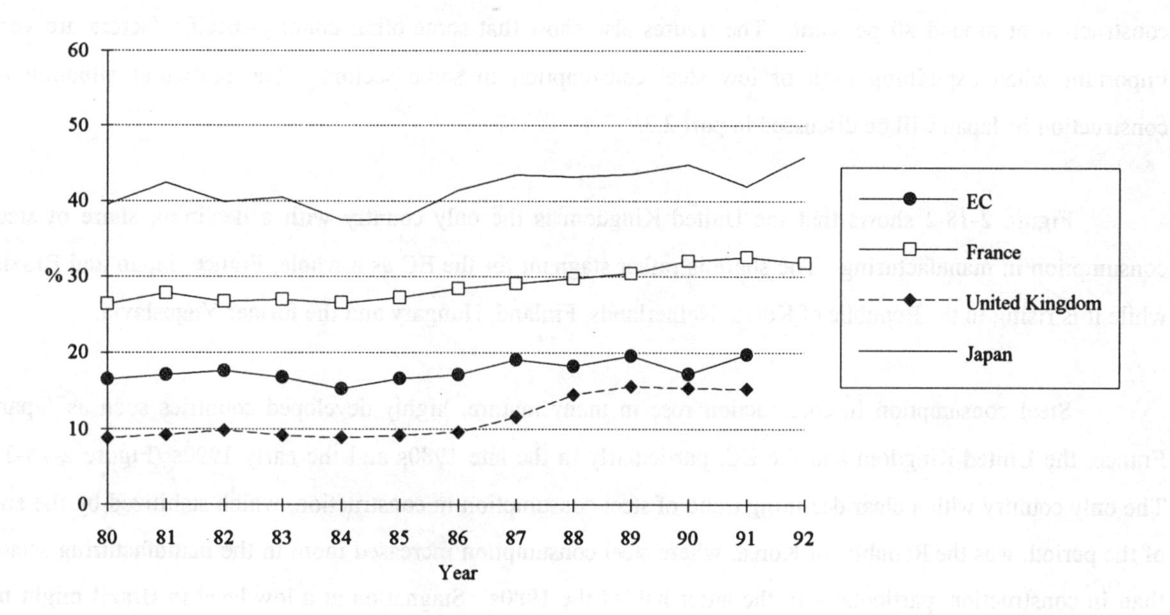

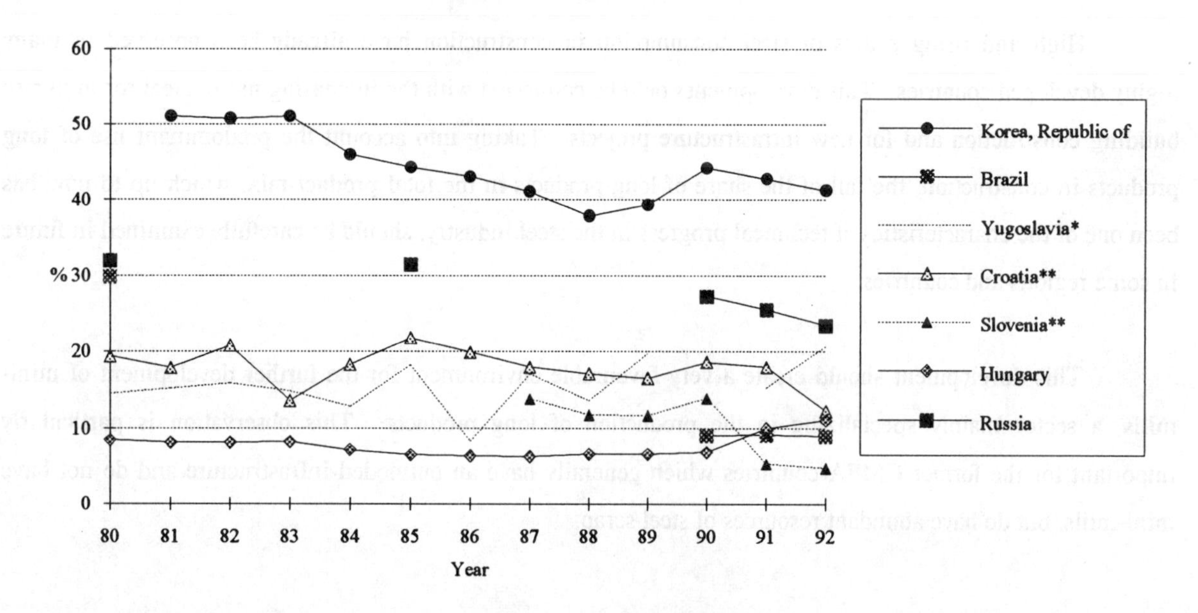

* Data for 1991 and 1992 are for the FR of Yugoslavia (Serbia and Montenegro). Data up to 1989 are for the former Yugoslavia.

** Croatia and Slovenia have reported data for years before they became independent States.

simple result of the high share of both gross fixed capital formation and construction in GDP in these two countries, as already discussed. On the other hand, other countries which are at very different levels of economic development such as the United Kingdom, the EC, Finland, the former Yugoslavia and Brazil have a very high share of steel consumption in manufacturing (50-70%) and a comparatively low share in construction (10-20%). It is interesting to note that both France and the Russian Federation have a relatively high share of construction at around 30 per cent. The figures also show that some other country-specific factors are very important when explaining high or low steel consumption in some sectors. The particular situation of construction in Japan will be discussed in part 2.3.

Figure 2-18-2 shows that the United Kingdom is the only country with a declining share of steel consumption in manufacturing. The share is rather stagnant for the EC as a whole, France, Japan and Brazil, while it is rising in the Republic of Korea, Netherlands, Finland, Hungary and the former Yugoslavia.

Steel consumption in construction rose in many mature, highly developed countries such as Japan, France, the United Kingdom and the EC, particularly in the late 1980s and the early 1990s (Figure 2-18-3). The only country with a clear declining trend of steel consumption in construction, which stabilized by the end of the period, was the Republic of Korea, where steel consumption increased more in the manufacturing sector than in construction, particularly in the latter half of the 1980s. Stagnation at a low level in Brazil might be explained by statistical inconsistency, particularly since Brazil reported much higher levels of steel use in construction in 1970.

High and rising shares of steel consumption in construction have already been observed in many highly developed countries. This development could be connected with the increasing use of steel for high-rise building construction and for new infrastructure projects. Taking into account the predominant use of long products in construction, the fall of the share of long products in the total product-mix, which up to now has been one of the characteristics of technical progress in the steel industry, should be carefully examined in future in some regions and countries.

This development should create a very favourable environment for the further development of mini-mills, a sector mainly specializing in the production of long products. This observation is particularly important for the former CMEA countries which generally have an outmoded infrastructure and do not have mini-mills, but do have abundant resources of steel scrap.

2.2.3. Steel intensity by industrial sectors

Figures 2-19-1 to 2-19-4 present developments in the sectoral steel intensities of selected countries in the last decade. It is important to note that the data are comparable among countries since they are all calculated in the same units (grams per constant 1980 United States dollar). They are set out in ascending order from the lowest in figure 2-19-1 (steel intensity of total GDP: range of 20-140 grams per United States dollar) to the highest in figure 2-19-4 (steel intensity of construction: range of 50-700 grams per United States dollar).

The countries in figure 2-19-1 are divided into two groups: those with high steel intensities over 40 grams per United States dollar, which include the Republic of Korea, Japan, and the former Yugoslavia and those with low intensities under 40 grams per United States dollar, which include the EC, France, the Netherlands, the United Kingdom, Finland, Sweden and Brazil. Only the Republic of Korea which has the highest intensity of GDP, is still rising on the intensity curve. The other countries such as Japan, the EC, France and the United Kingdom show a stagnant trend, but some did experience a slight increase again in the late 1980s. Finland, Sweden, Hungary, Brazil and the former Yugoslavia have declining intensities which, however, are not particularly pronounced.

Figure 2-19-2 presents developments in the steel intensities of the sum of manufacturing and construction. They are naturally higher than for total GDP. The absence of a clearly declining trend in the steel intensity of GDP observed in many highly developed countries is confirmed by similar developments in the intensities of this sector. Possible explanations for this observation are the following:

- First, technological progress in both steel and steel-using industries which brought about savings in steel consumption has reached certain limits and has stopped.
- Secondly, the decline of the steel intensity of some sectors has been compensated by increases in other sectors.

Further analysis confirms that the second factor is more common, while the first has also occurred in some countries.

Figures 2-19-3 and 2-19-4 show the development of steel intensities in manufacturing and construction, respectively. Construction is by far the most intensive steel consumer in the economy in many countries. This is the case of the Republic of Korea, Japan, some countries in the EC and in the former Yugoslavia in the most recent period. The steel intensity of construction was 350 grams per United States dollar in Japan, 5 times more than the steel intensity of manufacturing, and 700 grams per United States dollar in the Republic of Korea, 10 times more than manufacturing.

Manufacturing is the most intensive steel consumer only in very few countries. The steel intensity of manufacturing is rising in the Republic of Korea, while it is declining in the United Kingdom, Finland, Sweden and Japan, and is stagnant in France and the EC as a whole.

On the other hand, it is very important to be aware that no country shows a declining trend in the steel intensity of construction. It is clearly increasing in Japan and the Republic of Korea, while it is stagnant in some cases like the former Yugoslavia and Brazil. This implies that in general the decline of the steel intensity of manufacturing was compensated for by an increase in construction.

Figure 2-19-1 Steel intensity of industrial sectors in selected countries: ***Total GDP***

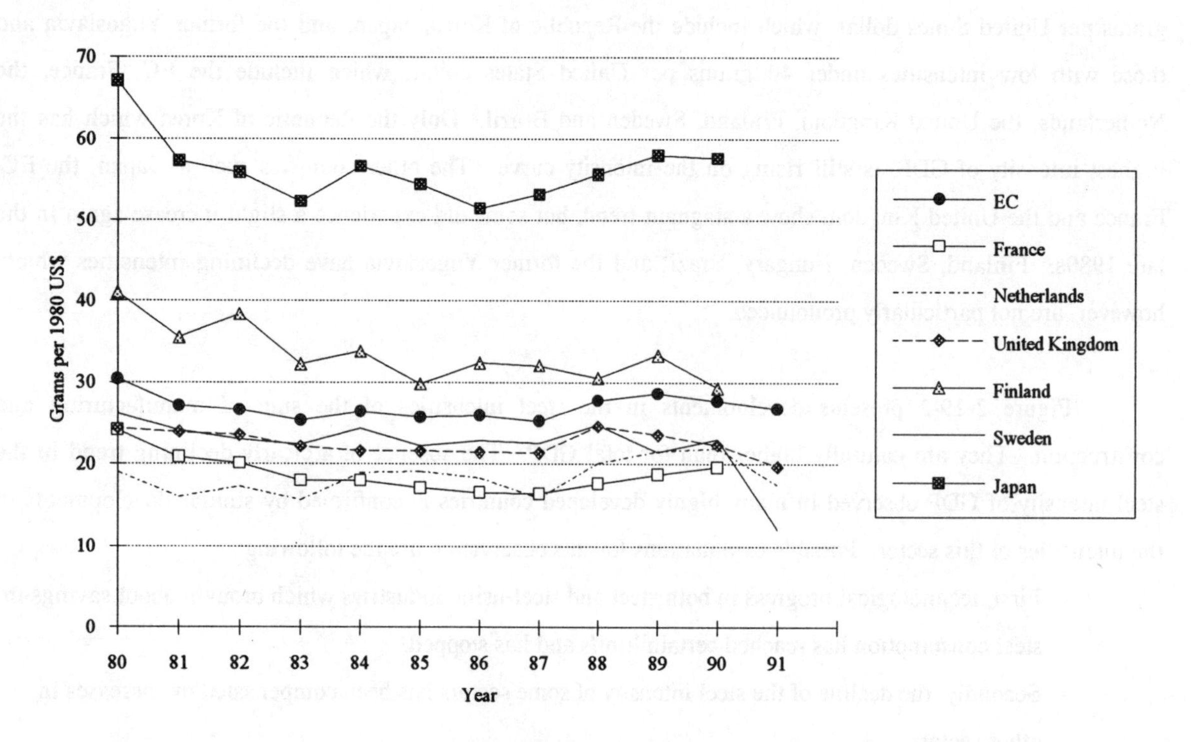

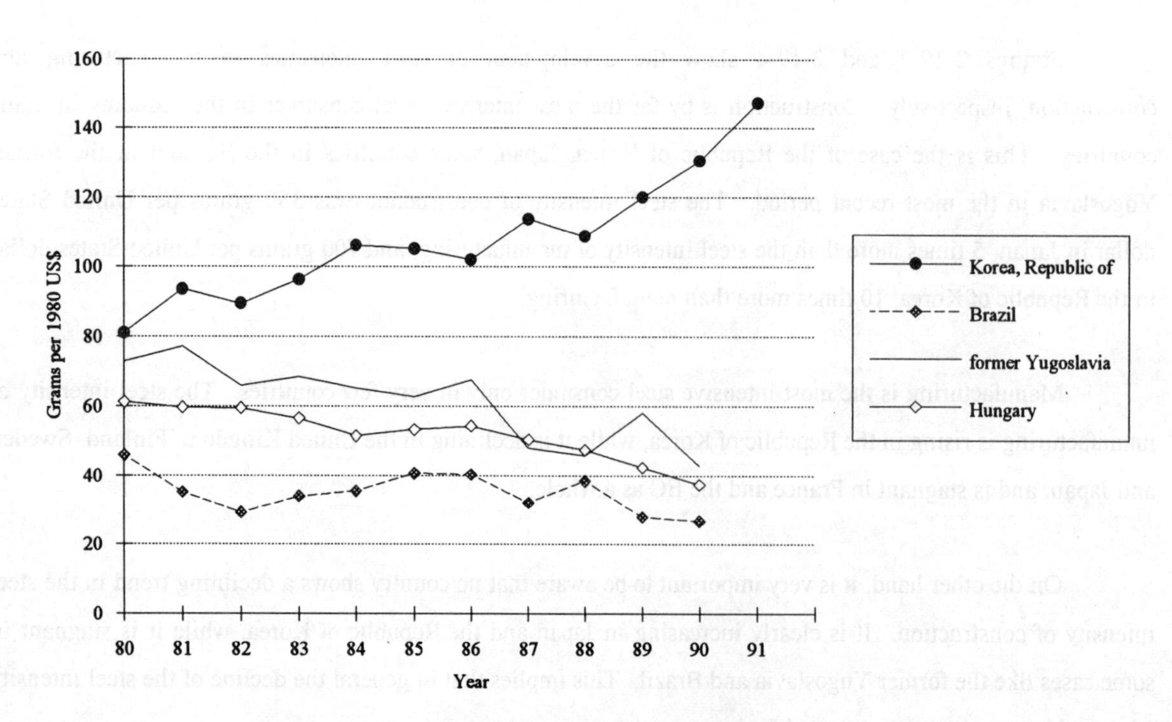

Figure 2-19-2 Steel intensity of industrial sectors in selected countries: ***Manufacturing plus construction***

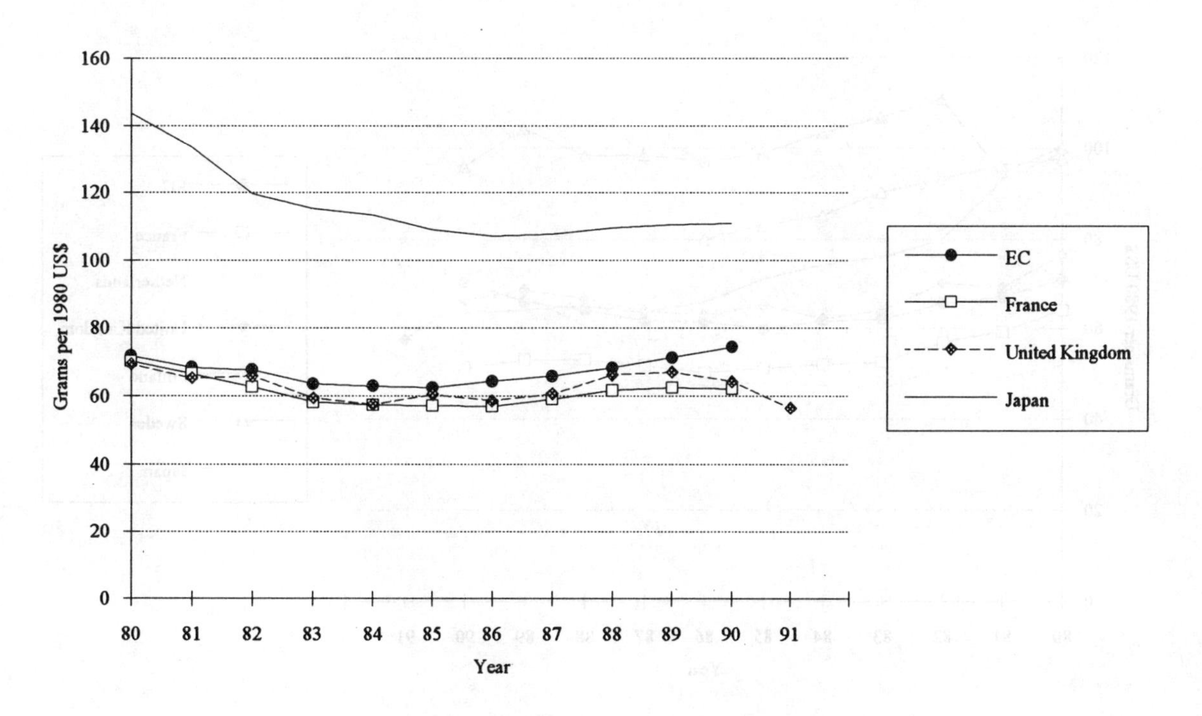

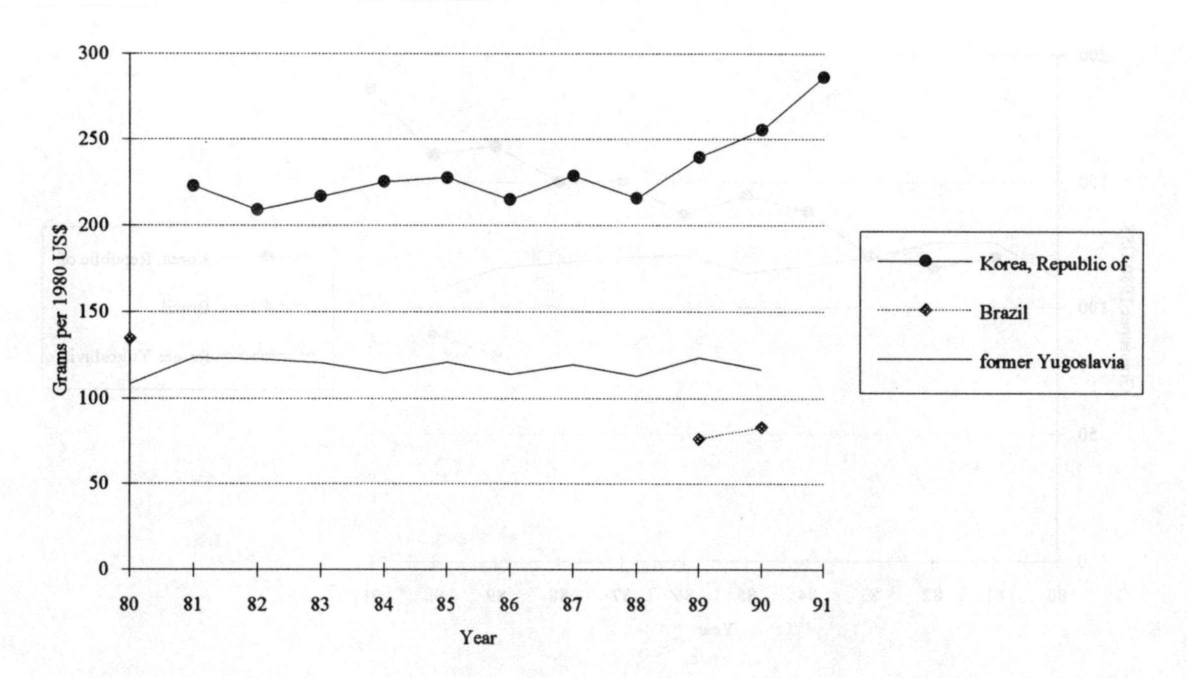

Figure 2-19-3 Steel intensity of industrial sectors in selected countries: *Manufacturing*

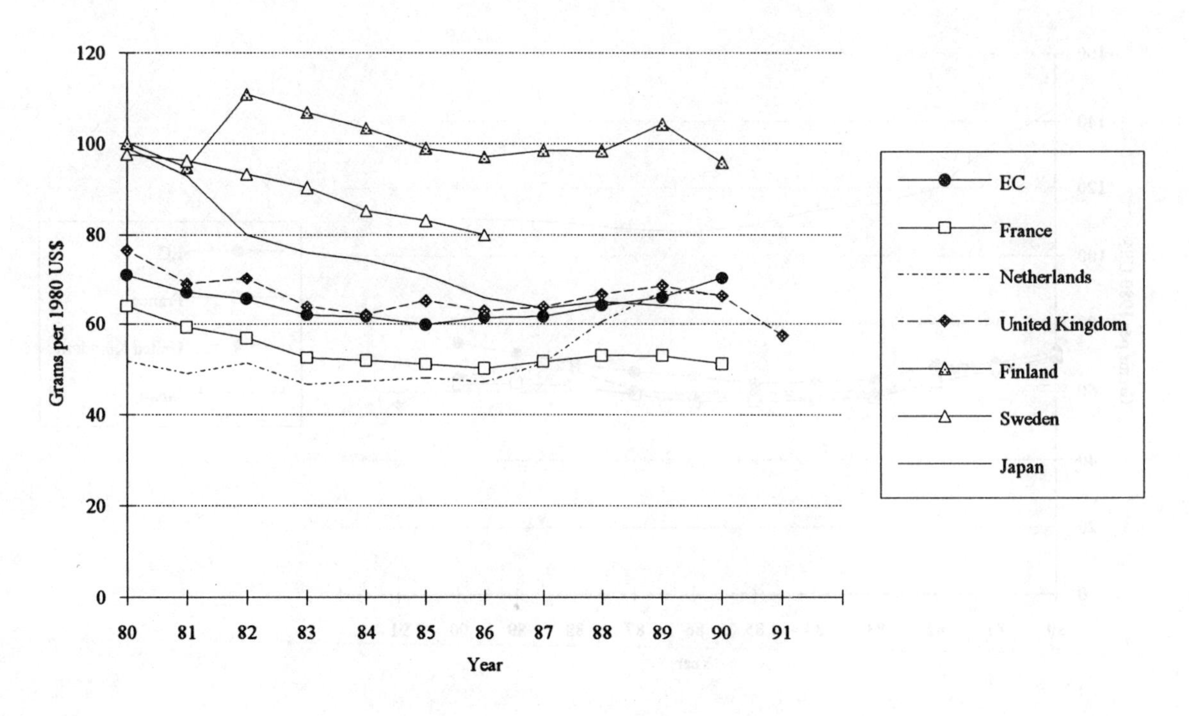

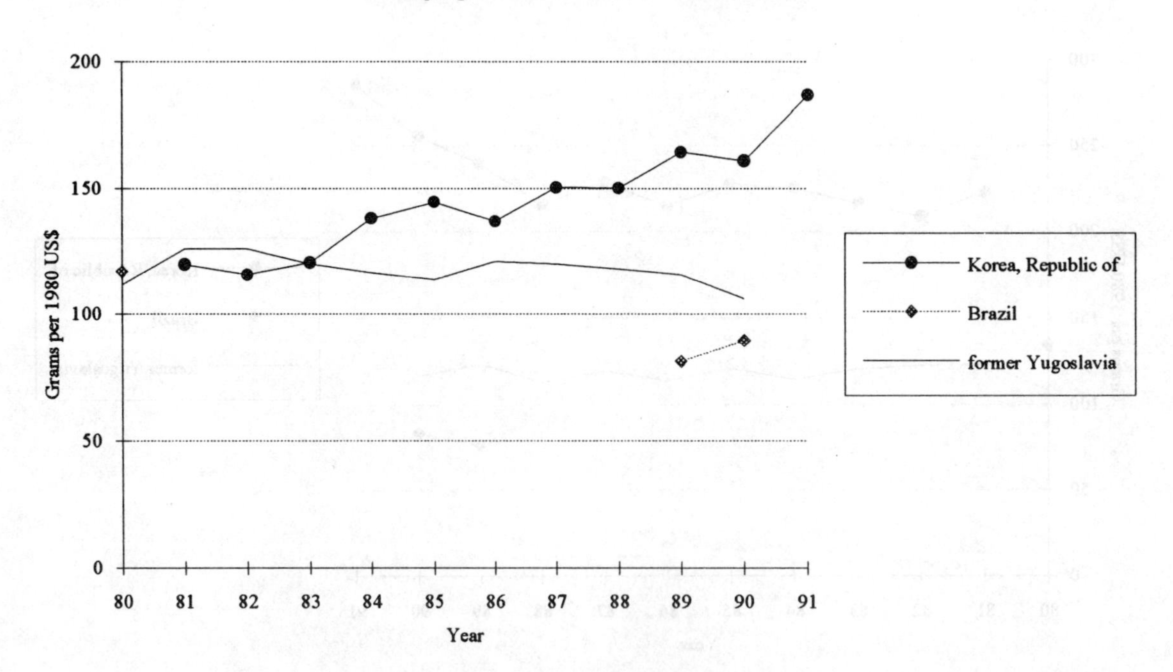

Figure 2-19-4 Steel intensity of industrial sectors in selected countries: ***Construction***

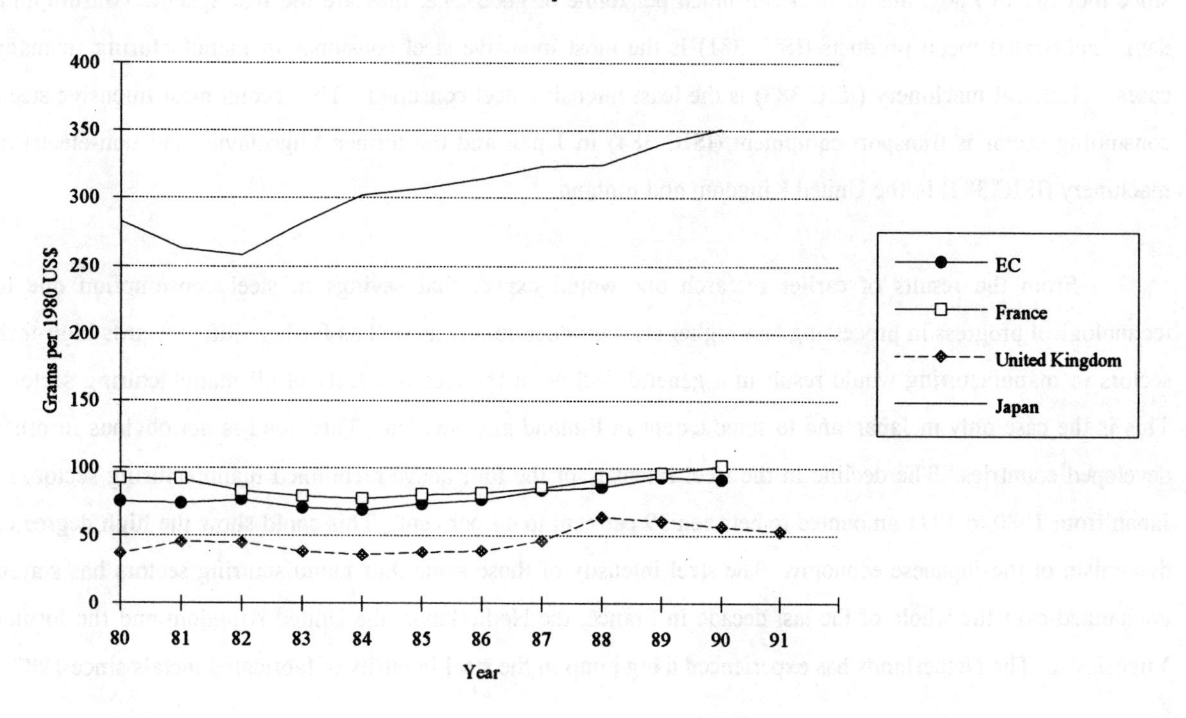

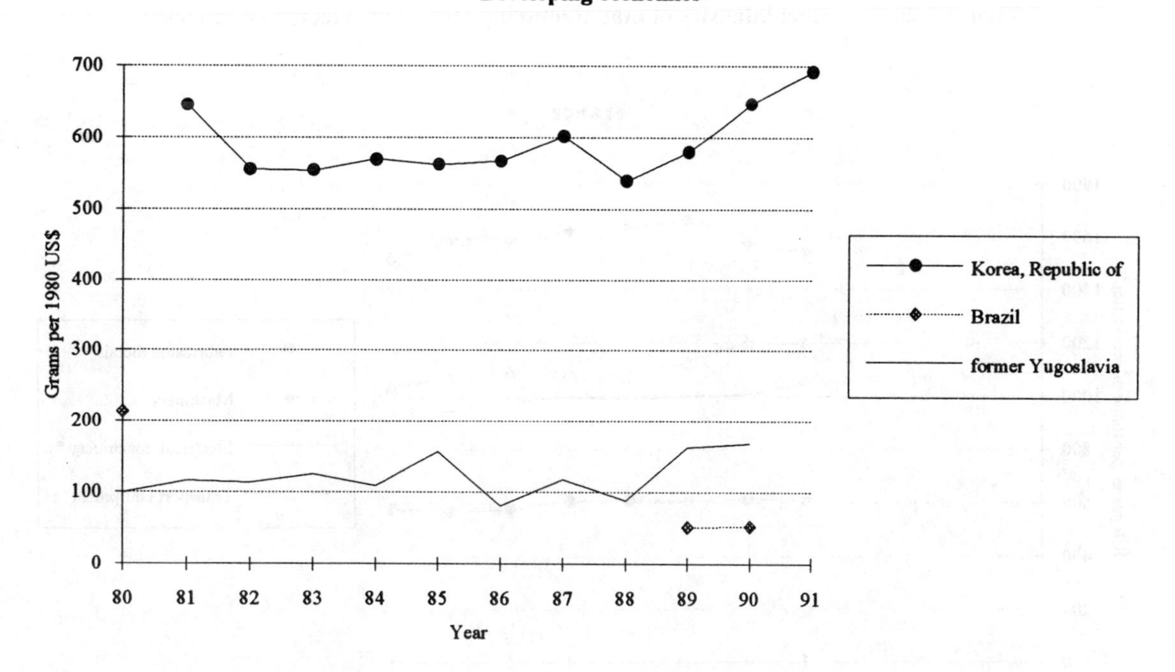

2.2.4. Steel intensity by manufacturing sectors

Several countries supplied data for the detailed analysis of the intensities of four manufacturing sub-sectors (ISIC 381 - 384). They are presented in Figure 2-20. The data for the United Kingdom are the best, since they are in kilograms of steel consumed per tonne of goods, i.e. they are the real specific consumption data. Fabricated metal products (ISIC 381) is the most intensive steel consumer in manufacturing in many cases. Electrical machinery (ISIC 383) is the least intensive steel consumer. The second most intensive steel-consuming sector is transport equipment (ISIC 384) in Japan and the former Yugoslavia, and non-electrical machinery (ISIC 382) in the United Kingdom and Finland.

From the results of earlier research one would expect that savings in steel consumption due to technological progress in processing and higher steel product quality as well as further shifts towards high-tech sectors in manufacturing would result in a general decline of the steel intensity of all manufacturing sectors. This is the case only in Japan and to some extent in Finland and Sweden. This trend is not obvious in other developed countries. The decline in the steel intensity of the four above-mentioned manufacturing sectors in Japan from 1980 to 1991 amounted to between 39 per cent to 44 per cent. This could show the high degree of dynamism of the Japanese economy. The steel intensity of those same four manufacturing sectors has stayed unchanged over the whole of the last decade in France, the Netherlands, the United Kingdom and the former Yugoslavia. The Netherlands has experienced a big jump in the steel intensity of fabricated metals since 1987.

Figure 2-20 Steel intensity of manufacturing sectors in selected countries:

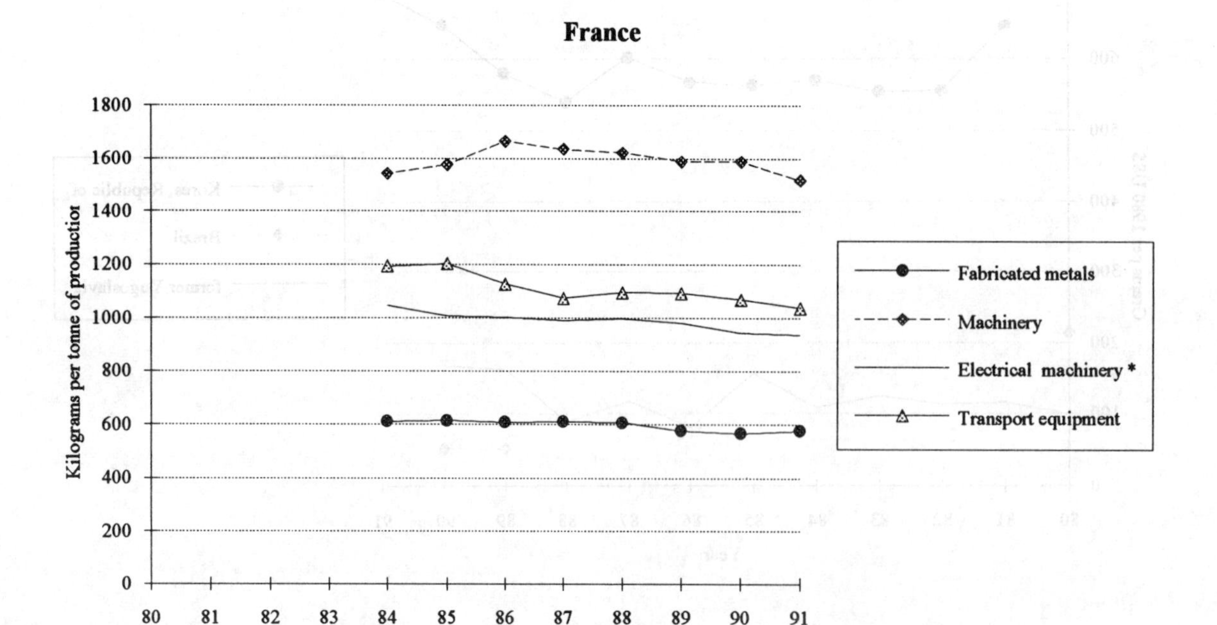

* Grams per hundred 1970 francs.

Figure 2-20 Steel intensity of manufacturing sectors in selected countries (continued):

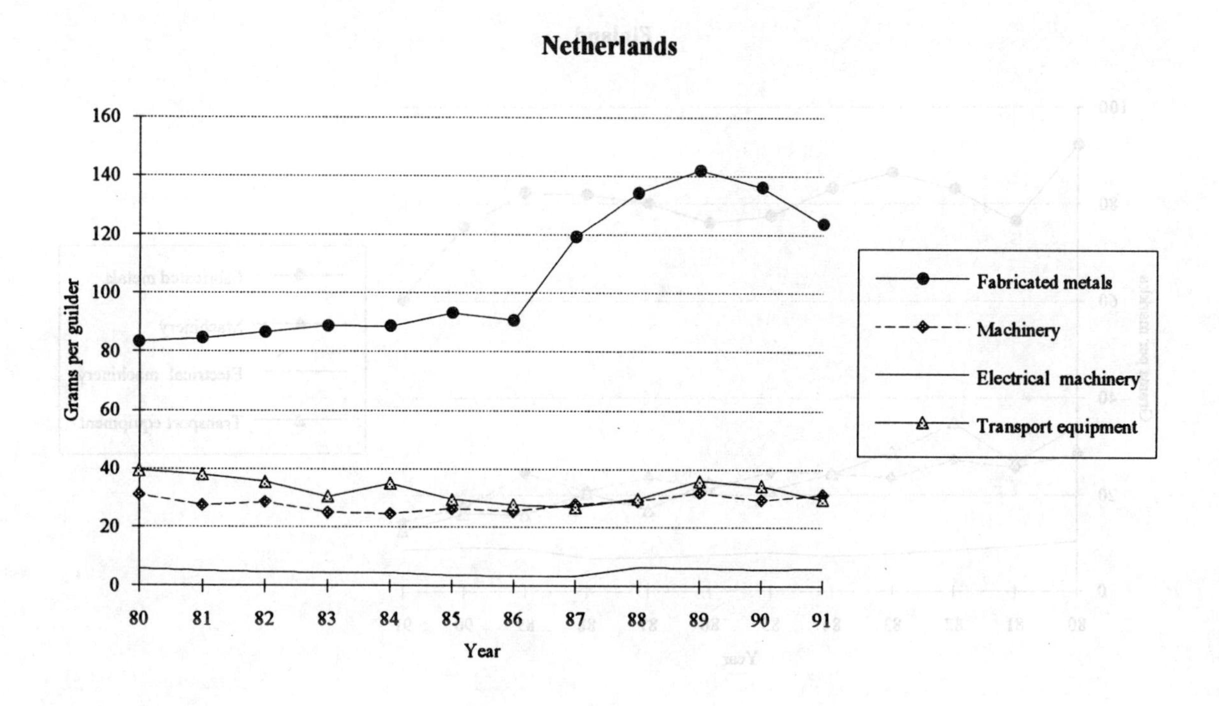

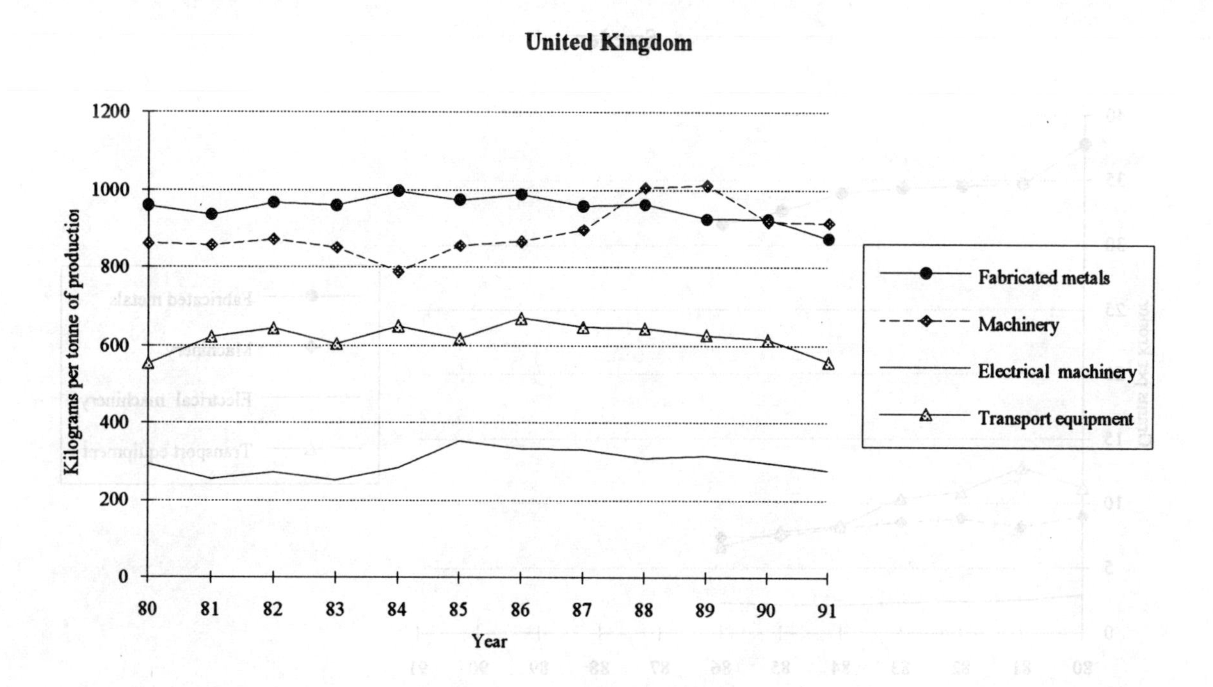

Figure 2-20 Steel intensity of manufacturing sectors in selected countries (continued):

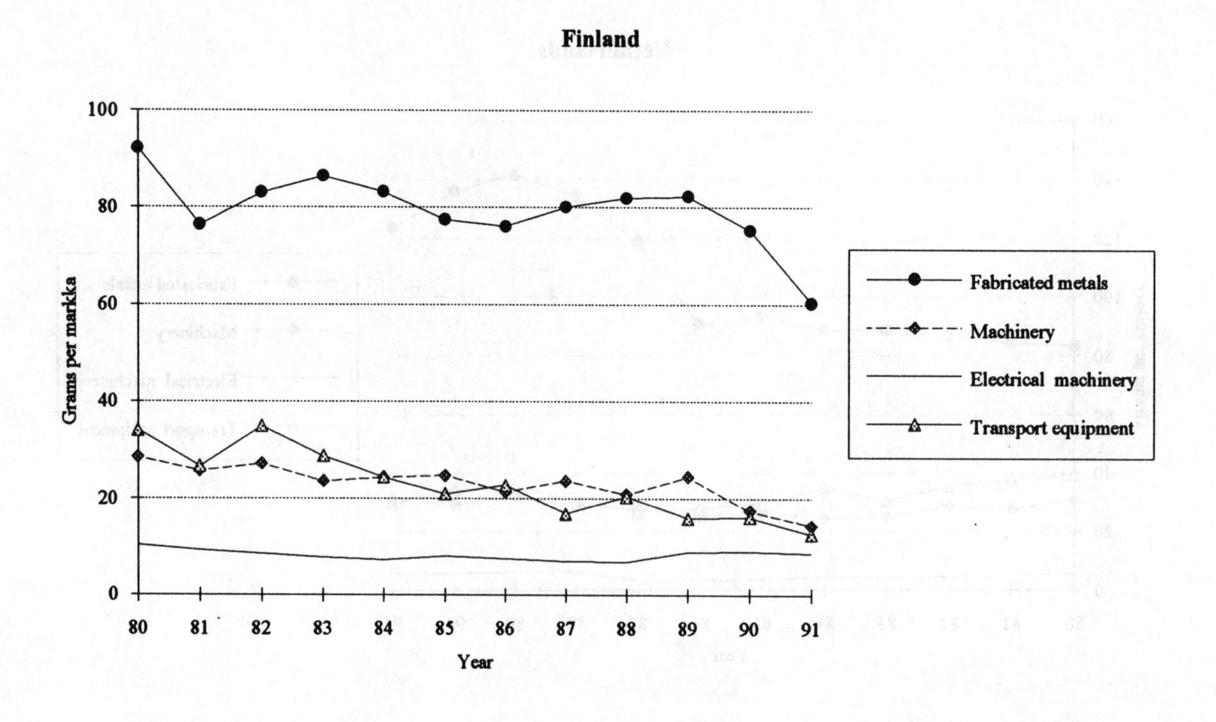

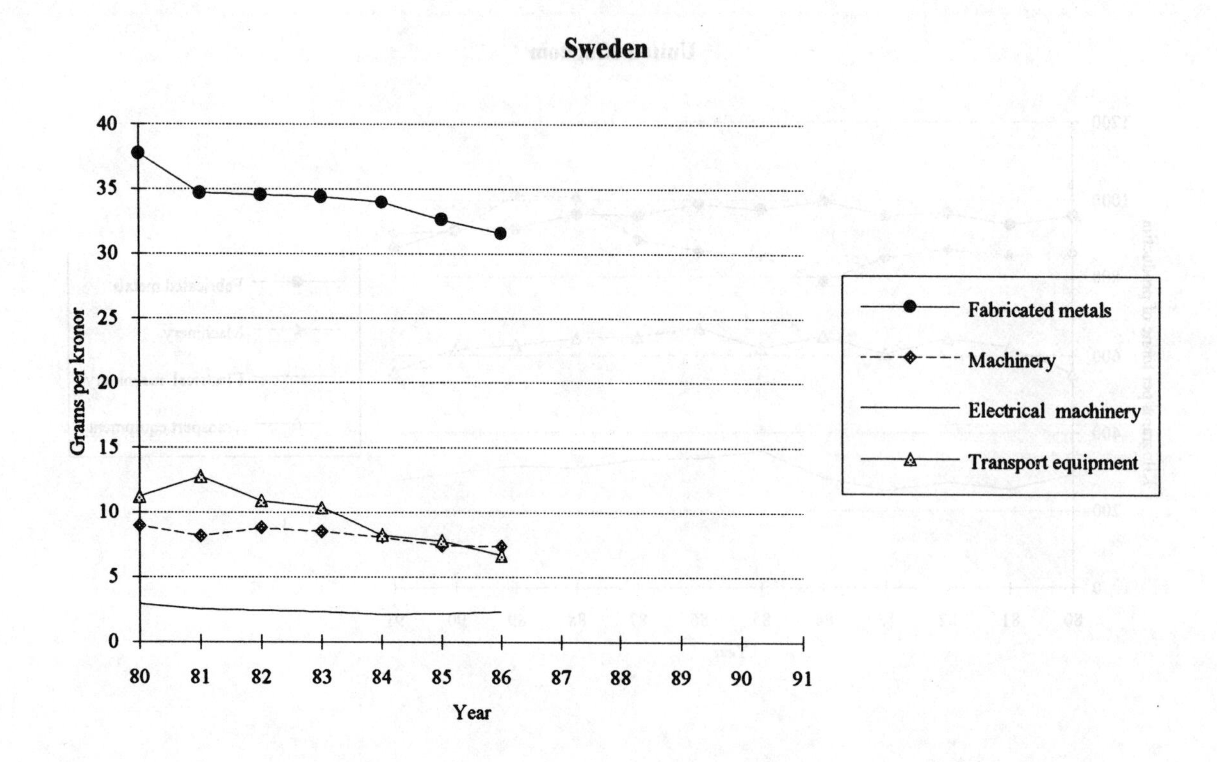

Figure 2-20 Steel intensity of manufacturing sectors in selected countries (continued):

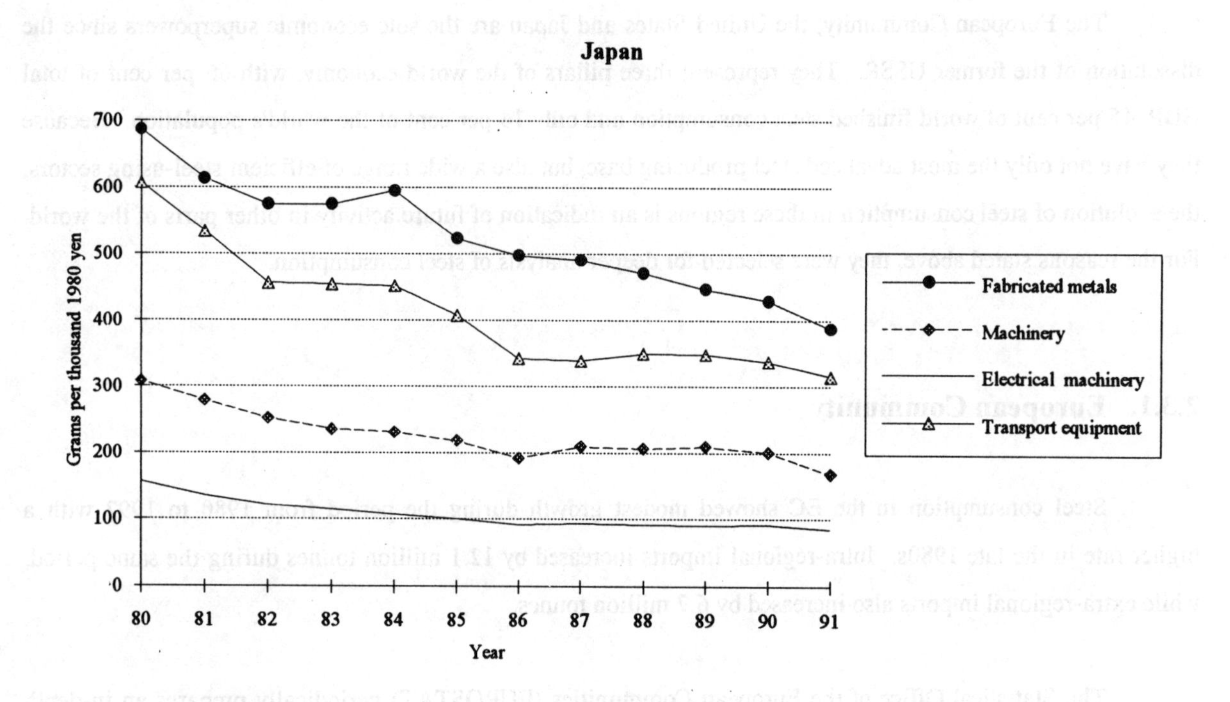

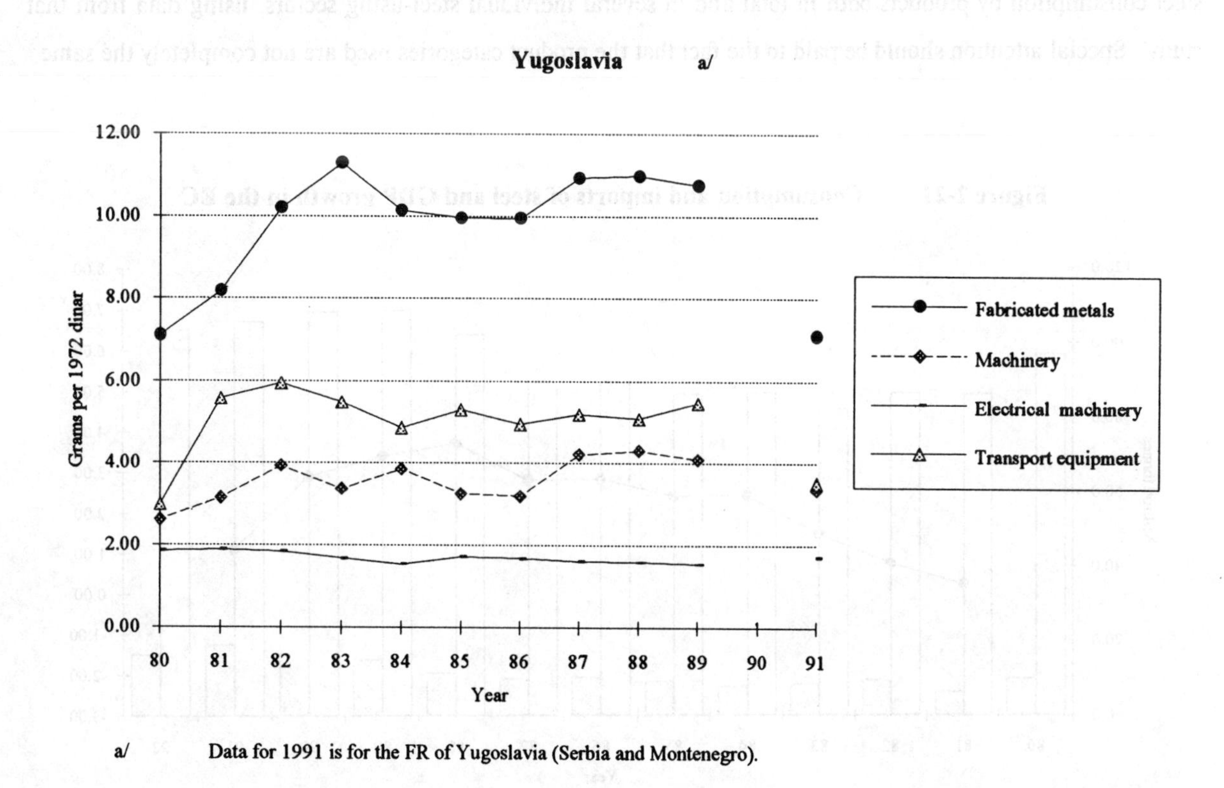

a/ Data for 1991 is for the FR of Yugoslavia (Serbia and Montenegro).

2.3. Changes in steel consumption in the European Community, the United States and Japan

The European Community, the United States and Japan are the sole economic superpowers since the dissolution of the former USSR. They represent three pillars of the world economy, with 66 per cent of total GDP, 45 per cent of world finished steel consumption and only 14 per cent of the world's population. Because they have not only the most advanced steel producing base, but also a wide range of efficient steel-using sectors, the evolution of steel consumption in these regions is an indication of future activity in other parts of the world. For the reasons stated above, they were selected for deeper analysis of steel consumption.

2.3.1. European Community

Steel consumption in the EC showed modest growth during the period from 1980 to 1992 with a higher rate in the late 1980s. Intra-regional imports increased by 12.1 million tonnes during the same period, while extra-regional imports also increased by 6.7 million tonnes.

The Statistical Office of the European Communities (EUROSTAT) periodically prepares an in-depth analysis of trends in steel consumption. The latest was published in 1993.[10] What follows are explanations for steel consumption by products both in total and in several individual steel-using sectors, using data from that study. Special attention should be paid to the fact that the product categories used are not completely the same

Figure 2-21 Consumption and imports of steel and GDP growth in the EC

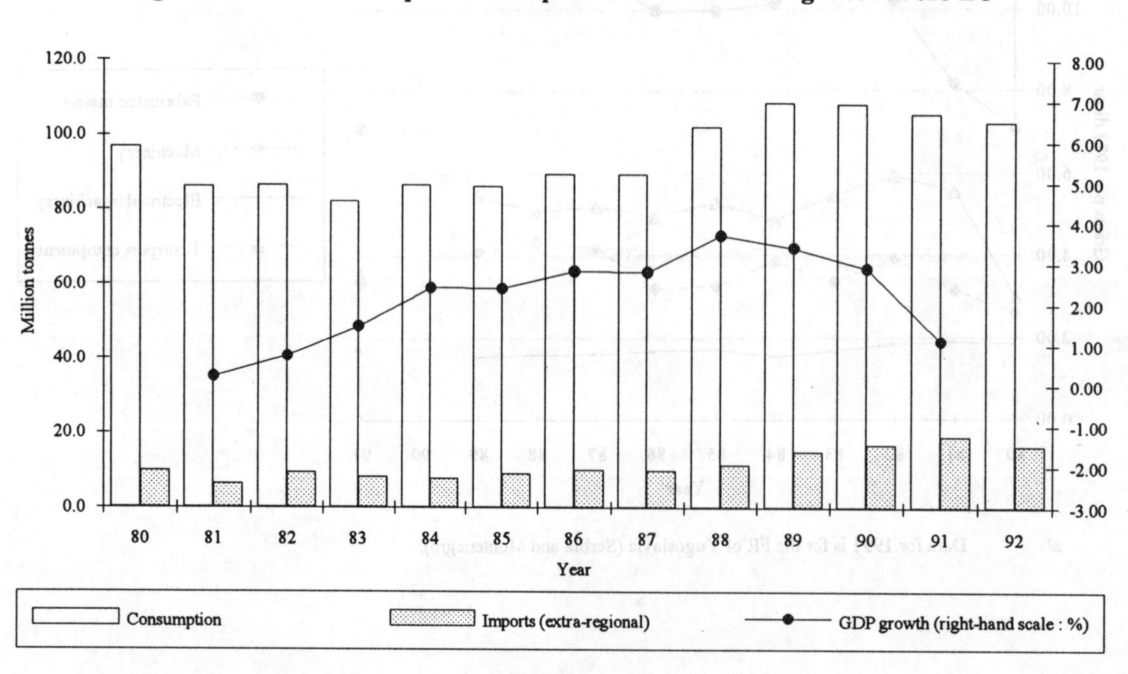

10/ *Steel Consumption by User Branch, 1970-1990*, Eurostat, 1993, ISBN 92-826-5584-9.

as those used for this study and that from 1985 membership in the EC increased from 10 to 12. Therefore, the coverage of the data is somewhat different.

Steel consumption by products

There has been a general increasing trend in the share of flat products, but a declining trend in both shares of long products and tubes and pipes. Merchant bars ranked as the most used material in 1990, representing 16.6 per cent of total consumption, followed by sheets, coated sheets and plates (Table 2-9).

During the period from 1980 to 1990, the most dramatic change in the structure of steel consumption has been the rapid increase in coated sheets. In 1985 the consumption of coated sheets was 9.3 million tonnes, or 11.5 per cent of total steel consumption, but in 1990 it reached 12.7 million tonnes, representing 12.8 per cent of total consumption.

Table 2-9 Steel consumption by products in the EC

Products	Amount (1000t)			Share (%)		
	1980	1985	1990	1980	1985	1990
Long products	25 722	23 714	28 764	29.9	29.4	28.9
Heavy sections	6 767	5 868	7 930	7.9	7.3	8.0
Merchant bars	15 486	14 009	16 496	18.0	17.4	16.6
Wire rods	3 469	3 837	4 338	4.0	4.8	4.4
Flat products	32 656	31 294	39 997	37.9	38.8	40.2
Narrow strip	2 108	2 100	2 068	2.4	2.6	2.1
Plate	10 017	8 562	11 185	11.6	10.6	11.3
Sheets	13 145	11 326	14 026	15.3	14.0	14.1
Coated sheets	7 386	9 306	12 718	8.6	11.5	12.8
Tubes and pipes	10 033	9 189	10 380	11.7	11.4	10.4
Others	17 700	16 545	20 250	20.6	20.5	20.4
Total	86 111	80 742	99 391	100.0	100.0	100.0

Consumption in fabricated metals

Fabricated metals was the largest steel-using sector with a consumption of 33.6 million tonnes of steel in 1990, or 33.8 per cent of the total (Table 2-10-1). The main products used by the branch include coated sheets, heavy sections, sheets and plates. While the consumption of long products increased by 2.0 million tonnes from 1985 to 1990, the consumption of flat products increased more by 3.6 million tonnes, resulting in an increase in the share of flat products to 48.0 per cent.

There was a general increase in the consumption of coated sheets among most sub-sectors of the fabricated metals sector such as structural steelwork, metal goods and drums, boilers and other vessels. There was only one sub-sector with decreased coated sheet consumption. Interestingly, this was cans and boxes which is the largest consumer of coated sheets (28.5 per cent of total coated sheet consumption in 1990) and uses coated sheets almost exclusively (97.5 per cent in 1990). Actually, steel consumption in this sub-sector declined from 1985 to 1992, while other sectors showed a steady growth in overall steel consumption. This may suggest that material substitution and other savings in steel consumption are being particularly pursued in the cans and boxes industry.

Table 2-10-1 Consumption by products in steel using sectors in the EC: *Fabricated metals (ISIC 381)*

Products	Amount (1000t)			Share (%)		
	1980	1985	1990	1980	1985	1990
Long products	7 949	7 695	9 646	29.6	29.3	28.7
Heavy sections	3 143	3 381	4 909	11.7	12.9	14.6
Merchant bars	2 381	2 061	2 103	8.9	7.8	6.3
Wire rods	2 425	2 253	2 634	9.0	8.6	7.8
Flat products	12 333	12 515	16 141	45.9	47.6	48.0
Narrow strip	486	336	475	1.8	1.3	1.4
Plate	3 631	3 381	4 422	13.5	12.9	13.1
Sheets	3 255	3 003	4 684	12.1	11.4	13.9
Coated sheets	4 961	5 795	6 560	18.5	22.0	19.5
Tubes and pipes	2 221	2 199	2 652	8.3	8.4	7.9
Others	4 373	3 887	5 193	16.3	14.8	15.4
Total	26 876	26 296	33 632	100.0	100.0	100.0

Consumption in machinery except electrical

The machinery industry except electrical consumed 13.3 million tonnes of steel in 1990, or 13.3 per cent of the total (Table 2-10-2). Consumption in this sector increased by 4.0 per cent annually from 1985 to 1990, but the growth pattern by products was almost the same and the shares of products did not change significantly. The main products used by the branch included plate, tubes and pipes, merchant bars and sheets.

Table 2-10-2 Consumption by products in steel using sectors in the EC: *Machinery except electrical (ISIC 382)*

Products	Amount (1000t)			Share (%)		
	1980	1985	1990	1980	1985	1990
Long products	2 501	2 128	2 774	20.6	19.6	20.9
Heavy sections	649	475	651	5.3	4.4	4.9
Merchant bars	1 722	1 533	1 949	14.2	14.1	14.7
Wire rods	130	120	174	1.1	1.1	1.3
Flat products	4 306	3 782	4 816	35.5	34.8	36.3
Narrow strip	202	213	274	1.7	2.0	2.1
Plate	2 487	2 136	2 483	20.5	19.6	18.7
Sheets	1 123	1 036	1 452	9.3	9.5	11.0
Coated sheets	494	397	607	4.1	3.7	4.6
Tubes and pipes	1 726	1 602	1 961	14.2	14.7	14.8
Others	3 601	3 362	3 706	29.7	30.9	28.0
Total	12 134	10 874	13 257	100.0	100.0	100.0

Consumption in electrical machinery

The electrical machinery industry consumed 4.7 million tonnes of steel in 1990, or 4.7 per cent of the total (Table 2-10-3). The main products used by the branch were sheets and coated sheets. Fifty-five per cent of the increase in consumption from 1985 to 1990 was accounted for by coated sheets. This is a reflection of a general increase in added-value brought about by more high-tech products in this sector.

Table 2-10-3 Consumption by products in steel using sectors in the EC: ***Electrical machinery (ISIC 383)***

Products	Amount (1000t)			Share (%)		
	1980	1985	1990	1980	1985	1990
Long products	240	220	214	6.0	5.8	4.6
Heavy sections	23	19	21	0.6	0.5	0.4
Merchant bars	181	171	186	4.6	4.5	4.0
Wire rods	36	30	7	0.9	0.8	0.1
Flat products	2 377	2 230	2 908	59.9	58.3	62.1
Narrow strip	98	77	71	2.5	2.0	1.5
Plate	220	178	225	5.5	4.7	4.8
Sheets	1 662	1 450	1 613	41.9	37.9	34.5
Coated sheets	397	525	999	10.0	13.7	21.3
Tubes and pipes	117	97	94	2.9	2.5	2.0
Others	1 233	1 275	1 466	31.1	33.4	31.3
Total	3 967	3 822	4 682	100.0	100.0	100.0

Consumption in automobiles and other vehicles

The automobile and other vehicles manufacturing sector was the third largest steel consuming branch with steel consumption of 17.5 million tonnes in 1990, representing 17.6 per cent of total consumption (Table 2-10-4). The main products used by the branch were sheets and coated sheets. Fifty-five per cent of the increase in consumption from 1985 to 1990 was accounted for by coated sheets as well as steel casting and forging. The most notable change in the sector was the rapid penetration of coated sheets into vehicle manufacturing, replacing sheets. While the consumption volume of sheets did not change at all, coated sheet consumption increased by 1.2 million tonnes from 1985 to 1990. As a result, the share of coated sheets in the total increased to 14.6 per cent in 1990. However, automobiles and other vehicles manufacturing is still the largest user of steel sheets among the industrial sectors, consuming 35 per cent of that product in 1990. The use of coated sheets has made great inroads into vehicle manufacturing. This branch was the second largest consumer of coated sheets in 1990 with a share of 20.0 per cent, only surpassed by cans and boxes which had a share of 28.5 per cent.

Consumption in construction

The construction sector used 18.7 million tonnes of steel in 1990, representing 18.8 per cent of total steel consumption. Steel consumption in this branch has been increasing constantly except for a decrease in

Table 2-10-4 Consumption by products in steel using sectors in the EC:
Autos and other vehicles (ISIC 384b)

Products	Amount (1000t)			Share (%)		
	1980	1985	1990	1980	1985	1990
Long products	1 511	1 263	1 548	10.1	8.7	8.8
Heavy sections	270	133	122	1.8	0.9	0.7
Merchant bars	1 048	928	1 280	7.0	6.4	7.3
Wire rods	193	202	146	1.3	1.4	0.8
Flat products	8 196	8 159	9 820	54.7	56.3	56.0
Narrow strip	956	1 028	1 064	6.4	7.1	6.1
Plate	1 037	863	1 305	6.9	6.0	7.4
Sheets	5 634	4 869	4 896	37.6	33.6	27.9
Coated sheets	569	1 399	2 555	3.8	9.6	14.6
Tubes and pipes	1 259	1 230	1 370	8.4	8.5	7.8
Others	4 030	3 848	4 803	26.9	26.5	27.4
Total	14 996	14 500	17 541	100.0	100.0	100.0

1990. EUROSTAT attributes the increase to the strong rise in specific steel consumption in this sector in spite of the decline in the production of the building industry.

The most important steel product used in construction is merchant bars, representing 53.4 per cent of the total steel consumed in the branch in 1990, though the use of wire rods and coated sheets has been increasing (Table 2-10-5). The construction sector is the most important consumer of merchant bars, using more than 60 per cent of that product.

Table 2-10-5 Consumption by products in steel using sectors in the EC:
Construction (ISIC 500)

Products	Amount (1000t)			Share (%)		
	1980	1985	1990	1980	1985	1990
Long products	10 397	9 779	12 503	71.6	68.2	67.0
Heavy sections	1 203	828	1 268	8.3	5.8	6.8
Merchant bars	8 765	7 991	9 969	60.3	55.8	53.4
Wire rods	429	960	1 266	3.0	6.7	6.8
Flat products	932	1 221	1 881	6.4	8.5	10.1
Narrow strip	31	9	9	0.2	0.1	0.0
Plate	297	488	496	2.0	3.4	2.7
Sheets	197	147	336	1.4	1.0	1.8
Coated sheets	407	577	1 040	2.8	4.0	5.6
Tubes and pipes	2 166	2 057	1 991	14.9	14.4	10.7
Others	1 035	1 272	2 287	7.1	8.9	12.3
Total	14 530	14 329	18 662	100.0	100.0	100.0

2.3.2. The United States

Steel consumption in the United States experienced huge swings related to fluctuations in the GDP growth rate from 1980 to 1992 (Figure 2-22). The lowest level was 69.3 million tonnes in 1982, which was 18.8 per cent below the average figure for the 1980s.

The first half of the 1980s was also a period of growing imports in the United States market. This was mainly caused by the appreciation of the dollar during that period (Figure 2-23). After a number of antidumping and countervailing cases filed by the United States steel industry, the Voluntary Restraint Agreement (VRA) was introduced in 1982 between the United States and the EC to limit imports from the EC as a replacement for the Trigger Price Mechanism (TPM), a system based on the import reference price.[11/] VRA was expanded to 19 countries and the EC in 1984 and was in effect until the end of March 1992.

Even though VRA succeeded in limiting import volumes of steel products from certain countries until 1986, because of the lower price competitiveness caused by the subsequent depreciation of the dollar, most exporting countries under VRA could not fill their limits in the latter half of the 1980s.[12/] Furthermore, globalization in the industry also appears to have diminished trade. As the steel industry globalized during the 1980s, in many cases foreign direct investment became an alternative to trade. For example, as steelmakers from the world's leading steel exporter, Japan, invested in the United States steel industry after 1984, they began to increasingly use steel produced in the United States as opposed to steel produced in Japan to meet the needs of their customers located in the United States.[13/] At the same time, VRA is also seen to have contributed to broadening the composition of import sources, making room for non-VRA countries to penetrate the United States steel market.[14/]

There were no significant changes in the structure of consumption in the period from 1980 to 1992 (Table 2-11). However, the share of flat products in total shipments increased steadily, while that of pipes and tubes declined (Table 2-12). The decline in tubes and pipes was due to stagnant oil prices after the OCTG boom in 1980 and 1981.

Another change was increasing demand for coated sheets. Net shipments of coated sheets increased from 6.0 million tonnes in 1980 to 11.9 million tonnes in 1992. Increased foreign direct investment in the United States in such major flat product and coated sheet users as the automobile industry might have had some effect by both increasing and diversifying demand for high-end flat products. For example, automobile production by Japanese manufacturers - so called transplants - increased to 1.3 million units in 1990 from

---

11/ *The Western U.S. Steel Market: Analysis of Market Conditions and Assessment of the Effects of Voluntary Restraint Agreements on Steel Producing and Steel Consuming Industries*, USITC, March 1989, USITC publication 2165.

12/ *Steel Industry Annual Report on Competitive Conditions in the Steel Industry and Industry Efforts to Adjust and Modernize*, USITC, September 1991, USITC publication 2436.

13/ Ibid., pp. 2-25.

14/ *The Western U.S. Steel Market: Analysis of Market Conditions and Assessment of the Effects of Voluntary Restraint Agreements on Steel Producing and Steel Consuming Industries*, pp. 5-3, loc. cit.

almost nothing in 1980.[15/] The trend towards higher value-added steel in the flat product market is obvious in the United States.

Figure 2-22 Consumption and imports of steel and GDP growth in the United States

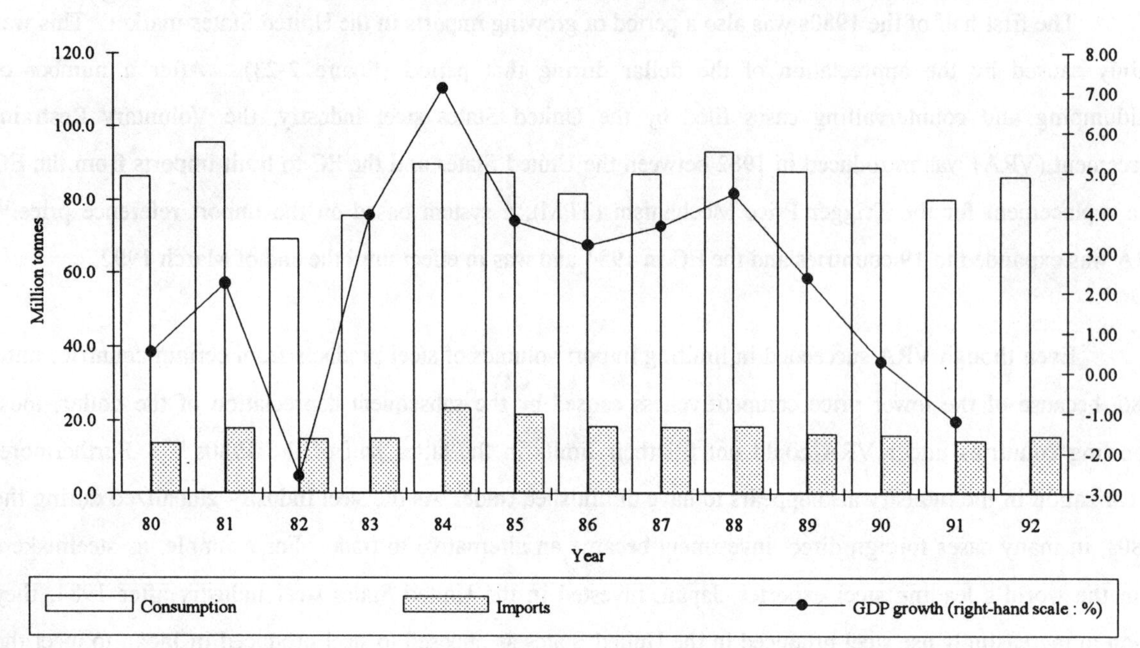

Table 2-11 Net shipments of steel products by market in the United States

Products	Amount (1000t)				Share (%)			
	1980	1985	1990	1992	1980	1985	1990	1992
Converting and service centres	21 812	24 017	31 883	31 845	26.0	32.9	37.5	38.7
Containers, etc.	5 551	4 089	4 474	3 974	6.6	5.6	5.3	4.8
Machinery except electrical	4 543	2 271	2 388	1 951	5.4	3.1	2.8	2.4
Electrical machinery	2 441	1 869	2 453	2 136	2.9	2.6	2.9	2.6
Automobile	12 124	12 950	11 100	11 092	14.5	17.7	13.1	13.5
Construction	11 890	11 230	12 115	12 230	14.2	15.4	14.3	14.9
Others	25 492	16 617	20 568	19 013	30.4	22.7	24.2	23.1
Total	83 853	73 043	84 981	82 241	100.0	100.0	100.0	100.0

Source: Annual Statistical Report, American Iron and Steel Institute.

15/ *Investment Behaviour for Plant and Equipment of the Japanese Transplants in the United States*, pp. 41, Nobuhisa Iwase, unpublished working paper at the Brooking Institution, 1992.

Figure 2-23 Steel imports and exchange rate in the United States

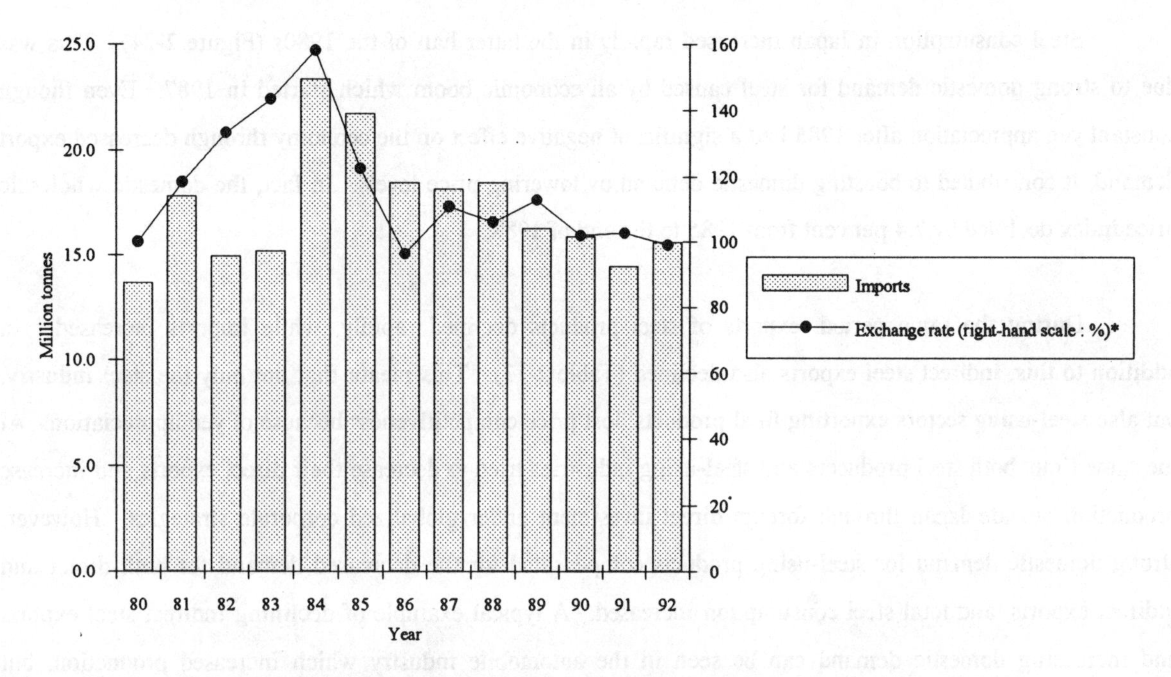

* Figures are an index of the multilateral trade-weighted value of the US dollar, whose value for 1980 is 100.
Sources: UN/ECE and *Economic Report of the President*, United States Government Printing Office.

Table 2-12 Net shipments by products in the United States

Products	Amount (1000t)				Share (%)			
	1980	1985	1990	1992	1980	1985	1990	1992
Ingots and semis	2 630	1 374	1 987	2 470	3.1	1.9	2.3	3.0
Long products	22 872	21 195	25 605	23 772	27.3	29.0	30.1	28.9
Rails and accessories	1 797	926	518	562	2.1	1.3	0.6	0.7
Steel pilings	346	326	423	454	0.4	0.4	0.5	0.6
Shapes	4 861	4 373	5 670	5 081	5.8	6.0	6.7	6.2
Bars	13 180	12 608	14 668	13 164	15.7	17.3	17.3	16.0
Wire rods	2 688	2 962	4 326	4 511	3.2	4.1	5.1	5.5
Flat products	45 841	43 808	50 502	49 625	54.7	60.0	59.4	60.3
Heavy and medium plate	8 080	4 327	7 945	7 102	9.6	5.9	9.3	8.6
Hot-rolled sheets and coils	12 785	13 538	14 080	13 911	15.2	18.5	16.6	16.9
Cold-rolled sheets and coils	14 243	14 250	14 015	13 524	17.0	19.5	16.5	16.4
Electric sheets	601	413	486	436	0.7	0.6	0.6	0.5
Tinplate	4 166	2 611	2 773	2 715	5.0	3.6	3.3	3.3
Coated sheets	5 966	8 669	11 203	11 937	7.1	11.9	13.2	14.5
Pipes and tubes	9 096	4 096	4 652	4 198	10.8	5.6	5.5	5.1
Others	3 414	2 570	2 235	2 176	4.1	3.5	2.6	2.6
Total	83 853	73 043	84 981	82 241	100.0	100.0	100.0	100.0

Source: Annual Statistical Report, American Iron and Steel Institute.

2.3.3. Japan

Steel consumption in Japan increased rapidly in the latter half of the 1980s (Figure 2-24). This was due to strong domestic demand for steel caused by an economic boom which started in 1987. Even though constant yen appreciation after 1985 had a significant negative effect on the economy through decreased export demand, it contributed to boosting domestic demand by lowering price levels. In fact, the domestic wholesale price index declined by 7.4 per cent from 1985 to the end of 1988.

During the same period, exports of steel products declined rapidly, while imports increased. In addition to this, indirect steel exports also declined (Table 2-13). This meant that not only the steel industry, but also steel-using sectors exporting final products, lost price competitiveness because of yen appreciation. At the same time, both steel producers and steel-using industries tried to decrease their direct exports and increase production outside Japan through foreign direct investment under globalized corporate strategies. However, strong domestic demand for steel-using products compensated for the decreased demand for both direct and indirect exports, and total steel consumption increased. A typical example of declining indirect steel exports and increasing domestic demand can be seen in the automobile industry which increased production, but decreased exports dramatically (Figure 2-25). In the case of the automobile industry, the shift of production bases to foreign countries, particularly the United States, accelerated this trend.

While steel consumption rapidly increased for all GDP components except indirect exports from 1985 to 1990, the largest increases were recorded by two private investment sectors: private housing investment and private investment in plant and equipment. The increases were 4.5 million tonnes (a growth rate of 66.2 per cent over five years) and 12.5 million tonnes (a growth rate of 55.6 per cent), respectively. As a result, the share of these two sectors in total domestic demand increased to 13.8 per cent and 42.8 per cent, respectively.

On the other hand, because of the slowing down of the Japanese economy steel consumption in all but one component of GDP declined from 1990 to 1992. Only public fixed capital spending showed increases. This component is used as an economic booster through infrastructure development.

Given the situation that investment activities were the major economic engine, the consumption of long products increased by 12 million tonnes from 1985 to 1990, while flat products increased by only 6.8 million tonnes (Table 2-14). This is an interesting observation, because the general perception is that steel consumption shifts from long products to flat products when economies get mature.

Bars and flat bars recorded the largest increase of a surprising 92.2 per cent from 1985 to 1990, and shapes also increased by 61.4 per cent in the same period. The share of long products in domestic orders increased from 37.9 per cent in 1980 to 43.5 per cent in 1990 and then dropped slightly to 41.5 per cent in 1992.

Figure 2-24 Consumption and imports of steel and GDP growth in Japan

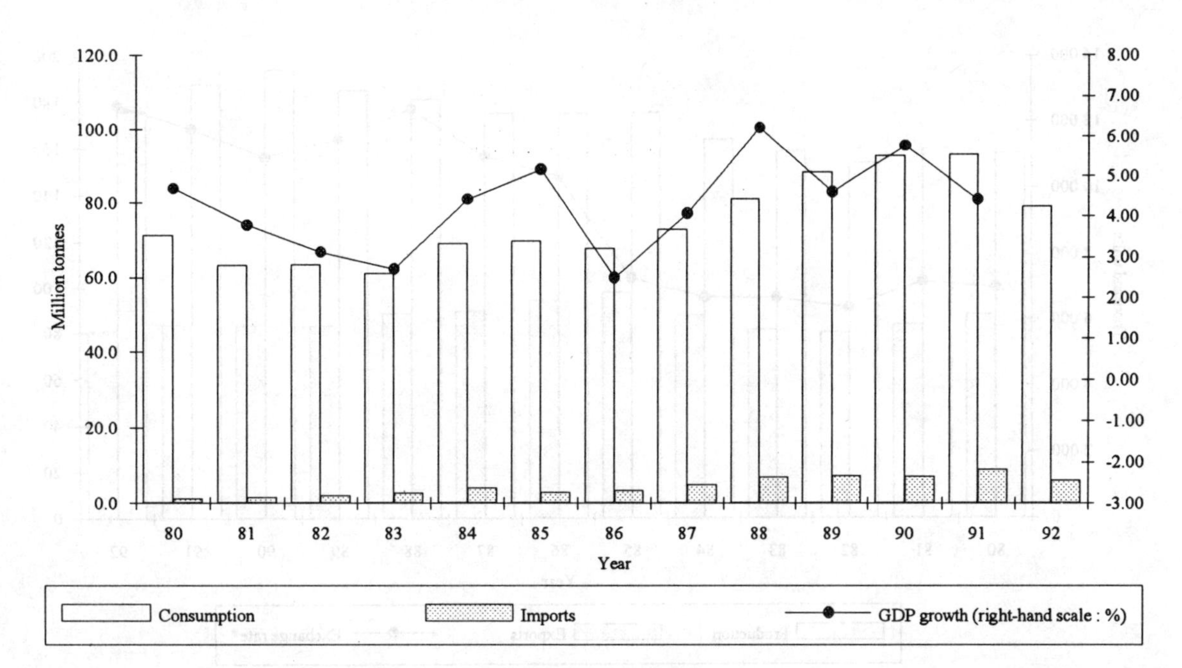

Table 2-13 Domestic consumption of steel products by components of GDP in Japan

GDP component	Amount (1000t)				Share (%)			
	1980	1985	1990	1992	1980	1985	1990	1992
Private consumption	4 448	4 823	7 146	6 465	7.4	8.1	8.8	9.0
Private housing investment	6 117	6 786	11 279	9 203	10.1	11.4	13.8	12.8
Private investment in plant and equipment	22 927	22 422	34 898	28 221	38.0	37.7	42.8	39.1
By manufacturing	7 495	8 272	12 432	8 879	12.4	13.9	15.3	12.3
By non-manufacturing	15 432	14 150	22 466	19 342	25.6	23.8	27.6	26.8
Public fixed capital spending	12 284	9 856	12 916	13 271	20.4	16.6	15.9	18.4
Indirect exports	14 495	15 586	15 232	15 000	24.0	26.2	18.7	20.8
Total	60 271	59 473	81 472	72 160	100.0	100.0	100.0	100.0

Source: *Handbook for Iron and Steel Statistics*, Japan Iron and Steel Federation.

Among flat products, the same trend occurred as in the EC and the United States, i.e. a shift from ordinary sheets to coated sheets. Coated sheets had a share of 17. 4 per cent in total domestic orders in 1992, up from only 8.6 per cent in 1980.

These observations confirm that steel intensity is declining in all manufacturing sectors in Japan, while in construction it is increasing, as seen in the previous section. Because of technological developments and the increasing need for higher value-added products, steel consumption in manufacturing is shifting to higher-end coated sheets with an accompanying declining trend in steel intensity.

Figure 2-25 Production and exports of automobiles and exchange rate in Japan

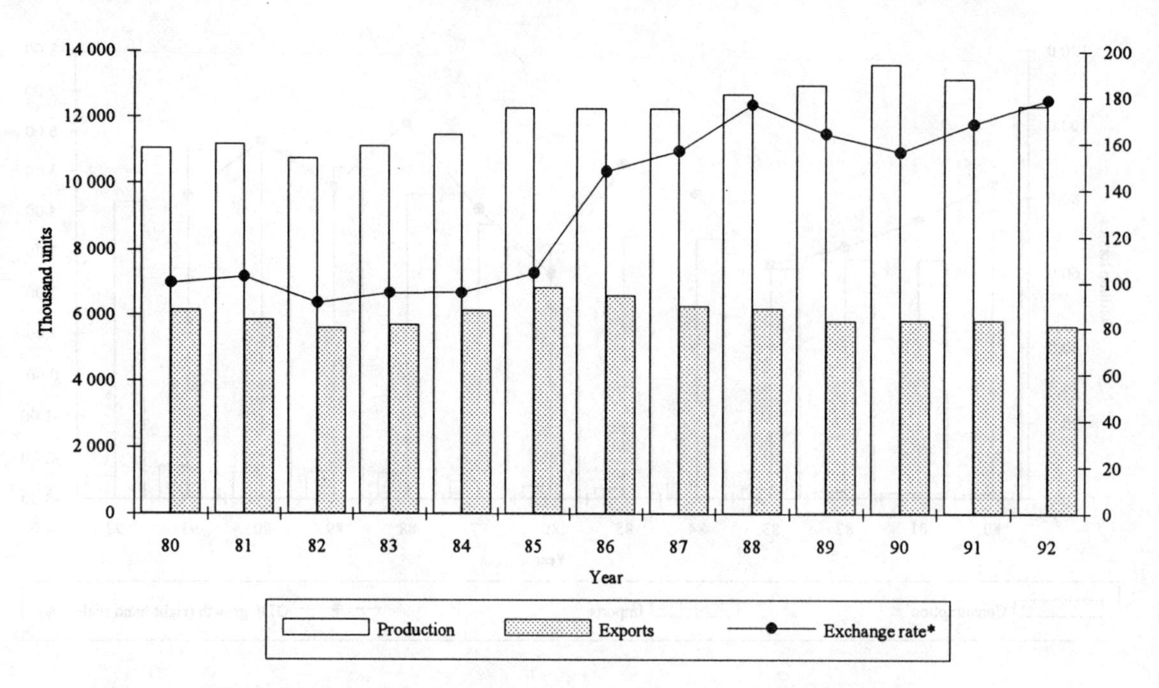

* Figures are an index of the exchange rate between the yen and the dollar, whose value for 1980 is 100.
Source: *Handbook for Iron and Steel Statistics*, Japan Iron and Steel Federation.

Table 2-14 Domestic orders for ordinary steel products in Japan

Products	Amount (1000t)				Share (%)			
	1980	1985	1990	1992	1980	1985	1990	1992
Long products	19 827	18 756	30 759	24 642	37.9	37.1	43.5	41.5
Rails and accessories	304	258	281	284	0.6	0.5	0.4	0.5
Sheet pilings	1 094	751	1 020	912	2.1	1.5	1.4	1.5
Shapes	6 852	7 218	11 652	9 583	13.1	14.3	16.5	16.1
Bars and flat bars	8 029	7 629	14 584	11 074	15.4	15.1	20.6	18.6
Wire rods	3 548	2 900	3 222	2 789	6.8	5.7	4.6	4.7
Flat products	27 614	27 525	34 337	29 866	52.8	54.4	48.5	50.3
Heavy and medium plate	9 138	6 825	8 045	6 898	17.5	13.5	11.4	11.6
Hot-rolled sheets and coils	5 392	5 894	7 093	6 595	10.3	11.6	10.0	11.1
Cold-rolled sheets and coils	7 082	6 105	5 655	4 367	13.5	12.1	8.0	7.4
Electric sheets	645	705	918	682	1.2	1.4	1.3	1.1
Tinplate	865	794	1 022	932	1.7	1.6	1.4	1.6
Zinc-coated sheets	1 075	1 280	1 705	1 513	2.1	2.5	2.4	2.5
Other coated sheets	3 417	5 922	9 899	8 879	6.5	11.7	14.0	14.9
Pipes and tubes	4 861	4 329	5 652	4 888	9.3	8.6	8.0	8.2
Total	52 302	50 610	70 748	59 396	100.0	100.0	100.0	100.0

Source: *Handbook for Iron and Steel Statistics*, Japan Iron and Steel Federation.

On the other hand, there is an emerging need for heavy and high-tensile steel products in construction in Japan for bridges and skyscrapers. Many big projects such as the super-long bridges which connect several main islands or both sides of Tokyo Bay are going on with the increased government spending on infrastructure. The Akashi Canal Bridge which is 3900 meters long used 100 thousand tonnes of steel. [16/] In addition, Japan is a country which has frequent big earthquakes, and this makes construction regulations stricter and tends to increase steel intensity.

However, it is worth pointing out that only recent technological developments in both the steel and construction industries have enabled the construction of super-long bridges and high-rise buildings which have strong resistance against winds and vibrations and can withstand earthquakes of a magnitude of more than 8.0. Technological innovation has created a new construction market and brought about an increasing need for steel, particularly long products.

16/ *Preparing the Future: Steel for Safer Living, Akashi Straits Suspension Bridge*, Manabu Ito, International Iron and Steel Institute, 1993, IISI/E/2710/0.

CHAPTER 3
STRUCTURAL CHANGES IN WORLD STEEL TRADE

3.1. General evolution of world trade in steel

3.1.1. General trends in world steel trade

World steel trade in terms of exports increased from 139.8 million tonnes in 1980 to 175.4 million tonnes in 1992, which represented an annual increase rate of 1.91 per cent. This rate was larger than the annual growth rate of steel consumption of 0.68 per cent during the same period, and resulted in an increase in the ratio of trade to steel consumption from 24.6 per cent in 1980 to 28.5 per cent in 1992 (Table 3-1 and Figure 3-1).

However, that increase in the ratio was not a linear, constant trend. In fact, after rising continually from 1980 to 1985, it experienced a constant decline to the same level as 1980 in 1989, and then began a rapid increase again. The export volume itself increased at quite a high pace from 1982 to 1985, but experienced sluggish growth from 1986 to 1989 with a drop in 1986. Because the period from 1986 to 1990 saw a boom in world steel consumption, the sluggish growth of trade decreased its ratio to steel consumption.

Two conclusions can be drawn from figure 3-1. First, there is a clear upward trend in trade in steel on a long-term basis, which means that trade is becoming more important than before in the structure of world steel consumption. Secondly, affected by some other short-term factors, world steel trade increased rapidly in the first half of the 1980s, but slowed down between 1985 and 1989. Two factors can be pointed out as major reasons for this short-term effect. One is that China increased its imports of steel products between 1983 and 1987 because of strong demand triggered by its economic boom during that period. Another is that the United States also increased its imports between 1984 and 1988, triggered by the dollar appreciation in the first half of the 1980s. Because these two countries have high import shares in total world trade, increased imports in these countries had a dramatic effect on world steel trade. In fact, the ratio of the combined import volume of these two countries to total steel consumption jumped from 3.3 per cent in 1980 to 7.0 per cent in 1985 (Figure 3-2). This trend is also seen in figure 3-1.

Table 3-1 World steel trade and ratio to total world steel consumption: *Exports*

(Million tonnes)

Item	1980	1981	1982	1983	1984	1985	1986	1987	1988	1989	1990	1991	1992
Exports	139.8	143.1	133.9	143.3	156.0	168.0	158.4	158.7	159.9	161.2	165.1	170.7	175.4
Consumption	568.0	553.3	524.4	537.7	578.7	588.5	593.1	617.3	654.8	666.0	650.7	623.5	614.8
Ratio (%)	24.6	25.9	25.5	26.7	27.0	28.5	26.7	25.7	24.4	24.2	25.4	27.4	28.5

Figure 3-1 World steel trade and ratio to total steel consumption

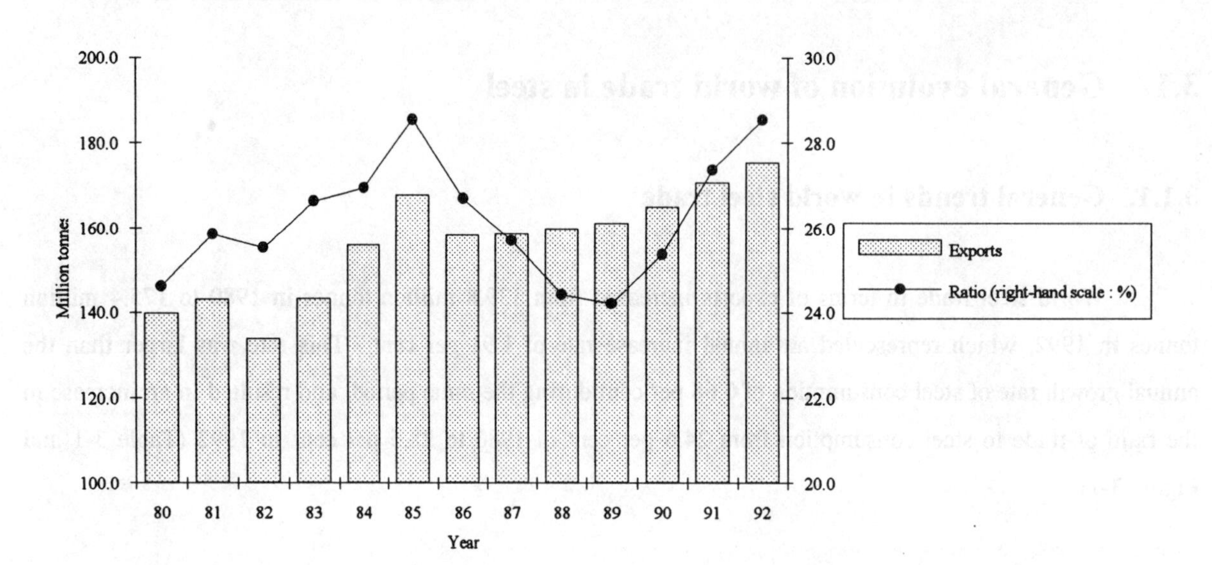

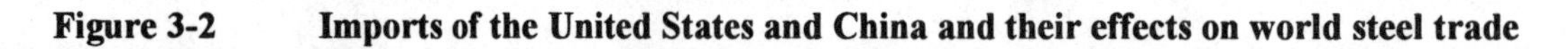
Figure 3-2 Imports of the United States and China and their effects on world steel trade

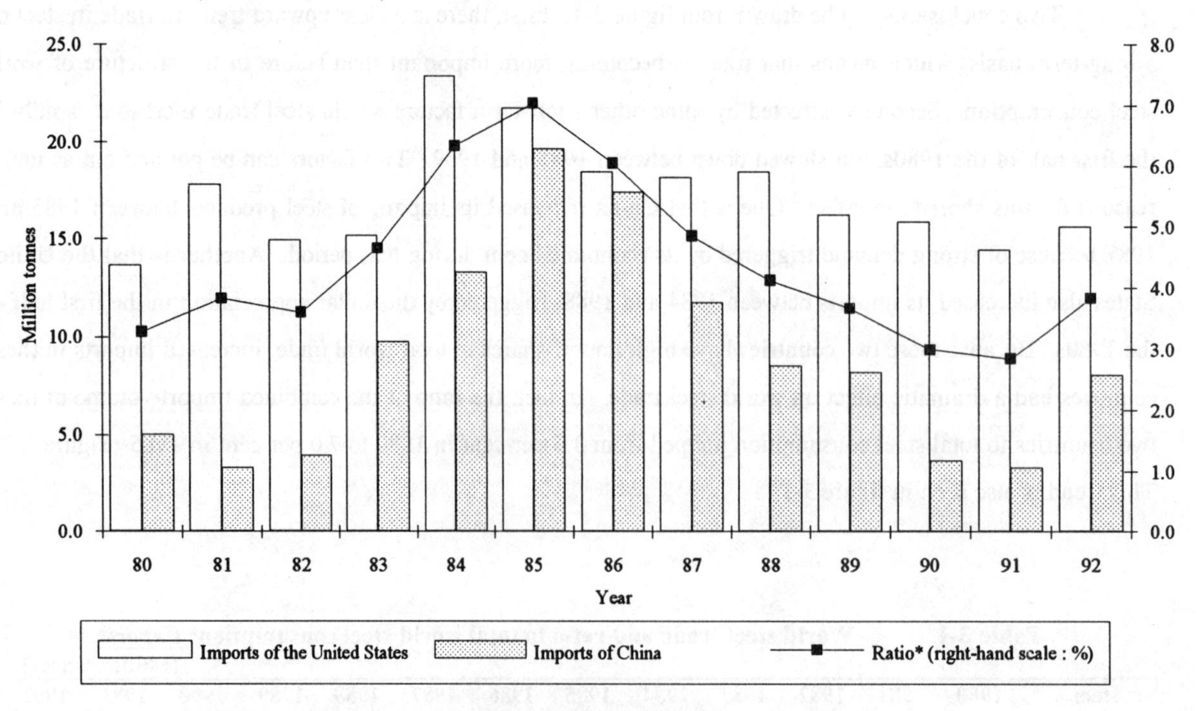

* Ratio of imports of the United States and China to total world steel consumption.

3.1.2. World steel exports by regions

Table 3-2-1 points to some interesting facts regarding the structure of world steel exports by regions and major exporting countries. First of all, relatively small producing regions and countries increased their exports and shares of total world trade. The Republic of Korea increased its exports by 7.2 million tonnes, or at an annual growth rate of 8.3 per cent, during the period from 1980 to 1992, boosting its export share of total world trade from 3.2 per cent in 1980 to 6.7 per cent in 1992. Brazil also increased its exports by 10.1 million tonnes at an annual growth rate of 18.6 per cent, and saw its export share of total world trade increase from only 1.1 per cent to 6.6 per cent. The combined share of these two countries increased from 5.5 per cent in 1980-1983 to 11.3 per cent in 1990-1992. In addition to these two emerging countries, Africa, the Middle East, Other America and Oceania increased their shares of total world exports.

Table 3-2-1 World steel exports by regions and major exporting countries a/

	Amount (1000t)				Share (%)			
	80-83	84-86	87-89	90-92	80-83	84-86	87-89	90-92
AFRICA	2 416	2 954	2 569	3 807	1.7	1.8	1.6	2.2
MIDDLE EAST	651	922	1 109	1 165	0.5	0.6	0.7	0.7
FAR EAST	37 288	39 895	33 403	33 879	26.6	24.8	20.9	19.9
Japan	29 368	30 616	22 674	17 006	21.0	19.0	14.2	10.0
Korea, Republic of	5 083	5 786	6 793	8 876	3.6	3.6	4.2	5.2
NORTH AMERICA	5 635	3 995	4 371	9 010	4.0	2.5	2.7	5.3
Canada	3 264	3 075	2 501	4 538	2.3	1.9	1.6	2.7
United States	2 371	921	1 870	4 472	1.7	0.6	1.2	2.6
OTHER AMERICA	4 316	9 612	12 872	15 562	3.1	6.0	8.0	9.1
Brazil	2 722	6 571	8 732	10 456	1.9	4.1	5.5	6.1
OCEANIA	1 402	1 275	1 161	2 459	1.0	0.8	0.7	1.4
EUROPE	88 304	102 144	104 454	104 518	63.1	63.5	65.3	61.3
EC	60 942	67 498	67 674	73 844	43.5	42.0	42.3	43.3
Belgium-Luxembourg	11 769	12 313	13 285	14 554	8.4	7.7	8.3	8.5
France	9 874	10 627	10 880	11 753	7.1	6.6	6.8	6.9
Germany	17 724	18 760	18 417	18 813	12.7	11.7	11.5	11.0
Italy	7 281	7 651	7 235	8 426	5.2	4.8	4.5	4.9
Netherlands	4 476	5 182	5 567	5 933	3.2	3.2	3.5	3.5
Spain	5 071	6 432	4 073	4 602	3.6	4.0	2.5	2.7
United Kingdom	3 572	4 780	6 528	7 844	2.6	3.0	4.1	4.6
EFTA	7 100	8 816	8 929	9 110	5.1	5.5	5.6	5.3
EASTERN EUROPE	12 614	14 810	15 693	12 221	9.0	9.2	9.8	7.2
former Czechoslovakia	3 382	3 866	4 061	4 506	2.4	2.4	2.5	2.6
Poland	1 889	2 219	2 333	3 545	1.3	1.4	1.5	2.1
Romania	2 036	2 924	3 194	1 215	1.5	1.8	2.0	0.7
former USSR	6 794	7 839	7 878	4 779	4.9	4.9	4.9	2.8
OTHER EUROPE	856	3 181	4 280	4 563	0.6	2.0	2.7	2.7
WORLD TOTAL	140 011	160 797	159 939	170 399	100.0	100.0	100.0	100.0

a/ Figures are annual average for period specified.

Secondly, the United States experienced a huge swing in export volumes during the period. With exports of 3.8 million tonnes in 1980 and 2.3 million tonnes in 1980-1983, volumes declined constantly in the first half of the 1980s and shrank to 0.9 million tonnes in 1985, but started to increase again to 4.5 million tonnes in 1990-1992. The changing international competitiveness of United States steel products affected by the fluctuation of the dollar during the period gave rise to this trend in exports.

Thirdly, Japan, a traditional large exporter, achieved its highest exports of 31.7 million tonnes in 1984, but experienced a sharp drop by 13.2 million to 18.5 million tonnes in 1992. Its share of total world trade declined from 21.0 per cent in 1980-1983 to 10.0 per cent in 1990-1992. The major reason for Japan's declining exports could be both the decreased price competitiveness of Japanese exports and increased joint-venture activity outside Japan necessitated by the constant appreciation of the yen. [1/]

Special attention should be paid when analysing the EC situation, because figures for the European Community also include intra-regional trade which is often regarded as internal trade. There was a general decline in the trade shares of member countries as well as for the EC as a whole, even though export volumes themselves increased in most member countries. Interestingly, among the member States, the United Kingdom recorded the largest increase in its share of total world trade by some 2 per cent. Germany became the world's largest steel exporter in 1992, replacing Japan by a small margin.

Finally, eastern Europe and the former USSR decreased their shares of total world trade with the exception of such countries as the former Czechoslovakia and Poland which increased both export volume and their shares. For many countries in transition, exports seem to be becoming an important factor in restructuring their economies, given the very weak steel demand in those countries.

Export figures, excluding intra-regional trade (Table 3-2-2), and figures only for intra-regional trade (Table 3-2-3) give us a somewhat different picture from those outlined above. Even though the definition and grouping of regions are sometimes difficult and not always meaningful depending on the regions and the degree of economic ties, this approach provides another angle to examine the structure of trade in steel.

The European Community had a high share of exports outside the region, 21.5 per cent of total world trade in 1980, but it decreased its share to 16.6 per cent in 1992. Most of the member States reduced their dependence on markets outside the EC with the exception of the United Kingdom and the Netherlands which increased their shares. Interestingly, the EC's intra-regional trade is so large that it accounts for almost one fourth of total world trade in steel, increasing from 33.3 million tonnes in 1980 to 45.4 million tonnes in 1992. This accelerated trade in steel inside the region can be said to be one of the important positive aspects of the efforts toward a single market in the region.

1/ "How Japanese industry is rebuilding the rust belt", *Technology Review*, Martin Kenny and Richard Florida, February-March, 1991.

Table 3-2-2 World steel exports by regions and major exporting countries (Extra-regional trade)

	Amount (1000t)				Share (%) a/			
	1980	1985	1990	1992	1980	1985	1990	1992
AFRICA	...	...	...	...	...	...	...	...
MIDDLE EAST	...	...	...	...	...	...	...	...
FAR EAST	...	...	...	...	...	...	...	...
Japan	16 691	13 588	6 215	6 105	11.9	8.1	3.8	3.5
Korea, Republic of	2 414	2 970	2 658	3 791	1.7	1.8	1.6	2.2
NORTH AMERICA	4 685	845	3 664	3 698	3.4	0.5	2.2	2.1
Canada	1 287	272	1 204	1 012	0.9	0.2	0.7	0.6
United States	3 398	573	2 460	2 686	2.4	0.3	1.5	1.5
OTHER AMERICA	...	...	...	...	...	...	...	...
Brazil	1 053	6 620	8 289	9 126	0.8	3.9	5.0	5.2
OCEANIA	...	...	...	...	...	...	...	...
EUROPE	...	...	...	...	...	...	...	...
EC	30 050	39 170	28 026	29 068	21.5	23.3	17.0	16.6
Belgium-Luxembourg	3 765	4 769	3 241	2 829	2.7	2.8	2.0	1.6
France	4 795	5 452	3 693	5 789	3.4	3.2	2.2	3.3
Germany	10 350	11 834	7 819	7 989	7.4	7.0	4.7	4.6
Italy	3 869	4 742	5 185	4 102	2.8	2.8	3.1	2.3
Netherlands	1 436	1 775	1 382	2 041	1.0	1.1	0.8	1.2
Spain	3 501	6 737	2 204	2 154	2.5	4.0	1.3	1.2
United Kingdom	1 585	2 833	3 804	3 257	1.1	1.7	2.3	1.9
EFTA	5 491	8 223	7 101	8 070	3.9	4.9	4.3	4.6
EASTERN EUROPE	...	...	...	...	...	...	...	...
former Czechoslovakia	2 209	2 468	2 806	4 197	1.6	1.5	1.7	2.4
Poland	1 254	1 007	2 931	3 316	0.9	0.6	1.8	1.9
Romania	...	...	1 335	714	...	...	0.8	0.4
former USSR	...	...	4 008	...	...	...	2.4	...
OTHER EUROPE	...	...	...	...	...	...	...	...
TOTAL	63 847	74 891	67 033	68 085	45.7	44.6	40.6	38.8
World total available b/	118 338	134 491	129 508	144 125	84.7	80.0	78.4	82.2

a/ Percentage share of total steel exports.
b/ Total steel trade by countries reporting extra- and intra-regional trade.
* Eastern Europe and the former USSR are treated as the same region.

Japan also had a high share of extra-regional exports of 11.9 per cent of total world trade in 1980, but it decreased its dependence on markets outside the Far East region to only 3.5 per cent in 1992. Its exports to the Far East region did not change dramatically throughout the period except for an increase in 1985 because of strong demand in China. However, as a result of the rapid decline in extra-regional exports, two thirds of Japan's exports were going to the Far East in 1992.

Brazil rapidly increased its steel exports to markets outside the Latin American region by 8.1 million tonnes from 1980 to 1992. Its share of extra-regional exports in total world trade increased from 0.8 per cent in 1980 to 5.2 per cent in 1992. On the other hand, the growth of intra-regional exports by Brazil increased by only 2.0 million tonnes. As a result, 79 per cent of exports by Brazil went outside Latin America in 1992.

Table 3-2-3 World steel exports by regions and major exporting countries (intra-regional trade)

		Amount (1000t)				Share (%) a/			
		1980	1985	1990	1992	1980	1985	1990	1992
AFRICA		...	...	...	...	...	...	...	...
MIDDLE EAST		...	...	...	...	...	...	...	...
FAR EAST		...	...	...	...	...	...	...	...
Japan		12 940	17 902	9 620	12 357	9.3	10.7	5.8	7.0
Korea, Republic of		2 068	2 706	4 600	7 906	1.5	1.6	2.8	4.5
NORTH AMERICA		2 679	2 949	4 603	5 265	1.9	1.8	2.8	3.0
Canada		2 235	2 653	2 772	3 951	1.6	1.6	1.7	2.3
United States		444	296	1 831	1 314	0.3	0.2	1.1	0.7
OTHER AMERICA		...	...	...	...	...	...	...	...
Brazil		449	491	651	2 458	0.3	0.3	0.4	1.4
OCEANIA		...	...	...	...	...	...	...	...
EUROPE		...	...	...	...	...	...	...	...
EC		33 276	31 942	43 422	45 379	23.8	19.0	26.3	25.9
Belgium-Luxembourg		9 887	7 842	11 840	10 877	7.1	4.7	7.2	6.2
France		5 912	5 390	7 751	6 008	4.2	3.2	4.7	3.4
Germany		8 684	8 211	10 275	10 847	6.2	4.9	6.2	6.2
Italy		2 877	3 229	3 028	4 860	2.1	1.9	1.8	2.8
Netherlands		3 180	3 414	4 211	3 943	2.3	2.0	2.6	2.2
Spain		1 032	1 055	1 975	2 609	0.7	0.6	1.2	1.5
United Kingdom		1 181	2 070	3 312	5 182	0.8	1.2	2.0	3.0
EFTA		1 186	1 100	1 500	1 555	0.8	0.7	0.9	0.9
EASTERN EUROPE		...	...	...	...	...	...	...	...
former Czechoslovakia		1 217	1 412	909	956	0.9	0.8	0.6	0.5
Poland		676	1 126	629	92	0.5	0.7	0.4	0.1
Romania		...	...	549	72	...	...	0.3	0.0
former USSR		...	...	924	...	...	...	0.6	...
OTHER EUROPE		...	...	...	...	...	...	...	...
TOTAL		54 491	59 628	67 407	76 040	39.0	35.5	40.8	43.3
World total available	b/	118 338	134 491	129 508	144 125	84.7	80.0	78.4	82.2

a/ Percentage share of total steel exports.
b/ Total steel trade by countries reporting extra- and intra-regional trade.
* Eastern Europe and the former USSR are treated as the same region.

Exports from the former Czechoslovakia and Poland to markets outside the former CMEA region increased to 2.4 per cent and 1.9 per cent of total world trade in 1992, respectively, while intra-regional exports declined sharply. This is a reflection of the very weak demand inside the former CMEA region, particularly after 1989, and their excessive capacity in steel production seemed to be compensated only by their export-drive outside the region. Another reason is that exports outside the CMEA region were the only way to get hard currencies and financing to pay for equipment bought from the west. At the same time, it is interesting to know that even in 1980 when the CMEA system was still functioning the dependence of the former Czechoslovakia and Poland on intra-regional exports was only 35.5 per cent and 35.0 per cent, respectively.

The ratio of intra- and extra-regional exports are summarized in table 3-3 and figure 3-4. Among the EC countries, Belgium-Luxembourg had the highest rate of intra-regional exports at close to 80 per cent. France had a relatively low dependency on intra-regional exports, but it still amounted to more than 50 per

Figure 3-3 Share of intra- and extra-regional trade by major exporting regions and countries: *Exports*
1980 (above) and 1992 (below)

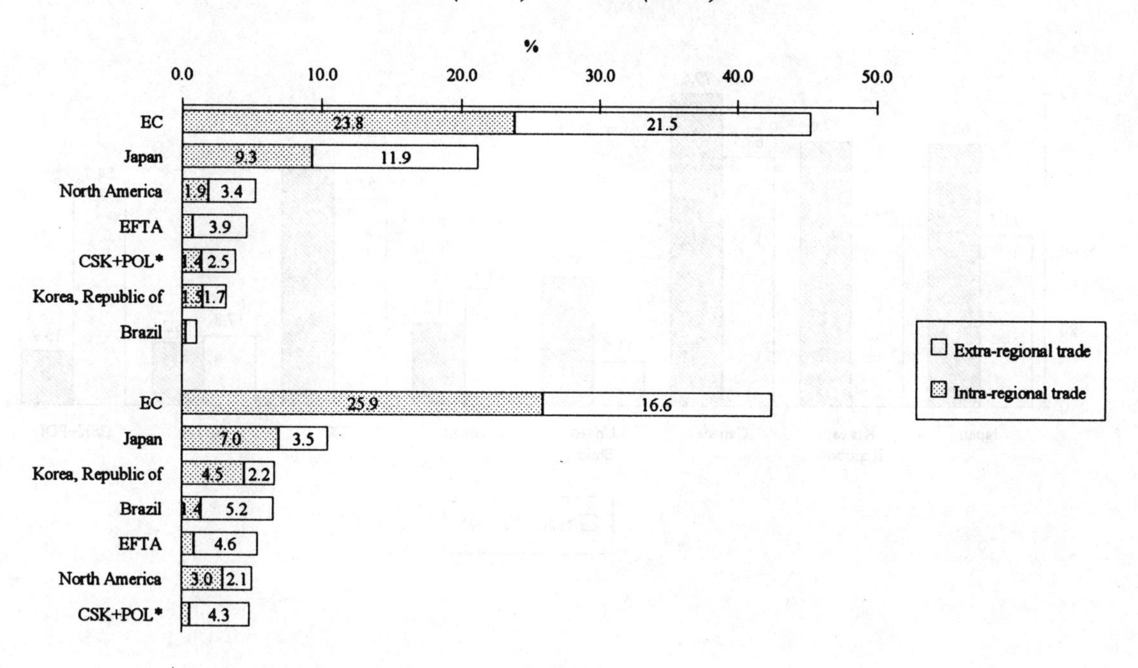

* CSK+POL is the former Czechoslovakia and Poland.

Table 3-3 Ratio of intra- and extra-regional trade: *Exports*

(%)

	Intra-regional trade				Extra-regional trade			
	1980	1985	1990	1992	1980	1985	1990	1992
AFRICA	...	...	...	...	...	...	...	...
MIDDLE EAST	...	...	...	...	...	...	...	...
FAR EAST	...	...	...	...	...	...	...	...
Japan	43.7	56.8	60.8	66.9	56.3	43.2	39.2	33.1
Korea, Republic of	46.1	47.9	63.4	67.6	53.9	52.1	36.6	32.4
NORTH AMERICA	36.4	77.7	55.7	58.7	63.6	22.3	44.3	41.3
Canada	63.5	90.7	69.7	79.6	36.5	9.3	30.3	20.4
United States	11.6	34.1	42.7	32.9	88.4	65.9	57.3	67.1
OTHER AMERICA	...	...	...	...	...	...	...	...
Brazil	29.9	6.9	7.3	21.2	70.1	93.1	92.7	78.8
OCEANIA	...	...	...	...	...	...	...	...
EUROPE	...	...	...	...	...	...	...	...
EC	52.5	44.9	60.8	61.0	47.5	55.1	39.2	39.0
Belgium-Luxembourg	72.4	62.2	78.5	79.4	27.6	37.8	21.5	20.6
France	55.2	49.7	67.7	50.9	44.8	50.3	32.3	49.1
Germany	45.6	41.0	56.8	57.6	54.4	59.0	43.2	42.4
Italy	42.6	40.5	36.9	54.2	57.4	59.5	63.1	45.8
Netherlands	68.9	65.8	75.3	65.9	31.1	34.2	24.7	34.1
Spain	22.8	13.5	47.3	54.8	77.2	86.5	52.7	45.2
United Kingdom	42.7	42.2	46.5	61.4	57.3	57.8	53.5	38.6
EFTA	17.8	11.8	17.4	16.2	82.2	88.2	82.6	83.8
EASTERN EUROPE	...	...	...	...	...	...	...	...
former Czechoslovakia	35.5	23.0	24.5	18.6	64.5	77.0	75.5	81.4
Poland	35.0	19.2	17.7	2.7	65.0	80.8	82.3	97.3
Romania	...	...	29.1	...	...	...	...	...
former USSR	...	...	18.7	...	...	...	...	...
OTHER EUROPE	...	...	...	...	...	...	...	...
WORLD TOTAL	46.0	44.3	52.0	52.8	54.0	55.7	48.0	47.2

Figure 3-4 Ratio of intra-regional trade: *Exports*

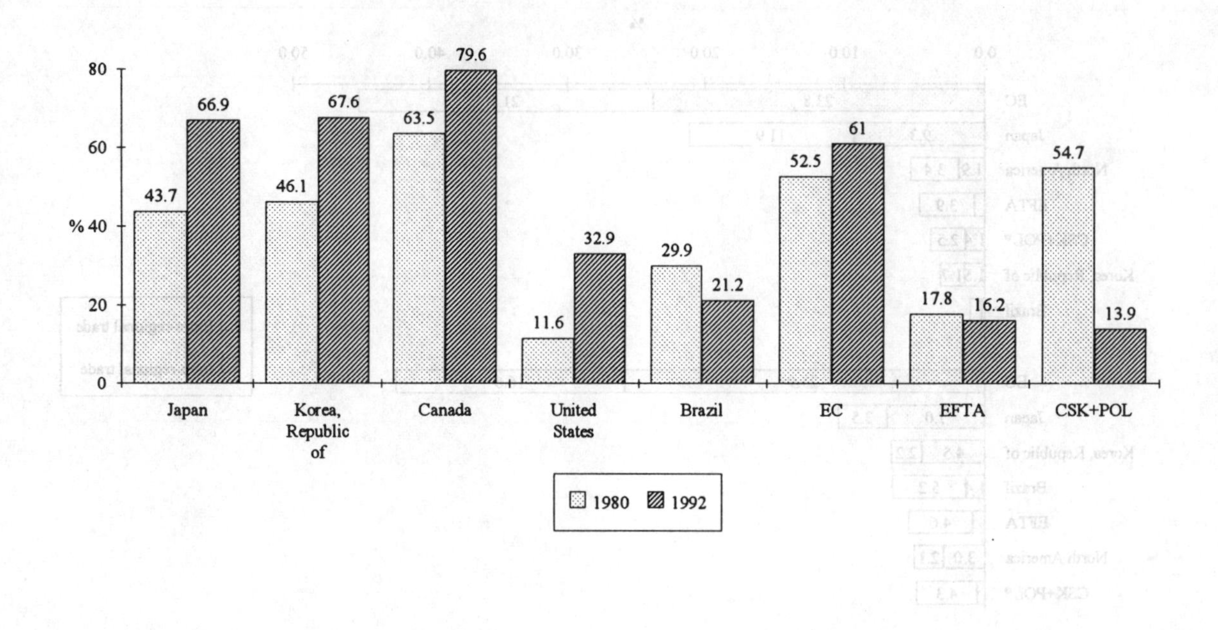

cent. The dependence on intra-regional trade of EFTA stayed low at less than 20 per cent because of the relatively low level of steel consumption in that region. Most of its extra-regional exports went to the EC. Figure 3-4 shows that all major exporting countries and regions are moving more towards their regional markets except Brazil and some of the former CMEA countries. In terms of steel trade, the integration of regional markets is creating much closer ties with neighbouring economic partners.

The world's largest exporters, including intra-regional trade, are shown in figure 3-5.

3.1.3. World steel imports by regions

Imports are the other side of trade. Therefore, they should show the same value as exports on a worldwide basis, i.e. a definite increasing trend in the past decade. However, the allocation of imports throughout the world shows a different pattern from that of exports. Table 3-4 and figure 3-6 summarize steel imports and the share of total world imports by regions and major importing countries.

Contrary to the case of exports, small steel producing regions and countries (Africa, Other America and Oceania) decreased their imports and their shares of total world imports. As seen in chapter 1, these regions suffered a decline in steel consumption. With weakness in demand and increasing production capacity, they naturally decreased their imports and increased their dependence on domestic sources.

Figure 3-5 Largest steel exporting countries in 1980 and 1992

1980

Million tonnes

0 5 10 15 20 25 30

Japan
former FR Germany
Belgium-Luxembourg
France
former USSR
Italy
Netherlands
Spain
Korea, Republic of
United States
Canada
former Czechoslovakia
United Kingdom
Austria
former GDR

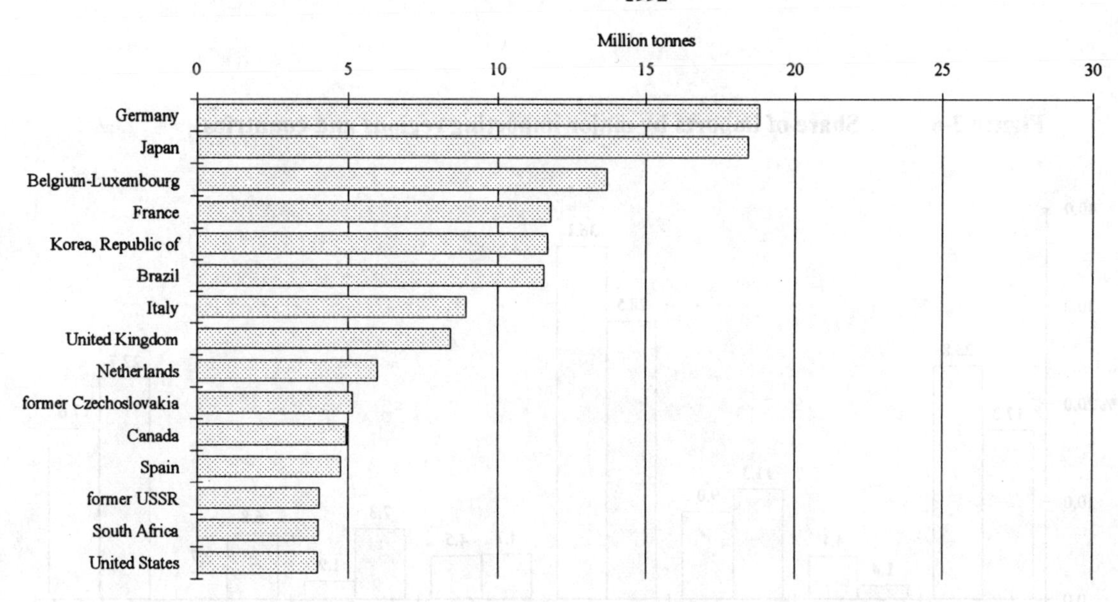

Table 3-4 World steel imports by regions and major importing countries a/

	Amount (1000t)				Share (%)			
	80-83	84-86	87-89	90-92	80-83	84-86	87-89	90-92
AFRICA	5 466	4 637	4 017	4 141	4.0	2.9	2.4	2.4
MIDDLE EAST	13 781	14 432	11 088	12 637	10.1	9.2	6.6	7.4
Iran	2 977	3 606	3 616	5 467	2.2	2.3	2.2	3.2
FAR EAST	25 500	37 790	38 998	47 903	18.7	24.0	23.4	28.2
China	5 511	16 791	9 486	5 020	4.0	10.7	5.7	3.0
Japan	1 855	3 405	6 365	7 408	1.4	2.2	3.8	4.4
Korea, Republic of	1 935	2 825	3 650	6 711	1.4	1.8	2.2	4.0
NORTH AMERICA	16 736	23 093	20 702	17 841	12.3	14.6	12.4	10.5
United States	15 399	21 189	17 631	15 310	11.3	13.4	10.6	9.0
OTHER AMERICA	6 375	4 302	4 481	6 045	4.7	2.7	2.7	3.6
OCEANIA	1 208	1 203	1 278	956	0.9	0.8	0.8	0.6
EUROPE	67 396	71 996	86 018	80 060	49.3	45.7	51.6	47.1
EC	38 914	41 528	51 726	61 359	28.5	26.3	31.0	36.1
Belgium-Luxembourg	2 848	3 374	4 468	4 735	2.1	2.1	2.7	2.8
France	7 063	6 715	8 184	10 292	5.2	4.3	4.9	6.1
Germany	11 111	11 616	12 928	4 952	8.1	7.4	7.7	2.9
Italy	5 487	6 399	9 038	10 423	4.0	4.1	5.4	6.1
Netherlands	3 490	3 969	4 449	5 084	2.6	2.5	2.7	3.0
United Kingdom	3 861	3 911	4 947	5 337	2.8	2.5	3.0	3.1
EFTA	6 380	6 826	7 663	7 582	4.7	4.3	4.6	4.5
EASTERN EUROPE	10 028	9 574	9 747	3 284	7.3	6.1	5.8	1.9
former USSR	9 343	10 109	10 233	4 557	6.8	6.4	6.1	2.7
OTHER EUROPE	2 731	3 960	6 648	3 279	2.0	2.5	4.0	1.9
WORLD TOTAL	136 608	157 639	166 818	169 805	100.0	100.0	100.0	100.0

a/ Figures are annual average for period specified.

Figure 3-6 Share of imports by major importing regions and countries a/

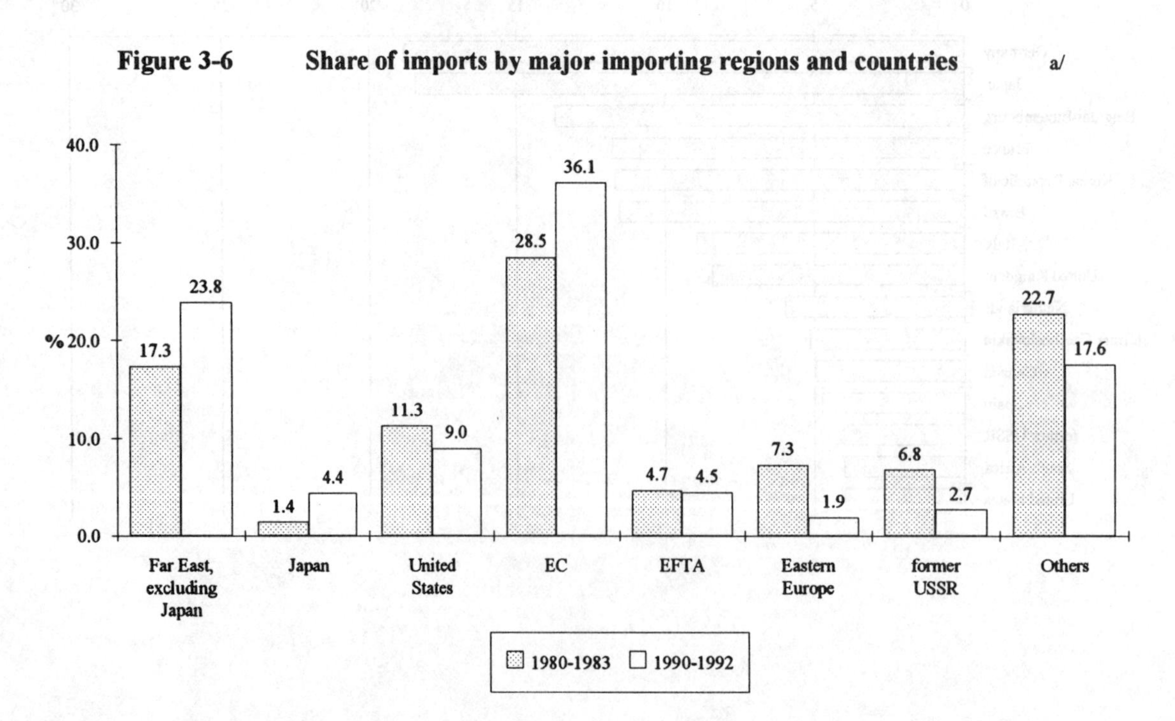

a/ Figures are annual average for period specified.

Table 3-5 Ratio of intra- and extra-regional trade in North America, the EC and EFTA: *Imports*

(%)

	Intra-regional trade				Extra-regional trade			
	1980	1985	1990	1992	1980	1985	1990	1992
NORTH AMERICA	18.3	12.5	24.6	29.4	81.7	87.5	75.4	70.6
Canada	46.9	15.2	64.3	58.5	53.1	84.8	35.7	41.5
United States	16.4	12.2	17.5	25.3	83.6	87.8	72.5	74.7
EC	77.4	77.8	72.1	73.4	22.6	22.2	27.9	26.6
Belgium-Luxembourg	91.7	90.7	98.1	99.8	8.3	9.3	1.9	0.2
France	103.5	101.0	92.1	92.1	-3.5	-1.0	7.9	7.9
Germany	69.3	65.1	49.8	50.3	30.7	34.9	50.2	49.7
Italy	62.7	70.2	59.6	63.5	37.3	29.8	40.4	36.5
Netherlands	81.0	78.0	88.7	88.4	19.0	22.0	11.3	11.6
Spain	75.3	103.8	89.8	88.8	24.7	-3.8	10.2	11.2
United Kingdom	69.0	75.9	78.2	82.7	31.0	24.1	21.8	17.3
EFTA	17.3	16.4	18.5	20.5	82.7	83.6	81.5	79.5

On the other hand, the Far East saw a dramatic increase in steel imports. Japan increased its imports by 5.6 million tonnes from 1980-1983 to 1990-1992, the Republic of Korea by 4.8 million tonnes and other Far East countries excluding China by 12.6 million tonnes. The total increase in the Far East during that period, 22.4 million tonnes, accounted for 67 per cent of the total increase in world steel imports. The share of the Far East, excluding Japan, and Japan increased from 17.3 per cent to 23.8 per cent, and 1.4 per cent to 4.4 per cent, respectively. This rapid increase was mainly caused by an eruption in steel consumption in the region with the exception of Japan. In the case of Japan, yen appreciation after 1985 was probably largely responsible, because Japan's increase in imports became much clearer after 1985. China had a huge swing in import volumes with the highest value of 19.6 million tonnes in 1985 and the lowest value of 3.3 million tonnes in 1981 (Figure 3-2). This is probably because the country still has an immature economic structure and experienced huge swings from economic boom to recession with a constant shortage in steel production capacity.

There was a significant increase in imports in the European Community of 22.4 million tonnes from 1980-1983 to 1990-1992 or 3 per cent annual growth during the period. Including its intra-regional trade, the EC was the biggest importer of steel products with a world share of 36.1 per cent in 1990-1992. Italy ranked at the top with Germany second and the United Kingdom third. EFTA experienced a modest growth in import volumes, but its share of total world imports declined. The former CMEA region sharply decreased its imports, resulting in a share of 4.6 per cent in 1990-1992, down from 14.1 per cent in 1980-1983.

The United States experienced a surge in imports in the middle of the 1980s, as already discussed. It accounted for 13.4 per cent of total world imports in 1984-1986, but its share declined to 9.0 per cent in 1990-1992. The relative importance of the United States as a traditional importer of steel products declined.

Figure 3-7 Share of intra- and extra-regional trade by North America, the EC and EFTA: *Imports*
1980 (above) and 1992 (below)

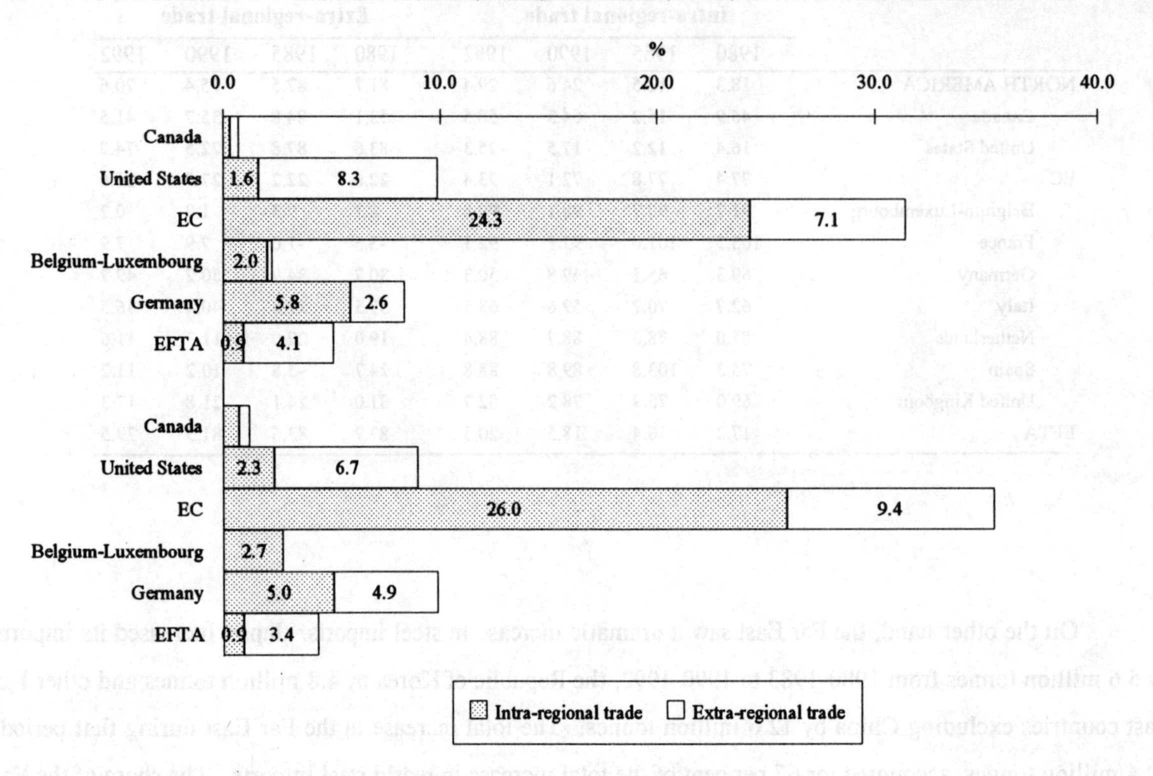

Because of the limited availability of import data, the distinction of imports between intra- and extra-regional trade can be seen only for North America, the EC and EFTA (Table 3-5, Figure 3-7 and Figure 3-8). There was a clear increasing trend in intra-regional trade for all these regions and countries with the exception of the EC, owing to a decline in Germany. Many member States of the EC have figures for intra-regional imports of more than 80 per cent. However, Germany's ratio of intra-regional trade declined from 69.3 per cent in 1980 to 50.3 per cent in 1992. This change could be partially explained by the unification of the Federal Republic of Germany and the former GDR which has a close relationship with other former CMEA countries. Excluding intra-regional trade, the EC had shares of total world imports of 7.1 per cent in 1980 and 9.4 per cent in 1992. The EC also increased its share of intra-regional trade in total world trade.

On the other hand, while the United States increased its ties with Canada in steel imports, it decreased its share of imports from outside North America from 8.3 per cent in 1980 to 6.7 per cent in 1992.

Of the list of largest steel importing countries in 1980 and 1992 (Figure 3-9), clear increasing trends in terms of volume can be seen. Japan, the Republic of Korea and Iran came into the top 10 in 1992 and the former USSR disappeared from the list.

Figure 3-8 Ratio of intra-regional trade: *Imports*

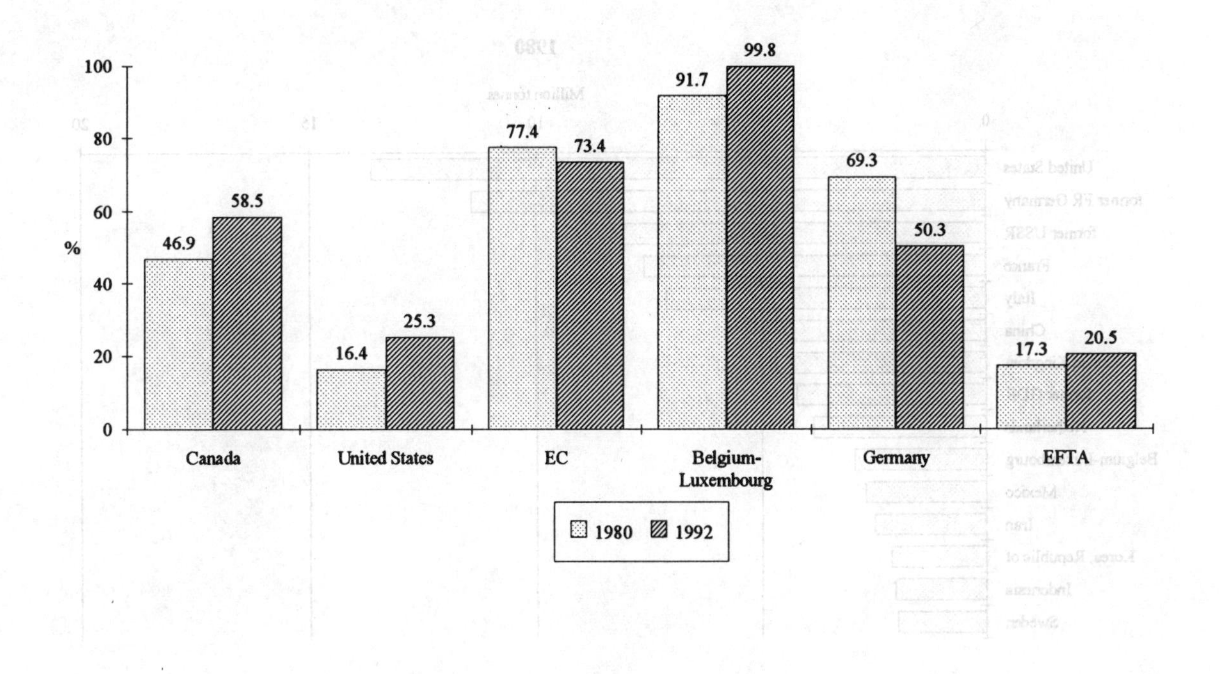

3.1.4. World trade balance by regions

As a result of the changes in exports and imports in steel trade described above, the structure of the world trade balance in steel changed (Table 3-6 and Figure 3-10). As already discussed in chapter 1 on the basis of crude steel equivalent, one general trend was the disappearance of huge net exporters such as Japan and the EC. However, it is interesting, because the trade balance in many regions shrank along with constant increases in both exports and imports on a worldwide basis.

Among the regions, the Far East, excluding Japan, rapidly increased its deficit in steel trade because the growth in exports could not catch up with the rapid pace of import growth to feed the increasing gap between rapidly growing consumption and production. The United States kept at almost the same level of net imports during the period with the exception of expanding net imports in the middle of the 1980s. Other America became a net exporter, owing to a rapid increase in exports by Brazil. The former CMEA countries increased their net exports, because of their sluggish consumption and decreased imports, even though their export volumes did not change dramatically.

Figure 3-9 Largest steel importing countries in 1980 and 1992

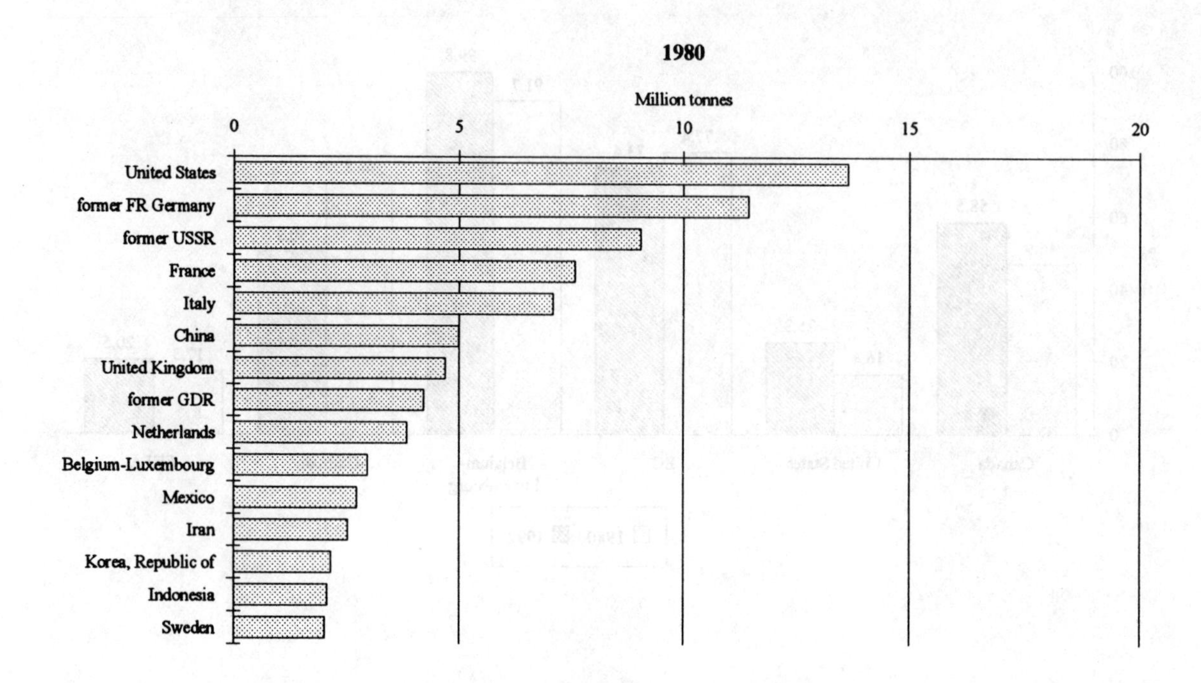

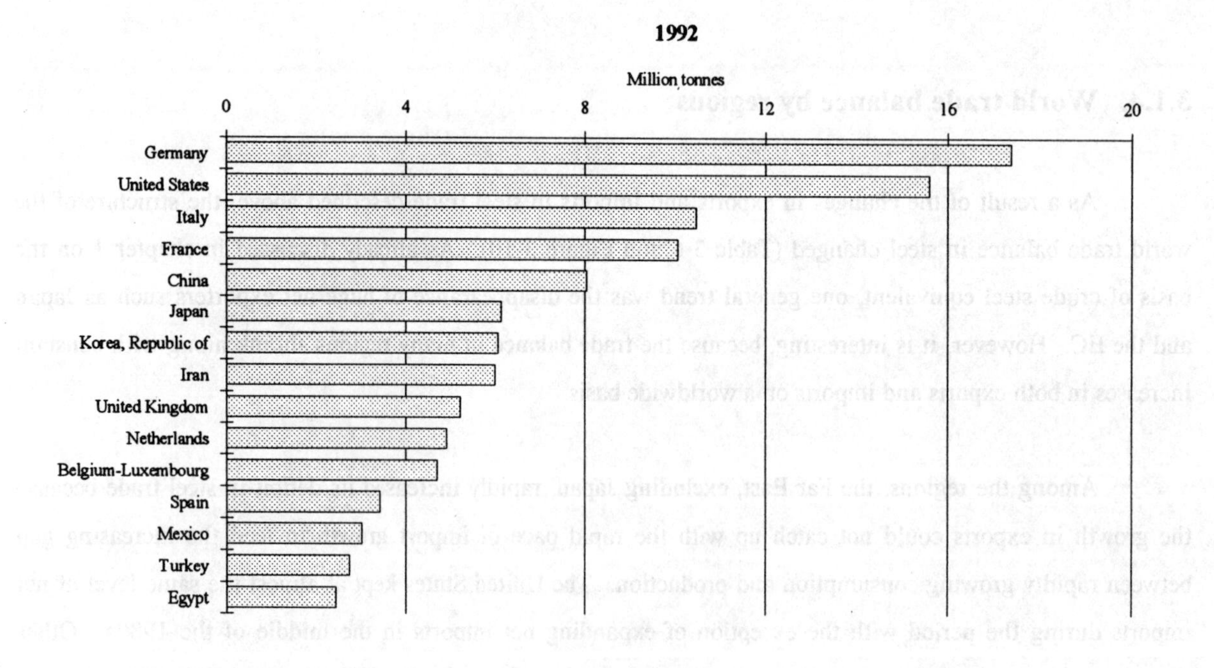

Table 3-6 World steel trade balance by region a/

Regions and countries	Amount (1000t)				Ratio (%) b/			
	80-83	84-86	87-89	90-92	80-83	84-86	87-89	90-92
AFRICA	-3 050	-1 683	-1 449	- 334	-2.2	-1.0	-0.9	-0.2
MIDDLE EAST	-13 130	-13 511	-9 979	-11 471	-9.4	-8.4	-6.2	-6.7
FAR EAST	11 788	2 105	-5 595	-14 024	8.4	1.3	-3.5	-8.2
Japan	27 513	27 211	16 309	9 598	19.7	16.9	10.2	5.6
NORTH AMERICA	-11 101	-19 098	-16 331	-8 831	-7.9	-11.9	-10.2	-5.2
United States	-13 028	-20 268	-15 761	-10 839	-9.3	-12.6	-9.9	-6.4
OTHER AMERICA	-2 059	5 311	8 391	9 518	-1.5	3.3	5.2	5.6
OCEANIA	193	71	- 117	1 503	0.1	0.0	-0.1	0.9
EUROPE	20 908	30 148	18 436	24 458	14.9	18.7	11.5	14.4
EC	22 028	25 971	15 948	12 485	15.7	16.2	10.0	7.3
EFTA	720	1 991	1 266	1 528	0.5	1.2	0.8	0.9
EASTERN EUROPE	2 586	5 235	5 946	8 937	1.8	3.3	3.7	5.2
former USSR	-2 551	-2 270	-2 355	223	-1.8	-1.4	-1.5	0.1
OTHER EUROPE	-1 874	- 779	-2 369	1 284	-1.3	-0.5	-1.5	0.8
WORLD TOTAL c/	3 404	3 158	-6 880	594	2.4	2.0	-4.3	0.3

a/ Figures are annual average for period specified.
b/ Percentage ratio of total exports.
c/ Because of discrepancies between export and import data, the WORLD TOTAL does not always equal zero.

Figure 3-10 Changes in steel trade balance a/

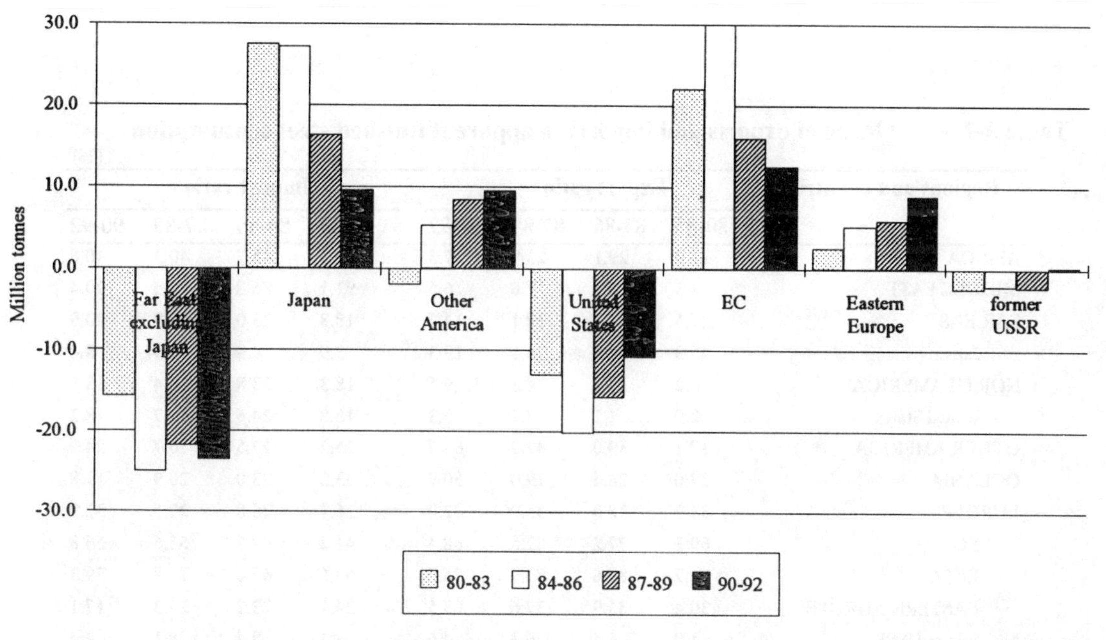

a/ Figures are annual average for period specified.

3.1.5. Ratio of imports and exports to steel consumption

The ratio of exports and imports to apparent finished steel consumption measures the relative importance of trade in steel consumption. Table 3-7 shows the general trend that both the export and import ratios to total world steel consumption have increased. As seen at the beginning of this chapter, the middle of the 1980s, specifically 1985, was an extraordinary boom period for steel trade, when both export and import ratios increased because of higher imports by the United States and China.

Smaller steel producing regions and countries such as Africa and the Middle East increased their export ratios, but decreased their import ratios. As a result, they increased their dependency on domestic sources. The United States did not see big changes in its export and import ratios with the exception of 1990-1992 for its export ratio and 1984-1986 for its import ratio. The European Community increased its import ratios mainly due to the expansion of intra-regional trade. EFTA countries also increased both ratios, contributing to more active world steel trade. The former CMEA countries experienced a sharp increase in export ratios with the opposite trend for import ratios.

Looking at figures for extra-regional trade alone (Table 3-8), the European Community decreased its export ratio by 17.4 per cent from 1985 to 1992, while it increased its import ratio by 5.4 per cent. This suggests that the region absorbed more imports from outside the EC than before, while its export dependence on outside the region declined. EFTA has a high degree of trade outside the region both in terms of exports and imports because of the relatively small size of its economies. As seen in figure 3-11, even though Japan increased the ratio of imports to steel consumption, it was still smaller than the world average.

Table 3-7 Ratio of exports and imports to apparent finished steel consumption a/

(%)

Regions and countries	Export ratio				Import ratio			
	80-83	84-86	87-89	90-92	80-83	84-86	87-89	90-92
AFRICA	21.9	29.1	25.9	37.5	49.5	45.7	40.5	40.8
MIDDLE EAST	4.3	5.5	7.6	6.5	91.1	86.8	76.4	70.4
FAR EAST	27.5	24.3	17.1	14.7	18.8	23.0	19.9	20.9
Japan	45.3	44.4	28.1	19.3	2.9	4.9	7.9	8.4
NORTH AMERICA	6.2	4.1	4.3	9.5	18.3	23.8	20.4	18.8
United States	2.9	1.1	2.1	5.3	18.8	24.6	19.7	18.2
OTHER AMERICA	17.1	39.0	47.9	63.7	25.2	17.5	16.7	24.7
OCEANIA	27.0	24.4	19.0	50.9	23.2	23.0	20.9	19.8
EUROPE	33.7	38.0	35.9	42.0	25.7	26.8	29.5	32.2
EC	69.3	77.2	67.5	68.4	44.2	47.5	51.6	56.8
EFTA	71.7	87.6	83.7	95.2	64.5	67.8	71.8	79.3
EASTERN EUROPE	30.4	35.9	37.6	63.5	24.1	23.2	23.3	17.1
former USSR	5.9	6.5	6.3	4.6	8.1	8.4	8.1	4.4
OTHER EUROPE	11.4	35.9	33.1	50.9	36.2	44.7	51.5	36.6
WORLD TOTAL	25.6	27.4	24.8	27.1	25.0	26.9	25.8	27.0

a/ Figures are annual average for period specified.

Table 3-8 Ratio of extra-regional trade to apparent finished steel consumption in North America, the EC and EFTA

(%)

Regions and countries	Export ratio				Import ratio			
	1980	1985	1990	1992	1980	1985	1990	1992
NORTH AMERICA	4.8	0.9	3.7	3.9	12.3	21.1	14.3	13.1
Canada	12.5	2.5	11.7	10.4	4.9	15.1	9.8	9.5
United States	3.9	0.7	2.8	3.1	13.2	21.8	14.8	13.5
EC	31.0	45.4	25.8	28.0	10.1	10.5	15.5	15.9
Belgium-Luxembourg	165.9	156.2	123.0	101.9	10.9	10.3	3.3	0.4
France	29.9	44.5	22.9	40.2	-1.7	-0.6	5.1	5.5
Germany	35.8	49.4	27.1	25.1	12.2	16.8	25.9	27.1
Italy	17.2	24.8	20.4	17.2	11.8	9.7	17.2	16.0
Netherlands	44.7	52.3	31.2	53.7	22.9	26.2	13.2	15.0
Spain	44.2	121.6	20.8	22.5	5.5	-1.0	3.2	4.0
United Kingdom	13.4	24.4	27.2	28.6	12.3	8.0	8.2	7.9
EFTA	50.5	85.2	67.7	83.4	52.2	58.0	62.9	62.1

Figure 3-11 Ratio of imports to apparent finished steel consumption a/

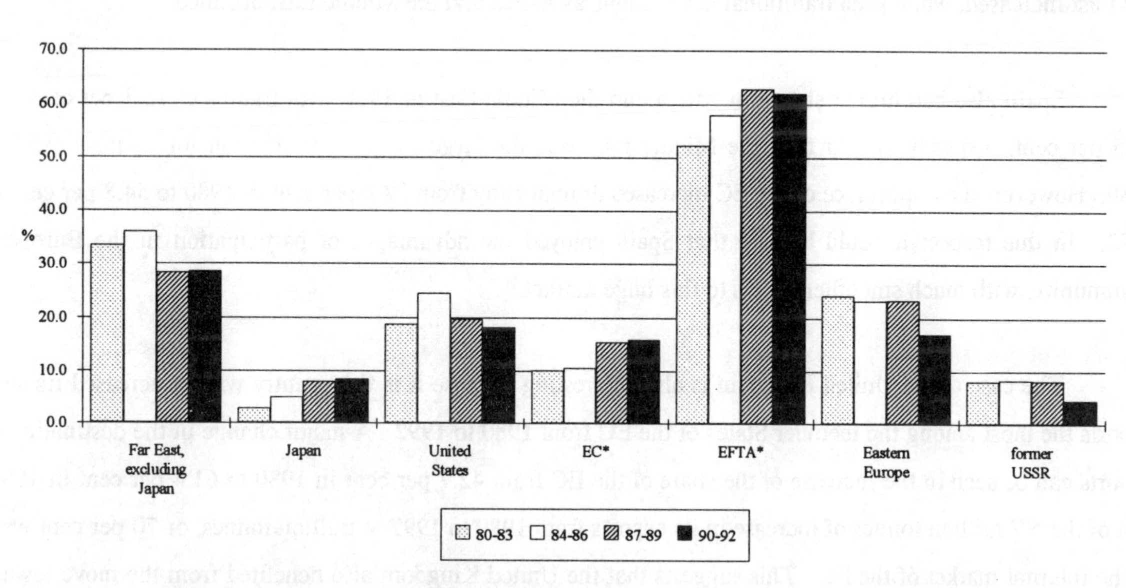

a/ Figures are annual average for period specified.

* Figures for EC and EFTA are ratio of extra-regional trade (imports) to apparent finished steel consumption.

3.2. Destination of exports and origin of imports

3.2.1. Destination of steel exports by major regions and countries

Tables 3-9-1 to 3-9-8 provide information on the destination and relative importance of steel exports by major regions and countries. Figures 3-12-1 to 3-12-8 show the relative importance of those destinations for steel exports by major regions and countries.

As already described, the internal market was the major destination of steel exports for the European Community. Its significance increased substantially over the period with the exception of 1985. In that year, the share of China and the United States jumped to 5.7 per cent and 9.0 per cent, respectively. Although China's importance declined after 1985, the share of the Far East as a whole increased from 3.9 per cent in 1980 to 7.0 per cent in 1992. The importance of all other regions for exports by the EC declined except for EFTA.

Italy has a relatively different structure of export destinations than other member States of the EC. Its dependence on intra-regional exports within the EC is smaller than the EC average. On the other hand, it has relatively high shares in Africa and the Middle East with figures of 11.7 per cent and 7.9 per cent in 1992, respectively. However, it is worth pointing out that the importance to Italy of the EC and EFTA as well as the Far East increased, while such traditional destinations as Africa and the Middle East declined.

Spain also had higher shares in Africa and the Middle East in 1980 with figures of 13.1 per cent and 24.5 per cent, respectively. In fact, the Middle East was the largest export destination for Spanish steel in 1980. However, the importance of the EC increased dramatically from 22.8 per cent in 1980 to 54.8 per cent in 1992. In this respect it could be said that Spain enjoyed the advantages of participation in the European Community, with much smoother access to this huge market.[2/]

The case of the United Kingdom is also interesting because it is the country which increased its steel exports the most among the member States of the EC from 1980 to 1992. A major change in the destination of exports can be seen in the increase of the share of the EC from 42.7 per cent in 1980 to 61.4 per cent in 1992. Out of the 5.7 million tonnes of increase in its exports from 1980 to 1992, 4 million tonnes, or 70 per cent went to the internal market of the EC. This suggests that the United Kingdom also benefited from the move toward the unification of the EC in terms of increased exports in steel.

EFTA also had increasing shares of exports to the EC and the Far East, while the importance of the United States was largest in 1985.

[2/] Gary Clyde Hufbauer and Jeffrey J. Schott, *North American Free Trade*, pp. 254, Institute for International Economics, 1992.

Among the eastern European countries, both the former Czechoslovakia and Poland rapidly increased their exports. The biggest increase in the destination of exports by the former Czechoslovakia was recorded for the EC. The increase from 1980 to 1992 amounted to almost 1 million tonnes, and its share increased from 25.3 per cent to 35.7 per cent in 1992. Big increases were also made in exports to the Far East (0.8 million tonnes) and to EFTA (0.3 million tonnes). The shares of these two regions in 1992 were 18.4 per cent and 9.5 per cent, respectively. The importance of the former CMEA countries declined from 35.5 per cent in 1980 to 18.6 per cent in 1992.

Poland also made a shift in export destinations from the former CMEA to new markets. However, the largest increases were to the Far East, not to its European partners. Its exports to the Far East rose from only 0.13 million tonnes in 1980 to 1.29 million tonnes in 1992, an increase from 6.9 per cent to 37.9 per cent in 1992. Poland also increased its exports to the EC by 0.58 million tonnes during the same period, or from 17.0 per cent to 26.7 per cent. Interestingly, Poland also increased its share of exports to Africa and the Middle East, while its share of exports to the former CMEA countries declined from 36 per cent in 1980 to only 8.6 per cent in 1992.

Exports from North America increased from 7.4 million tonnes in 1980 to 9.0 million tonnes in 1992 with a sharp decline to 3.8 million tonnes in 1985. However, the increase was completely due to an increase of trade inside the region, between Canada and the United States. The export volume and the importance of all other regions declined. This was also the case for Other America which was the second largest destination for North America. However, Other America is still the biggest destination for the United States with a 41.7 per cent share in 1992, down from 47.8 per cent in 1980. In 1985, even exports to this region dropped to only 0.2 million tonnes, an 88 per cent decline from 1980. This sharp drop may have been caused by a combination of currency fluctuations and the large foreign debts of the Latin American countries. But on a more long-term basis, sluggish steel consumption with slow economic growth in the region and an increase in the steel production of Brazil were the major factors for the declining importance of exports to this region from the United States.

An increase in steel exports through strengthened ties with Canada was the only factor behind United States export growth. Given the recent agreement of NAFTA (North American Free Trade Agreement), the United States and Canada might now increase their trade in steel with the new member, Mexico. One study suggests that steel trade within NAFTA could at least double and perhaps triple from the 1992 level, taking into account the experience of the European Community.[3]

Japan's exports declined from 1980 to 1992 for all destinations. Because exports to the Far East declined at a slower rate, the importance of this region increased relatively from 43.7 per cent in 1980 to 66.9 per cent in 1992.

---

[3] Ibid., pp. 254.

The Republic of Korea increased its exports mainly to the Far East, including Japan, and the United States. The increases from 1980 to 1992 amounted to 1.9 million tonnes for Japan, 3.9 million tonnes for Other Far East and 1.7 million tonnes for the United States. Exports to other regions declined except for Oceania. As a result, the importance of Japan, Other Far East and the United States increased to 24.0 per cent, 43.6 per cent and 22.3 per cent, respectively.

On the contrary, Brazil, another emerging steel exporter, increased its steel exports to all destinations. Among them the major increases were 5 million tonnes to the Far East, 2 million tonnes to Other America, 0.9 million tonnes to Europe and 0.9 million tonnes to the United States. Brazil increased its exports to major consumers such as the EC and Japan by 0.4 million tonnes each. As a result, the importance of the Far East rose dramatically from 4.8 per cent to 49.4 per cent. On the other hand, even though export volumes increased, the importance of Other America, the United States and Europe declined.

Table 3-9-1 Destination and relative importance of steel exports: EC

Destination	Amount (1000 t)				Share (%)			
	1980	1985	1990	1992	1980	1985	1990	1992
AFRICA	4 387	3 244	2 361	3 178	6.9	4.6	3.3	4.3
MIDDLE EAST	4 540	4 826	2 651	3 046	7.2	6.8	3.7	4.1
FAR EAST	2 485	6 792	3 950	5 243	3.9	9.6	5.5	7.0
Japan	38	118	277	185	0.1	0.2	0.4	0.2
NORTH AMERICA	4 148	7 166	5 963	4 826	6.6	10.1	8.3	6.5
United States	3 832	6 304	5 318	4 238	6.1	8.9	7.4	5.7
OTHER AMERICA	2 510	1 299	1 204	1 262	4.0	1.8	1.7	1.7
OCEANIA	68	94	133	100	0.1	0.1	0.2	0.1
EUROPE	45 110	46 182	52 239	52 857	71.2	64.9	73.1	71.0
EC	33 276	31 942	43 422	45 379	52.5	44.9	60.8	61.0
EFTA	5 013	4 873	5 954	6 034	7.9	6.9	8.3	8.1
EASTERN EUROPE	1 195	716	400	498	1.9	1.0	0.6	0.7
former USSR	4 638	6 537	1 324	8	7.3	9.2	1.9	0.0
OTHER EUROPE	989	2 113	1 138	938	1.6	3.0	1.6	1.3
OTHERS	79	1 510	2 947	3 934	0.1	2.1	4.1	5.3
WORLD TOTAL	63 327	71 112	71 449	74 446	100.0	100.0	100.0	100.0

Table 3-9-2 Destination and relative importance of steel exports: EFTA

Destination	Amount (1000 t)				Share (%)			
	1980	1985	1990	1992	1980	1985	1990	1992
AFRICA	69	64	81	120	1.0	0.7	0.9	1.2
MIDDLE EAST	147	218	91	129	2.2	2.3	1.1	1.3
FAR EAST	173	382	359	675	2.6	4.1	4.2	7.0
Japan	90	7	35	118	1.3	0.1	0.4	1.2
NORTH AMERICA	165	1 052	501	452	2.5	11.3	5.8	4.7
United States	141	1 000	464	426	2.1	10.7	5.4	4.4
OTHER AMERICA	84	47	45	56	1.3	0.5	0.5	0.6
OCEANIA	11	15	26	43	0.2	0.2	0.3	0.5
EUROPE	6 023	7 520	5 231	7 867	90.2	80.7	60.8	81.7
EC	3 896	5 086	3 135	6 109	58.3	54.6	36.4	63.5
EFTA	1 186	1 100	1 500	1 555	17.8	11.8	17.4	16.2
EASTERN EUROPE	177	300	69	154	2.7	3.2	0.8	1.6
former USSR	561	825	352	0	8.4	8.8	4.1	0.0
OTHER EUROPE	202	209	176	49	3.0	2.2	2.0	0.5
OTHERS	5	25	2 266	284	0.1	0.3	26.3	2.9
WORLD TOTAL	6 677	9 323	8 601	9 625	100.0	100.0	100.0	100.0

Table 3-9-3 Destination and relative importance of steel exports: *Former Czechoslovakia*

Destination	Amount (1000 t)				Share (%)			
	1980	1985	1990	1992	1980	1985	1990	1992
AFRICA	59	6	97	138	1.7	0.2	2.6	2.7
MIDDLE EAST	428	442	247	448	12.5	11.4	6.7	8.7
FAR EAST	157	314	398	949	4.6	8.1	10.7	18.4
Japan	...	...	...	57	...	...	...	1.1
NORTH AMERICA	18	37	25	23	0.5	1.0	0.7	0.5
United States	5	26	13	22	0.2	0.7	0.4	0.4
OTHER AMERICA	30	31	26	46	0.9	0.8	0.7	0.9
OCEANIA	...	...	...	1	...	...	...	0.0
EUROPE	2 728	2 998	2 910	3 461	79.6	77.3	78.3	67.2
EC	867	834	1 078	1 838	25.3	21.5	29.0	35.7
EFTA	182	265	291	489	5.3	6.8	7.8	9.5
EASTERN EUROPE	730	894	324	581	21.3	23.0	8.7	11.3
former USSR	487	518	585	375	14.2	13.4	15.8	7.3
OTHER EUROPE	463	487	632	177	13.5	12.5	17.0	3.4
OTHERS	7	52	14	86	0.2	1.3	0.4	1.7
WORLD TOTAL	3 426	3 880	3 715	5 153	100.0	100.0	100.0	100.0

Table 3-9-4 Destination and relative importance of steel exports: *Poland*

Destination	Amount (1000 t)				Share (%)			
	1980	1985	1990	1992	1980	1985	1990	1992
AFRICA	9	1	2	227	0.5	0.0	0.1	6.7
MIDDLE EAST	188	138	241	456	9.8	6.5	6.8	13.4
FAR EAST	133	251	86	1 290	6.9	11.8	2.4	37.9
Japan	...	...	...	19	...	...	...	0.6
NORTH AMERICA	22	54	63	28	1.1	2.5	1.8	0.8
United States	19	43	61	28	1.0	2.0	1.7	0.8
OTHER AMERICA	70	26	3	127	3.6	1.2	0.1	3.7
OCEANIA	0	...	0	0	0.0	...	0.0	0.0
EUROPE	1 474	1 661	3 002	1 244	76.3	77.9	84.3	36.5
EC	328	283	1 393	911	17.0	13.3	39.1	26.7
EFTA	285	78	906	218	14.8	3.7	25.4	6.4
EASTERN EUROPE	467	409	70	77	24.2	19.2	2.0	2.2
former USSR	209	717	559	15	10.8	33.6	15.7	0.4
OTHER EUROPE	185	173	74	23	9.6	8.1	2.1	0.7
OTHERS	36	1	163	36	1.8	0.0	4.6	1.1
WORLD TOTAL	1 930	2 133	3 560	3 408	100.0	100.0	100.0	100.0

Table 3-9-5 Destination and relative importance of steel exports: *North America*

Destination	Amount (1000 t)				Share (%)			
	1980	1985	1990	1992	1980	1985	1990	1992
AFRICA	162	52	136	105	2.2	1.4	1.6	1.2
MIDDLE EAST	328	47	352	119	4.5	1.2	4.3	1.3
FAR EAST	1 232	236	1 358	975	16.7	6.2	16.4	10.9
Japan	76	8	456	131	1.0	0.2	5.5	1.5
NORTH AMERICA	2 678	2 949	4 605	5 266	36.4	77.7	55.7	58.8
United States	2 234	2 652	2 772	3 951	30.3	69.9	33.5	44.1
OTHER AMERICA	2 154	312	1 068	1 850	29.2	8.2	12.9	20.6
OCEANIA	20	11	21	16	0.3	0.3	0.2	0.2
EUROPE	789	179	673	567	10.7	4.7	8.1	6.3
EC	733	141	580	538	10.0	3.7	7.0	6.0
EFTA	19	4	20	12	0.3	0.1	0.2	0.1
EASTERN EUROPE	18	0	1	3	0.2	0.0	0.0	0.0
former USSR	0	...	57	...	0.0	...	0.7	...
OTHER EUROPE	18	34	16	14	0.2	0.9	0.2	0.2
OTHERS	...	8	54	65	...	0.2	0.6	0.7
WORLD TOTAL	7 363	3 794	8 267	8 963	100.0	100.0	100.0	100.0

Table 3-9-6 Destination and relative importance of steel exports: *Japan*

Destination	Amount (1000 t)				Share (%)			
	1980	1985	1990	1992	1980	1985	1990	1992
AFRICA	1 342	471	302	499	4.5	1.5	1.9	2.7
MIDDLE EAST	3 996	3 410	1 104	1 528	13.5	10.8	7.0	8.3
FAR EAST	12 940	17 902	9 620	12 357	43.7	56.8	60.8	66.9
Japan	.	.	.	.	.	.	.	.
NORTH AMERICA	5 268	5 288	3 114	2 706	17.8	16.8	19.7	14.7
United States	4 843	4 966	2 927	2 540	16.3	15.8	18.5	13.8
OTHER AMERICA	2 548	962	437	372	8.6	3.1	2.8	2.0
OCEANIA	646	737	300	302	2.2	2.3	1.9	1.6
EUROPE	2 889	2 719	898	547	9.7	8.6	5.7	3.0
EC	921	383	369	299	3.1	1.2	2.3	1.6
EFTA	147	128	123	200	0.5	0.4	0.8	1.1
EASTERN EUROPE	80	25	10	6	0.3	0.1	0.1	0.0
former USSR	1 651	2 114	355	...	5.6	...	2.2	...
OTHER EUROPE	90	69	42	42	0.3	0.2	0.3	0.2
OTHERS	...	1	59	151	...	0.0	0.4	0.8
WORLD TOTAL	29 629	31 490	15 835	18 462	100.0	100.0	100.0	100.0

Table 3-9-7 Destination and relative importance of steel exports: *Korea, Republic of*

Destination	Amount (1000 t)				Share (%)			
	1980	1985	1990	1992	1980	1985	1990	1992
AFRICA	31	18	12	28	0.7	0.3	0.2	0.2
MIDDLE EAST	915	861	230	563	20.4	15.2	3.2	4.8
FAR EAST	2 068	2 706	4 600	7 906	46.1	47.9	63.4	67.6
Japan	903	1 623	2 986	2 805	20.1	28.7	41.1	24.0
NORTH AMERICA	947	1 614	1 452	2 645	21.1	28.6	20.0	22.6
United States	907	1 532	1 415	2 604	20.2	27.1	19.5	22.3
OTHER AMERICA	125	73	82	127	2.8	1.3	1.1	1.1
OCEANIA	62	106	146	179	1.4	1.9	2.0	1.5
EUROPE	388	55	182	240	8.7	1.0	2.5	2.1
EC	386	55	134	207	8.6	1.0	1.8	1.8
EFTA	1	0	9	6	0.0	0.0	0.1	0.1
EASTERN EUROPE	0	...	...	0	0.0	...	...	0.0
former USSR	...	...	...	...	...	...	...	...
OTHER EUROPE	...	0	39	27	...	0.0	0.5	0.2
OTHERS	- 53	217	553	10	...	3.8	7.6	0.1
WORLD TOTAL	4 482	5 650	7 258	11 697	100.0	100.0	100.0	100.0

Table 3-9-8 Destination and relative importance of steel exports: *Brazil*

Destination	Amount (1000 t)				Share (%)			
	1980	1985	1990	1992	1980	1985	1990	1992
AFRICA	101	206	335	228	6.7	2.9	3.8	2.0
MIDDLE EAST	51	552	687	439	3.4	7.8	7.7	3.8
FAR EAST	73	3 052	4 996	5 726	4.8	42.9	55.9	49.4
Japan	6	581	687	444	0.4	8.2	7.7	3.8
NORTH AMERICA	571	1 480	1 528	1 488	38.0	20.8	17.1	12.8
United States	540	1 377	1 394	1 406	35.9	19.4	15.6	12.1
OTHER AMERICA	449	491	651	2 458	29.9	6.9	7.3	21.2
OCEANIA	5	61	58	74	0.3	0.9	0.6	0.6
EUROPE	296	1 268	876	1 170	19.7	17.8	9.8	10.1
EC	290	791	537	666	19.3	11.1	6.0	5.7
EFTA	2	9	29	9	0.1	0.1	0.3	0.1
EASTERN EUROPE	...	2	2	7	...	0.0	0.0	0.1
former USSR	0	126	6	...	0.0	...	0.1	...
OTHER EUROPE	4	341	303	488	0.2	4.8	3.4	4.2
OTHERS	- 42	- 2	- 192	...	...	0.0	-2.1	...
WORLD TOTAL	1 502	7 109	8 940	11 584	100.0	100.0	100.0	100.0

Figure 3-12-1 Destination and relative importance of steel exports: *EC*

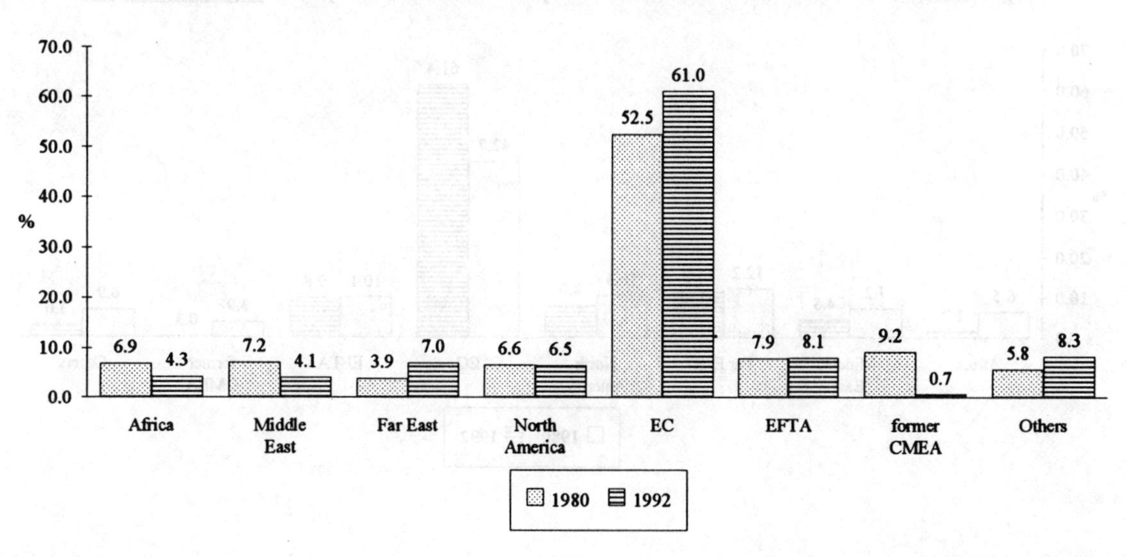

Figure 3-12-2 Destination and relative importance of steel exports: *Italy*

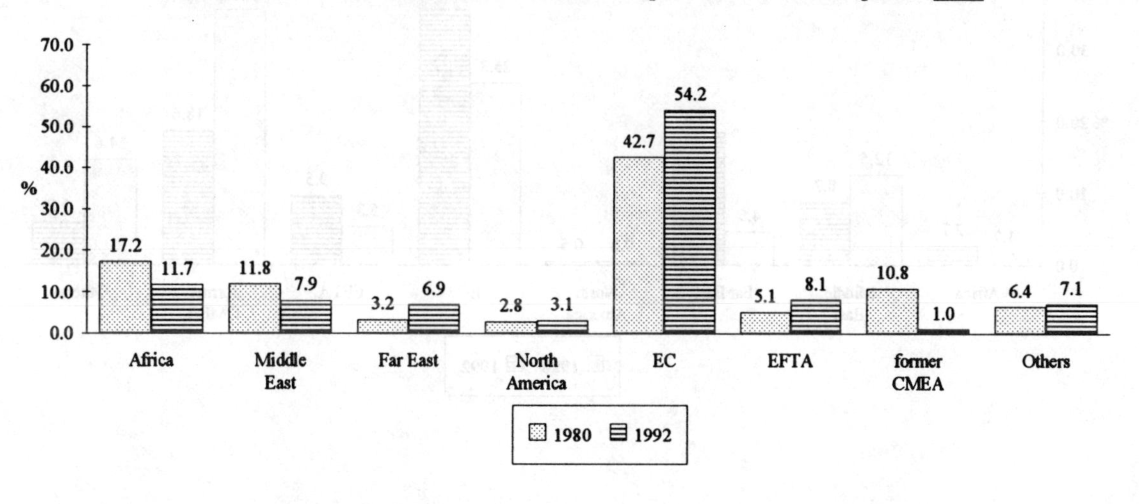

Figure 3-12-3 Destination and relative importance of steel exports: *Spain*

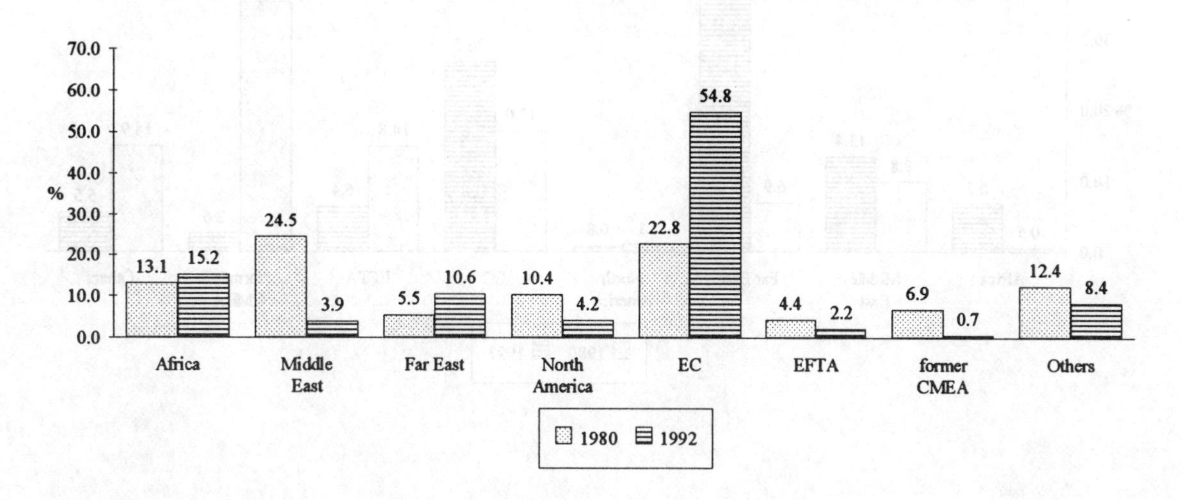

Figure 3-12-4 Destination and relative importance of steel exports: *United Kingdom*

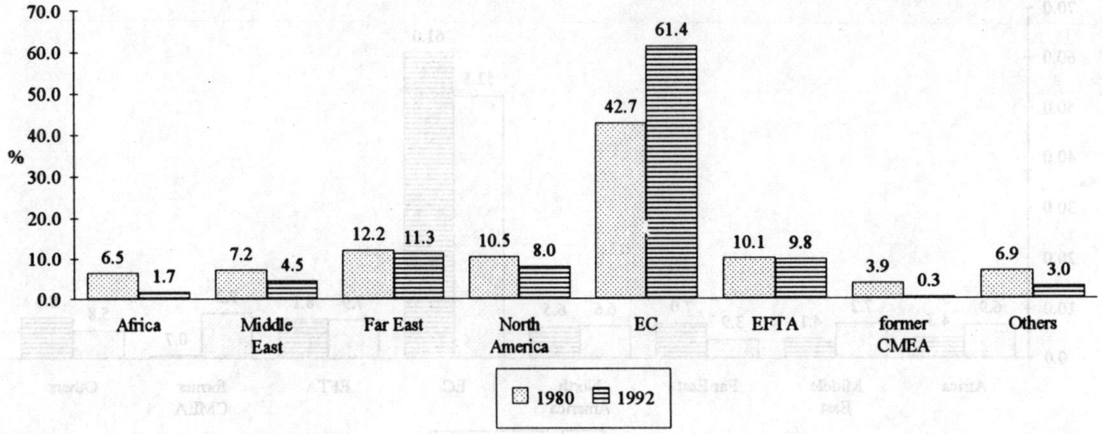

%	Africa	Middle East	Far East	North America	EC	EFTA	former CMEA	Others
1980	6.5	7.2	12.2	10.5	42.7	10.1	3.9	6.9
1992	1.7	4.5	11.3	8.0	61.4	9.8	0.3	3.0

Figure 3-12-5 Destination and relative importance of steel exports: *former Czechoslovakia*

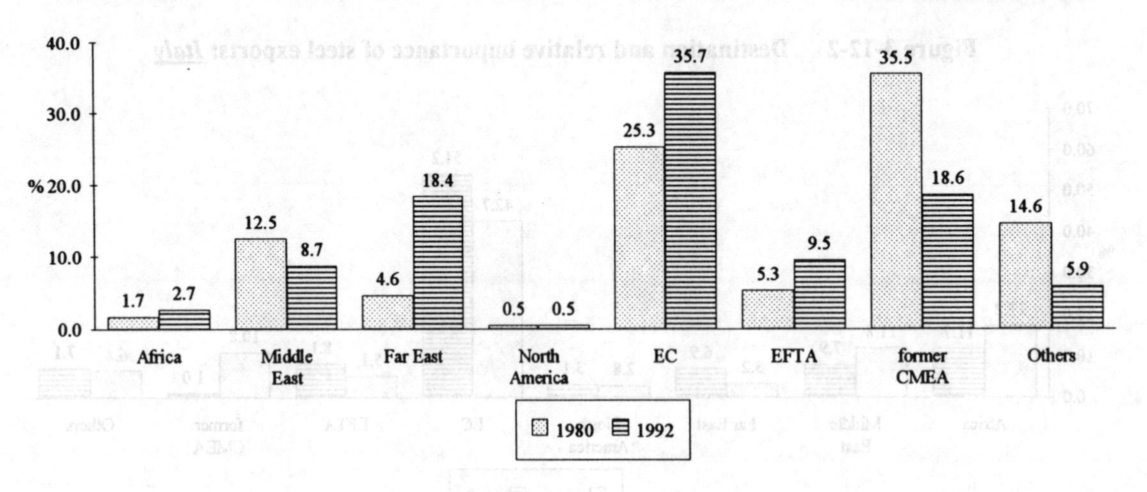

Figure 3-12-6 Destination and relative importance of steel exports: *Poland*

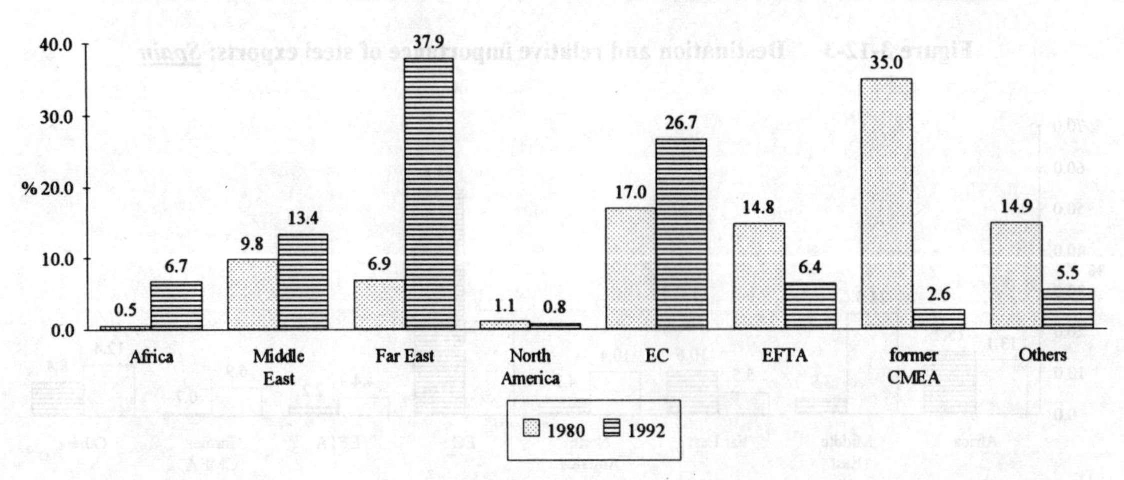

Figure 3-12-7 Destination and relative importance of steel exports: *United States*

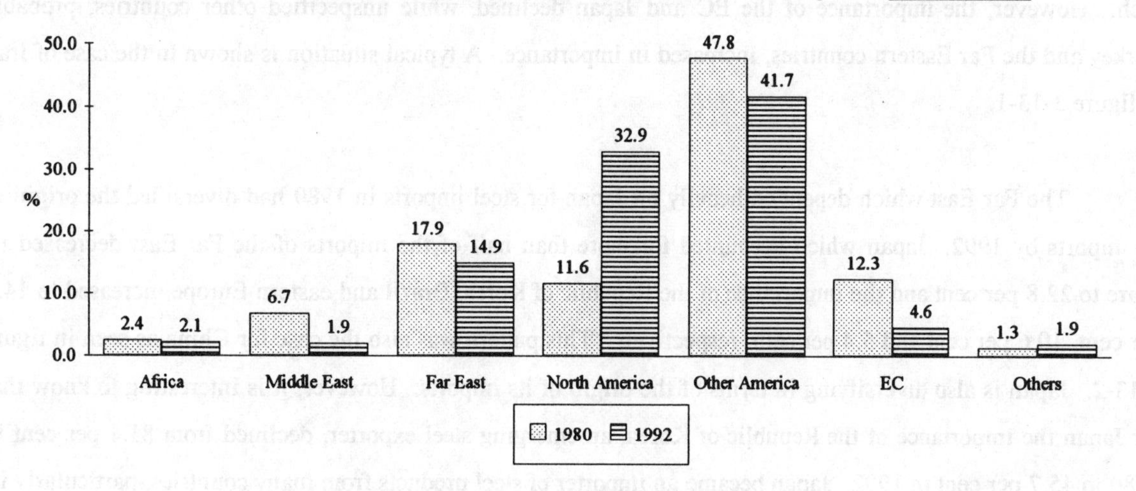

Figure 3-12-8 Destination and relative importance of steel exports: *Japan*

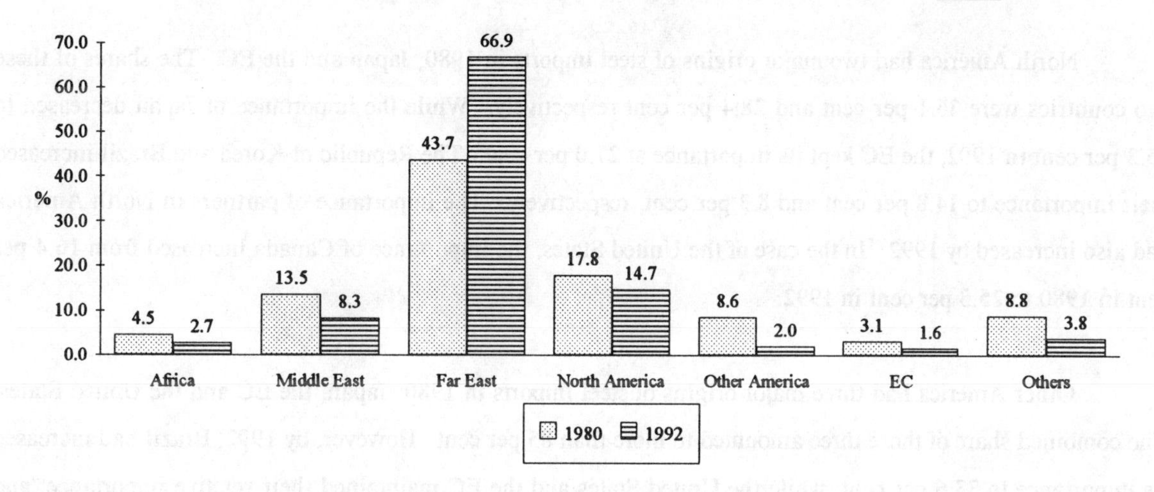

3.2.2. Origin of steel imports by major regions and countries

Tables 3-10-1 and 3-10-2 summarize the origin and relative importance of steel imports by region in 1980 and 1992. Figures 3-13-1 to 3-13-8 show changes in the relative importance of steel imports by region.

73.3 per cent of the steel imports of Africa originated from the EC countries in 1980. Italy had the highest share among the member States. Belgium-Luxembourg, France, Germany and Spain also had high shares. Japan had the highest share of 22.4 per cent as a single country. In 1992, the situation had not changed dramatically. Even though the total imports of Africa declined from 6.0 million tonnes in 1980 to 4.4 million tonnes in 1992, the EC kept its importance with a share of 71.8 per cent. Interestingly, the eastern European countries increased both volume and importance, while Japan's importance slipped to 11.3 per cent.

The Middle East imported mainly from the EC and Japan, which had shares of around 35 per cent each. However, the importance of the EC and Japan declined, while unspecified other countries, probably Turkey and the Far Eastern countries, increased in importance. A typical situation is shown in the case of Iran in figure 3-13-1.

The Far East which depended heavily on Japan for steel imports in 1980 had diversified the origin of its imports by 1992. Japan which accounted for more than half of the imports of the Far East decreased its share to 22.8 per cent and the importance of the Republic of Korea, Brazil and eastern Europe increased to 14.6 per cent, 10.6 per cent and 5.4 per cent, respectively. This pattern was also the case for China as seen in figure 3-13-2. Japan is also diversifying in terms of the origin of its imports. However, it is interesting to know that for Japan the importance of the Republic of Korea, an emerging steel exporter, declined from 81.4 per cent in 1980 to 45.7 per cent in 1992. Japan became an importer of steel products from many countries, particularly its Asian partners. In the case of the Republic of Korea, the importance of Japan declined sharply from 85.1 per cent in 1980 to 23.8 per cent in 1992 and the share of Brazil and unspecified countries increased.

North America had two major origins of steel imports in 1980, Japan and the EC. The shares of these two countries were 36.1 per cent and 28.4 per cent respectively. While the importance of Japan decreased to 15.2 per cent in 1992, the EC kept its importance at 27.0 per cent. The Republic of Korea and Brazil increased their importance to 14.8 per cent and 8.3 per cent, respectively. The importance of partners in North America had also increased by 1992. In the case of the United States, the importance of Canada increased from 16.4 per cent in 1980 to 25.3 per cent in 1992.

Other America had three major origins of steel imports in 1980: Japan, the EC and the United States. The combined share of these three amounted to more than 85 per cent. However, by 1992, Brazil had increased its importance to 33.6 per cent, while the United States and the EC maintained their relative importance, and Japan's rapidly decreased to only 5.1 per cent.

As stated previously, the EC relied heavily on internal trade and this situation did not change from 1980 to 1992, even though imports increased by 18.8 million tonnes during the period. The share of internal trade amounted to 73.4 per cent in 1992, higher than the figure for exports, and increases in imports from, for example, EFTA and some eastern European countries were marginal. Among the member States, Germany, interestingly, decreased its dependence on EC steel imports. The share of intra-regional trade was 50.3 per cent in 1992, 20 per cent less than the EC average. In the case of the United Kingdom, the importance of intra-regional trade increased to 82.7 per cent in 1992.

EFTA depended more on trade with the EC than its intra-regional trade throughout the period. Because eastern Europe and the former USSR decreased their imports very rapidly to marginal amounts and because of the lack of data, it is not meaningful to point out structural changes in the import origins for these regions.

Table 3-10-1 Origin and relative importance of steel imports by region: *1980*

Origin of imports	Africa		Middle East		Far East		North America		Other America	
	Amount (1000t)	Share (%)	Amount (1000t)	Share (%)	Amount (1000t)	Share (%)	Amount (1000t)	Share (%)	Amount (1000t)	Share (%)
EC	4 387	73.3	4 540	38.6	2 485	10.5	4 148	28.4	2 510	30.9
Belgium-Luxembourg	680	11.4	469	4.0	204	0.9	825	5.7	202	2.5
France	819	13.7	467	4.0	405	1.7	883	6.0	658	8.1
Germany	721	12.0	1 180	10.0	924	3.9	1 184	8.1	670	8.3
Italy	1 160	19.4	796	6.8	215	0.9	192	1.3	247	3.0
Netherlands	149	2.5	99	0.8	145	0.6	249	1.7	216	2.7
Spain	596	10.0	1 109	9.4	251	1.1	474	3.2	398	4.9
United Kingdom	180	3.0	200	1.7	338	1.4	289	2.0	113	1.4
EFTA	69	1.2	147	1.2	173	0.7	165	1.1	84	1.0
EASTERN EUROPE	...	...	...	...	...	...	...	...	...	...
former Czechoslovakia	59	1.0	428	3.6	157	0.7	18	0.1	30	0.4
Poland	9	0.2	188	1.6	133	0.6	22	0.2	70	0.9
Romania	...	...	...	...	...	...	...	...	...	...
former USSR	...	...	...	...	...	...	...	...	...	...
Canada	71	1.2	72	0.6	547	2.3	2 235	15.3	317	3.9
United States	91	1.5	257	2.2	686	2.9	444	3.0	1 836	22.6
Japan	1 342	22.4	3 996	34.0	12 940	54.8	5 268	36.1	2 548	31.4
Korea, Republic of	31	0.5	915	7.8	2 068	8.8	947	6.5	125	1.5
Brazil	101	1.7	51	0.4	73	0.3	571	3.9	449	5.5
Others	- 105	-1.8	1 785	15.2	4 628	19.6	823	5.6	249	3.1
WORLD TOTAL	5 987	100.0	11 761	100.0	23 599	100.0	14 601	100.0	8 118	100.0

Origin of imports	EC		EFTA		Eastern Europe		former USSR		Other Europe	
	Amount (1000t)	Share (%)	Amount (1000t)	Share (%)	Amount (1000t)	Share (%)	Amount (1000t)	Share (%)	Amount (1000t)	Share (%)
EC	33 276	77.4	5 013	73.1	1 195	11.9	4 638	51.2	989	37.1
Belgium-Luxembourg	9 887	23.0	691	10.1	147	1.5	444	4.9	82	3.1
France	5 912	13.7	653	9.5	168	1.7	631	7.0	83	3.1
Germany	8 684	20.2	2 183	31.8	549	5.5	2 552	28.2	369	13.9
Italy	2 877	6.7	344	5.0	146	1.5	579	6.4	158	5.9
Netherlands	3 180	7.4	411	6.0	18	0.2	117	1.3	30	1.1
Spain	1 032	2.4	198	2.9	91	0.9	221	2.4	151	5.7
United Kingdom	1 181	2.7	279	4.1	16	0.2	92	1.0	48	1.8
EFTA	3 896	9.1	1 186	17.3	177	1.8	561	6.2	202	7.6
EASTERN EUROPE	...	...	...	...	...	...	...	...	...	...
former Czechoslovakia	867	2.0	182	2.7	730	7.3	487	5.4	463	17.4
Poland	328	0.8	285	4.2	467	4.6	209	2.3	185	6.9
Romania	...	...	...	...	...	...	...	...	...	...
former USSR	...	...	...	...	...	...	.	.	...	...
Canada	262	0.6	4	0.1	0	0.0	0	0.0	7	0.3
United States	471	1.1	15	0.2	17	0.2	0	0.0	11	0.4
Japan	921	2.1	147	2.1	80	0.8	1 651	18.2	90	3.4
Korea, Republic of	386	0.9	1	0.0	0	0.0	...	...	...	...
Brazil	290	0.7	2	0.0	...	...	0	0.0	4	0.1
Others	3 512	8.2	488	7.1	8 581	85.4	2 214	24.4	1 360	51.1
WORLD TOTAL	43 014	100.0	6 856	100.0	10 050	100.0	9 064	100.0	2 663	100.0

Table 3-10-2 Origin and relative importance of steel imports by region: *1992*

Origin of imports	Africa		Middle East		Far East		North America		Other America	
	Amount (1000t)	Share (%)	Amount (1000t)	Share (%)	Amount (1000t)	Share (%)	Amount (1000t)	Share (%)	Amount (1000t)	Share (%)
EC	3 178	71.8	3 046	22.9	5 243	9.7	4 826	27.0	1 262	17.2
Belgium-Luxembourg	231	5.2	318	2.4	613	1.1	520	2.9	135	1.8
France	359	8.1	327	2.5	670	1.2	910	5.1	168	2.3
Germany	505	11.4	852	6.4	1 579	2.9	1 501	8.4	326	4.5
Italy	1 048	23.7	712	5.4	617	1.1	278	1.6	157	2.1
Netherlands	95	2.1	146	1.1	265	0.5	656	3.7	95	1.3
Spain	725	16.4	187	1.4	507	0.9	200	1.1	255	3.5
United Kingdom	143	3.2	378	2.8	953	1.8	673	3.8	123	1.7
EFTA	120	2.7	129	1.0	675	1.2	452	2.5	56	0.8
EASTERN EUROPE	398	9.0	1 244	9.4	2 927	5.4	76	0.4	187	2.6
former Czechoslovakia	138	3.1	448	3.4	949	1.8	23	0.1	46	0.6
Poland	227	5.1	456	3.4	1 290	2.4	28	0.2	127	1.7
Romania	9	0.2	59	0.4	322	0.6	2	0.0	6	0.1
former USSR	...	...	...	...	...	...	...	...	...	...
Canada	19	0.4	42	0.3	379	0.7	3 951	22.1	183	2.5
United States	85	1.9	77	0.6	596	1.1	1 314	7.4	1 667	22.8
Japan	499	11.3	1 528	11.5	12 357	22.8	2 706	15.2	372	5.1
Korea, Republic of	28	0.6	563	4.2	7 906	14.6	2 645	14.8	127	1.7
Brazil	228	5.2	439	3.3	5 726	10.6	1 488	8.3	2 458	33.6
Others	- 126	-2.8	6 218	46.8	18 403	33.9	385	2.2	1 007	13.8
WORLD TOTAL	4 429	100.0	13 285	100.0	54 212	100.0	17 842	100.0	7 319	100.0

Origin of imports	EC		EFTA		Eastern Europe		former USSR		Other Europe	
	Amount (1000t)	Share (%)	Amount (1000t)	Share (%)	Amount (1000t)	Share (%)	Amount (1000t)	Share (%)	Amount (1000t)	Share (%)
EC	45 379	73.4	6 034	79.7	498	26.1	8	0.4	938	27.7
Belgium-Luxembourg	10 877	17.6	852	11.3	11	0.6	...	...	133	3.9
France	6 008	9.7	626	8.3	23	1.2	...	...	114	3.4
Germany	10 847	17.5	2 083	27.5	280	14.7	...	...	147	4.3
Italy	4 860	7.9	725	9.6	92	4.8	...	...	268	7.9
Netherlands	3 943	6.4	569	7.5	5	0.3	1	0.1	82	2.4
Spain	2 609	4.2	106	1.4	35	1.8	...	...	60	1.8
United Kingdom	5 182	8.4	824	10.9	27	1.4	...	...	108	3.2
EFTA	6 109	9.9	1 555	20.5	154	8.1	...	...	49	1.4
EASTERN EUROPE	3 500	5.7	831	11.0	724	37.9	509	25.5	612	18.1
former Czechoslovakia	1 838	3.0	489	6.5	581	30.4	375	18.8	177	5.2
Poland	911	1.5	218	2.9	77	4.0	15	0.8	23	0.7
Romania	122	0.2	32	0.4	16	0.8	56	2.8	144	4.3
former USSR	...	...	...	...	...	...	.	.	...	...
Canada	355	0.6	1	0.0	0	0.0	...	...	8	0.2
United States	183	0.3	11	0.1	3	0.2	...	...	6	0.2
Japan	299	0.5	200	2.6	6	0.3	...	...	42	1.2
Korea, Republic of	207	0.3	6	0.1	0	0.0	...	...	27	0.8
Brazil	666	1.1	9	0.1	7	0.4	...	...	488	14.4
Others	5 150	8.3	-1 080	-14.3	518	27.1	1 483	74.2	1 216	35.9
WORLD TOTAL	61 848	100.0	7 567	100.0	1 910	100.0	2 000	100.0	3 387	100.0

Figure 3-13-1 Origin and relative importance of steel imports: *Iran*

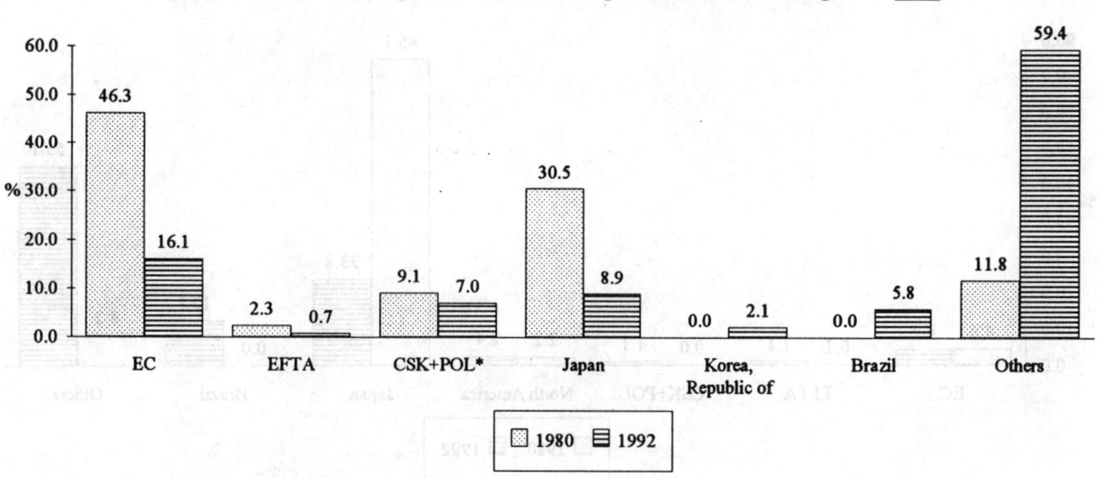

* CSK+POL is the former Czechoslovakia and Poland.

Figure 3-13-2 Origin and relative importance of steel imports: *China*

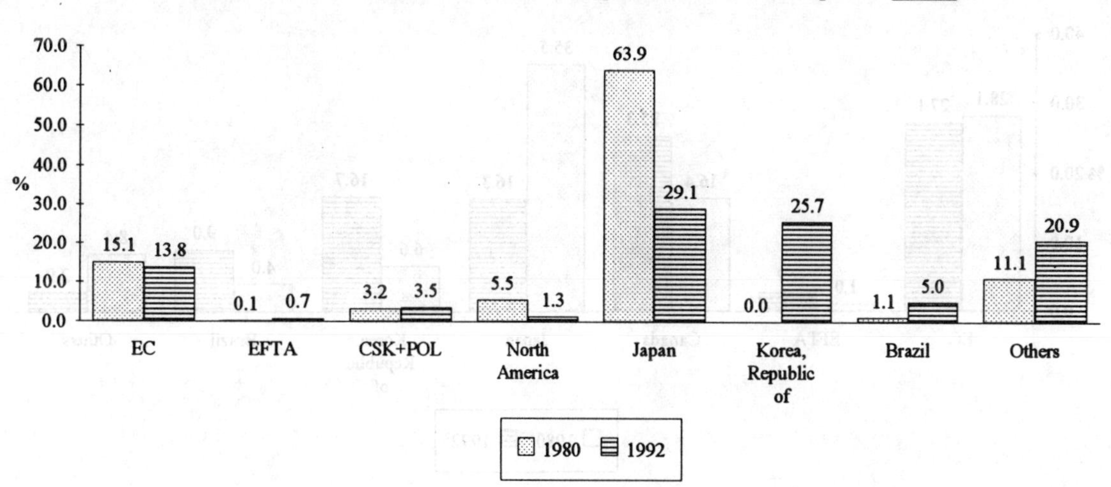

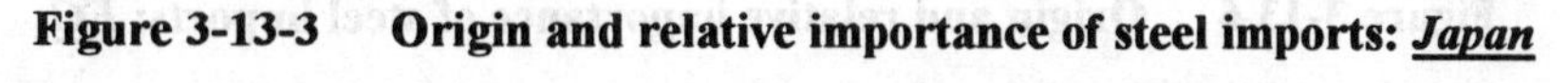

Figure 3-13-3 Origin and relative importance of steel imports: *Japan*

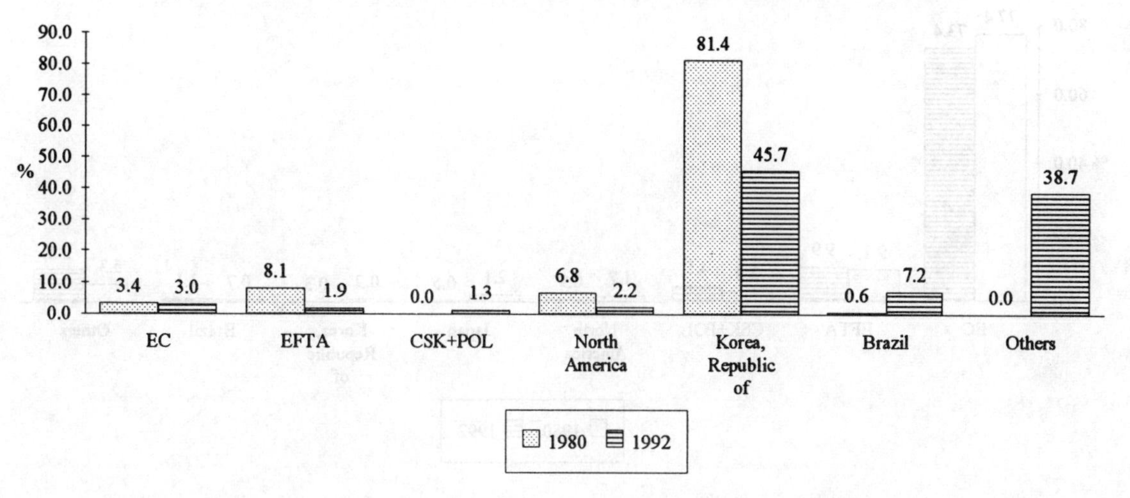

Figure 3-13-4 Origin and relative importance of steel imports: *Korea, Republic of*

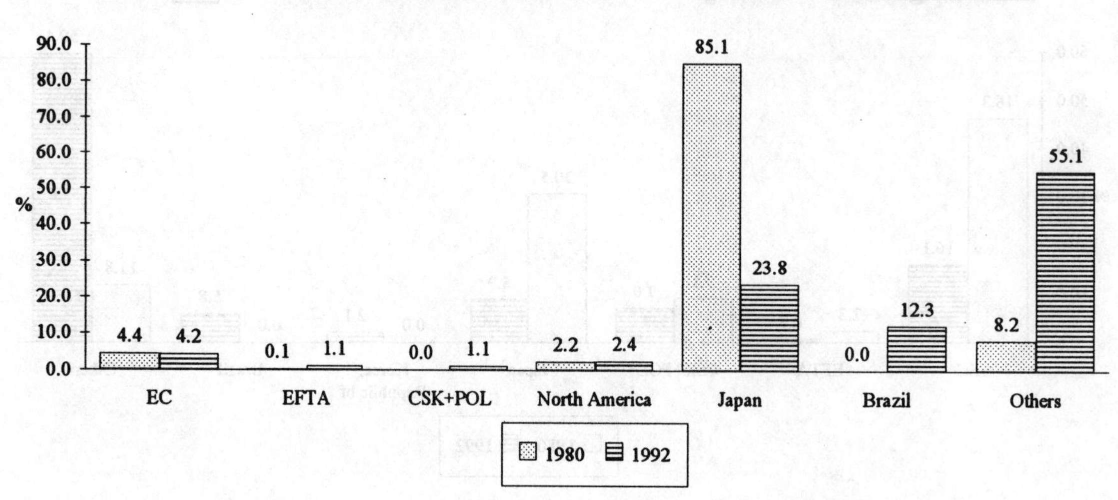

Figure 3-13-5 Origin and relative importance of steel imports: *United States*

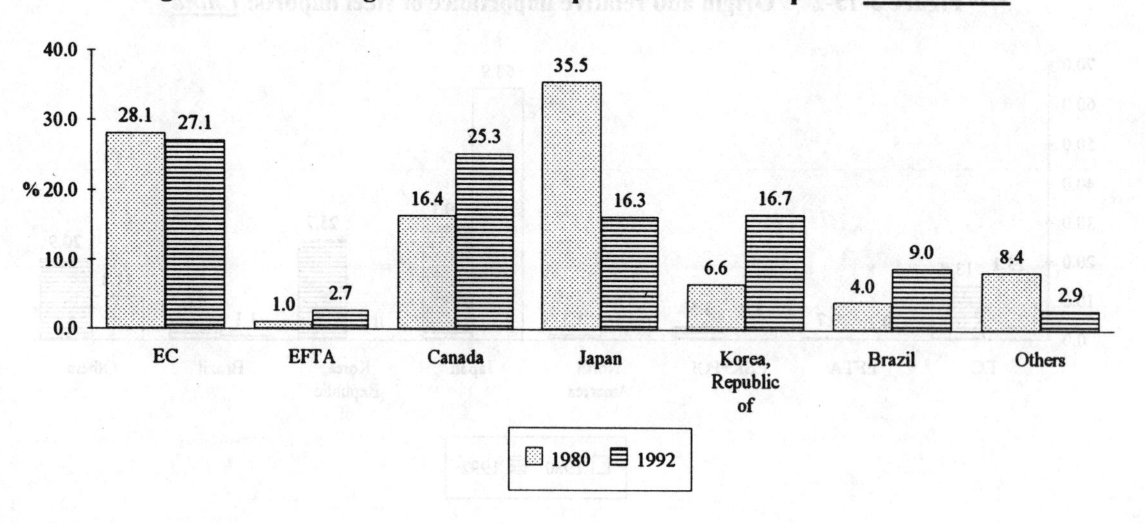

Figure 3-13-6 Origin and relative importance of steel imports: *EC*

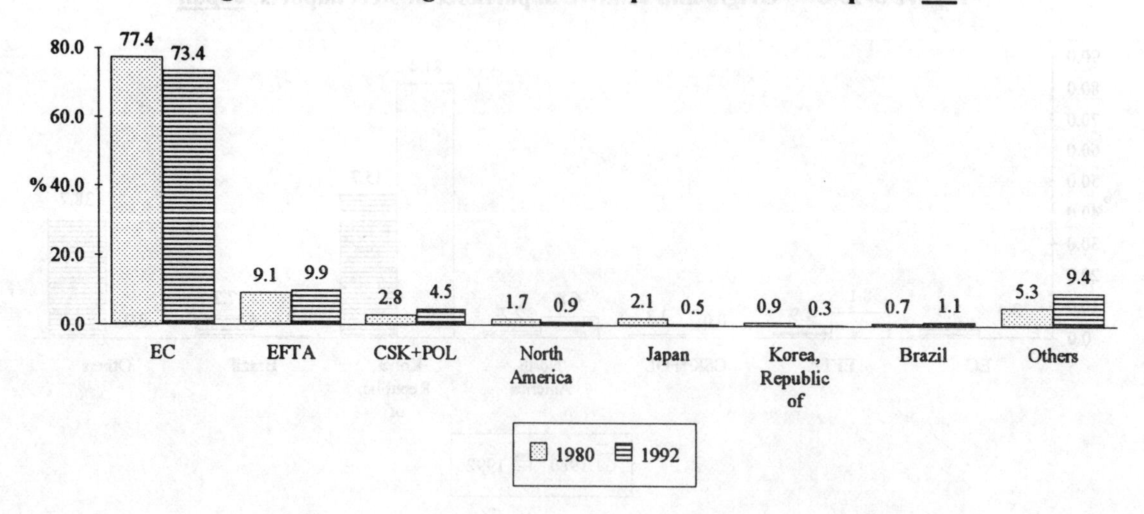

Figure 3-13-7 Origin and relative importance of steel imports: *Germany*

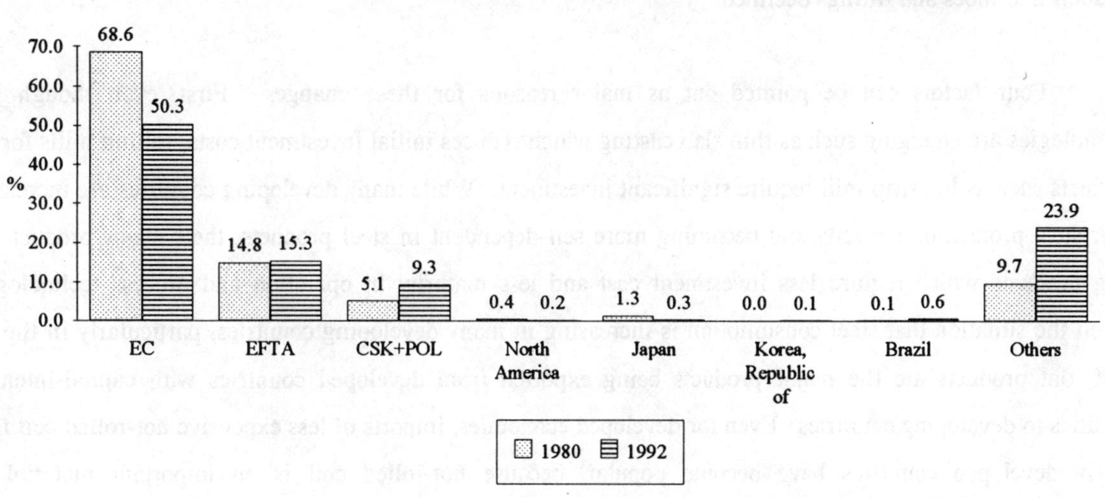

Figure 3-13-8 Origin and relative importance of steel imports: *United Kingdom*

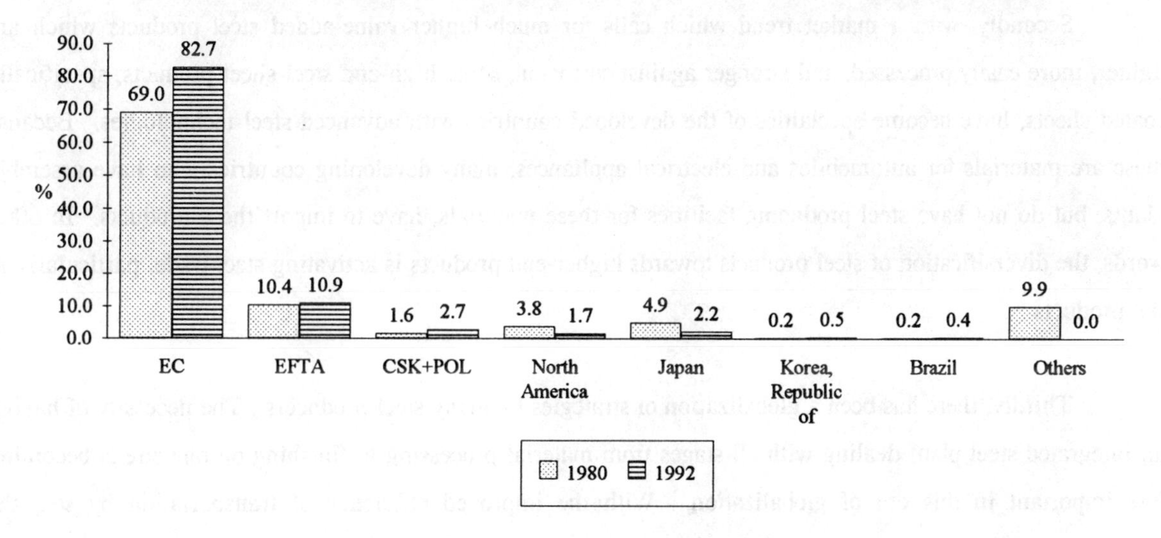

3.3. Changes in product structure of world steel trade

3.3.1. Evolution in product structure of steel exports

Table 3-11 and Figure 3-14 summarize changes in the product structure of steel trade by major exporters, covering 89.3 per cent of total world trade in 1980 and 83.2 per cent in 1992. Among various interesting observations, the first is that the share of flat products increased. Trade in flat products increased by almost 20 million tonnes over the 12 years. The share of flat products increased from 46.6 per cent in 1980 to 53.6 per cent in 1992. Particularly after 1985, this was the sole engine behind increases in world trade. Secondly, trade in ingots and semis increased by 5.9 million tonnes and its share increased from 7.2 per cent in

1980 to 10.2 per cent in 1992. Finally, contrary to the trend in flat products and ingots and semis, trade in long products and tubes and fittings declined.

Four factors can be pointed out as major reasons for these changes. First, even though new technologies are emerging such as thin slab casting which reduces initial investment costs, rolling mills for flat products such as hot strip mill require significant investment. While many developing countries are increasing their steel production capacity and becoming more self-dependent in steel products, their major products are long products which require less investment cost and less maturity in operation and process technologies. Given the situation that steel consumption is increasing in many developing countries, particularly in the Far East, flat products are the major products being exported from developed countries with capital-intensive facilities to developing countries. Even for developed economies, imports of less expensive hot-rolled coil from nearly developed countries have become popular, because hot-rolled coil is an important material for construction, tube manufacturing and other uses, and does not require so much maturity in technologies or quality.

Secondly, with a market trend which calls for much higher value-added steel products which are lighter, more easily processed, and stronger against corrosion, some high-end steel sheet products, specifically coated sheets, have become specialties of the developed countries with advanced steel technologies. Because these are materials for automobiles and electrical appliances, many developing countries who have assembly plants, but do not have steel producing facilities for these materials, have to import these products. In other words, the diversification of steel products towards higher-end products is activating steel trade, particularly in flat products.

Thirdly, there has been a globalization of strategies by many steel producers. The necessity of having an integrated steel plant dealing with all stages from material processing to finishing on one site is becoming less important in this era of globalization. With the improved efficiency of transportation by sea, the transportation costs of heavy steel products have declined. Many integrated steel producers have been trying to set up joint activities with foreign producers for the exchange of steel products and semis.

Table 3-11 Structure of world exports by product group a/

Products	Amount (1000t)				Share (%)			
	1980	1985	1990	1992	1980	1985	1990	1992
Ingots and semis	9 041	13 870	13 075	14 959	7.2	9.7	9.7	10.2
Long products	39 633	44 561	36 554	37 603	31.7	31.1	27.1	25.8
Flat products	58 263	63 737	70 504	78 246	46.6	44.5	52.3	53.6
Tubes and fittings	18 009	20 923	14 542	15 067	14.4	14.6	10.8	10.3
Total	124 945	143 091	134 686	145 955	100.0	100.0	100.0	100.0

a/ Figures are sum of EC, EFTA, EASTERN EUROPE, Canada, USA, Japan, Republic of Korea and Brazil.
Total exports by these regions and countries amounted to 89.3% of total world exports in 1980 and 83.2% in 1992.

One example is a joint venture started in 1986 between USS, a subsidiary of USX Corporation in the United States, and Pohang Iron and Steel Ltd. (POSCO) in the Republic of Korea. POSCO exports hot-rolled coil to California where USS carries out cold-rolling and successive finishing processes. The amount of exports of hot-rolled coil varied from 600 thousand tonnes to 850 thousand tonnes per year from 1990 and 1992. Another example is a new joint venture between a Thai investor and Japan's NKK to produce electro-galvanized sheets in Thailand. In this joint venture, it is reported that the company is currently sourcing most of the cold-rolled sheets required to feed its 135 thousand tonne per year electro-galvanizing line from the Republic of Korea and Taiwan Province of China, but is also planning to import them from Brazil and China in the future. In the European Community, the same kind of exchange involving hot-rolled coil is becoming

Figure 3-14 Structure of world exports by product group a/

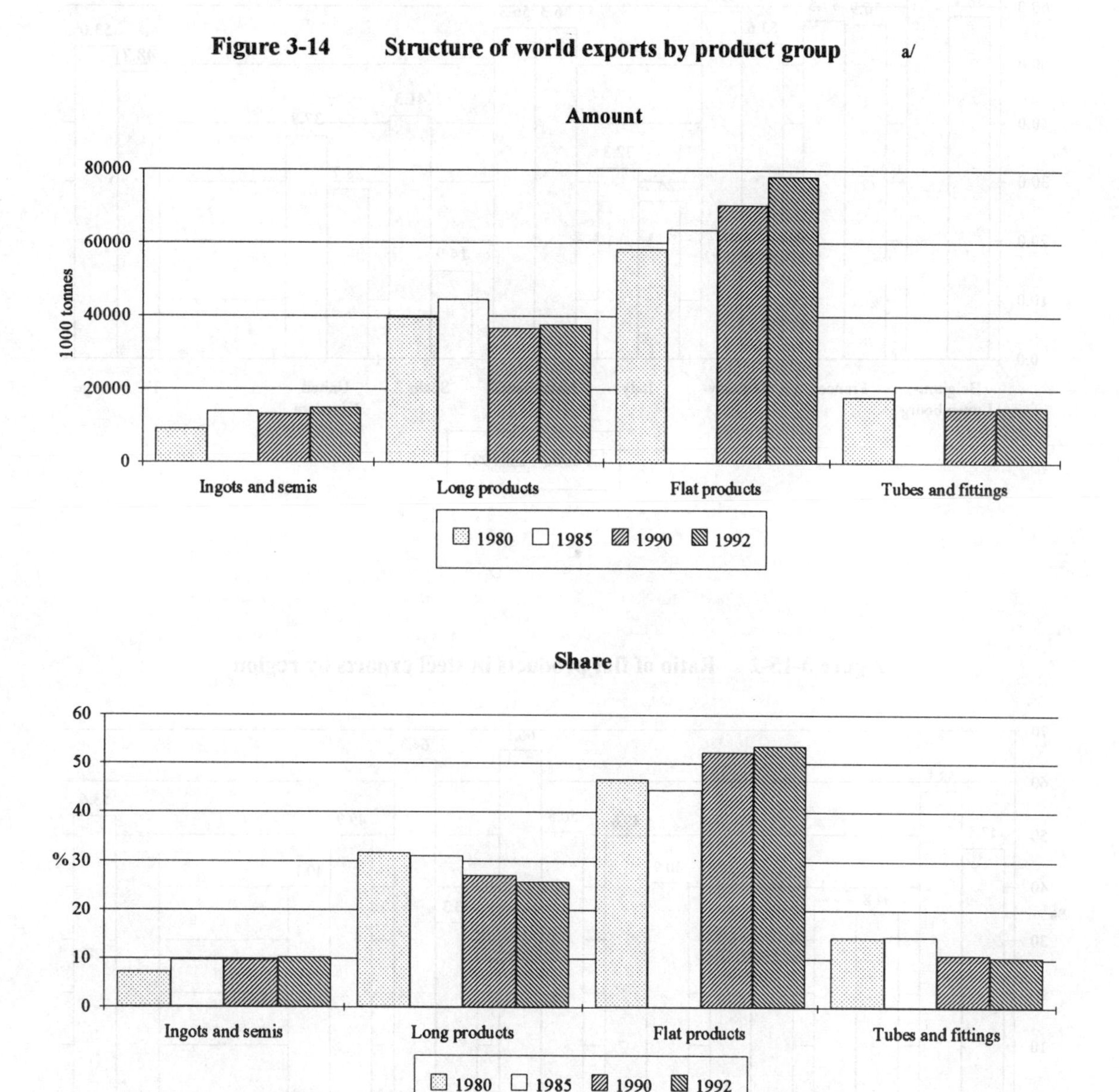

a/ Figures are sum of EC, EFTA, EASTERN EUROPE, Canada, USA, Japan, Republic of Korea and Brazil. Total exports by these regions and countries amounted to 89.3% of total world exports in 1980 and 83.2% in 1992.

common. The increasing trade in ingots and semis is a result of the same factor. Brazil has been exporting slabs to the United States, which are then rolled and further processed in that country.

Finally, regarding the sluggish tubes and fittings trade, the continuing low price of crude oil is one of the factors. After the oil drilling boom at the beginning of the 1980s, oil prices have been low and stable. This

Figure 3-15-1 Ratio of flat products in steel exports in the EC

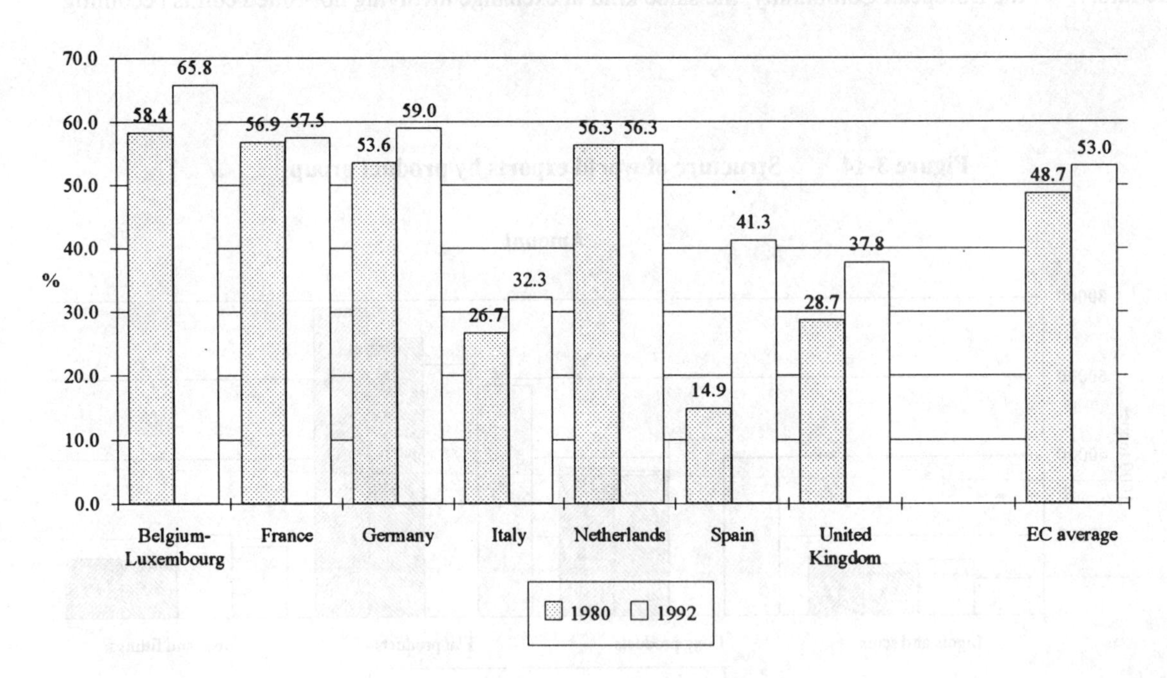

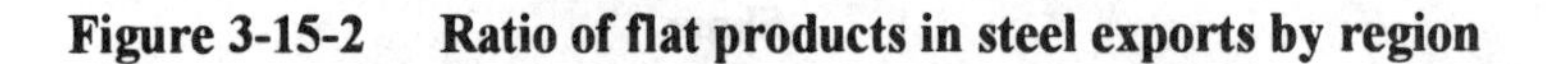

Figure 3-15-2 Ratio of flat products in steel exports by region

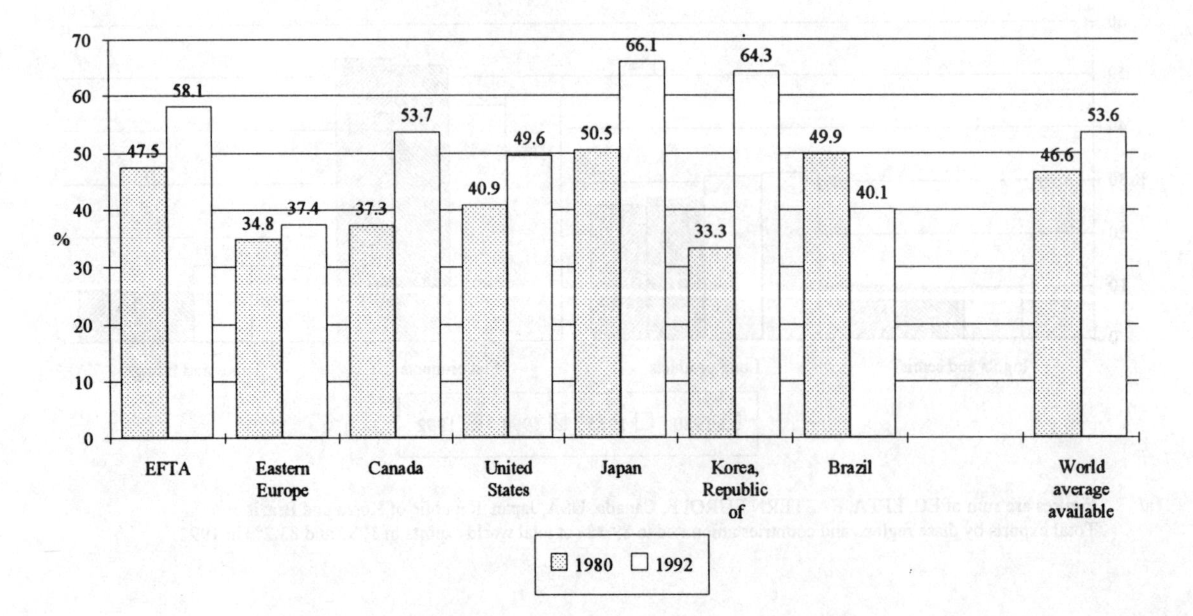

brought about a depressed market for the oil drilling business and cooled demand for OCTG products. In addition to this, the breakup of the former USSR and the difficult economic situation in the successor States reduced demand for pipes and tubes, these countries being traditional large importers of tube and pipe products.

As seen in tables 3-12-1 and 3-12-2, there is great diversity of products among the major exporters. Japan and Belgium-Luxembourg had the highest share of flat products in their exports with figures of 66.1

Table 3-12-1 Structure of exports by product group in 1980

Regions and countries	Amount (1000t)				Share (%)			
	Ingots and semis	Long products	Flat products	Tubes and fittings	Ingots and semis	Long products	Flat products	Tubes and fittings
EC	4 358	20 464	30 840	7 665	6.9	32.3	48.7	12.1
Belgium-Luxembourg	704	4 623	7 966	359	5.2	33.9	58.4	2.6
France	360	3 075	6 087	1 185	3.4	28.7	56.9	11.1
Germany	1 487	4 317	10 199	3 031	7.8	22.7	53.6	15.9
Italy	189	3 151	1 800	1 606	2.8	46.7	26.7	23.8
Netherlands	878	650	2 600	488	19.0	14.1	56.3	10.6
Spain	545	2 846	675	467	12.0	62.8	14.9	10.3
United Kingdom	185	1 429	793	359	6.7	51.7	28.7	13.0
EFTA	776	2 041	3 170	690	11.6	30.6	47.5	10.3
EASTERN EUROPE	1 066	5 311	4 089	1 276	9.1	45.2	34.8	10.9
former Czechoslovakia	341	1 369	1 191	525	10.0	40.0	34.8	15.3
Poland	87	1 240	555	48	4.5	64.2	28.8	2.5
Romania	34	810	707	433	1.7	40.8	35.6	21.8
former USSR	...	...	...	...	...	...	...	...
Canada	327	1 492	1 314	389	9.3	42.4	37.3	11.0
United States	830	953	1 571	488	21.6	24.8	40.9	12.7
Japan	189	8 032	14 957	6 453	0.6	27.1	50.5	21.8
Korea, Republic of	1 209	1 108	1 551	792	25.9	23.8	33.3	17.0
Brazil	285	232	771	256	18.5	15.0	49.9	16.6
World Total Available	9 040	39 633	58 263	18 009	7.2	31.7	46.6	14.4

Table 3-12-2 Structure of exports by product group in 1992

Regions and countries	Amount (1000t)				Share (%)			
	Ingots and semis	Long products	Flat products	Tubes and fittings	Ingots and semis	Long products	Flat products	Tubes and fittings
EC	5 901	21 762	39 452	7 262	7.9	29.3	53.0	9.8
Belgium-Luxembourg	487	3 698	9 018	503	3.6	27.0	65.8	3.7
France	1 000	3 013	6 789	995	8.5	25.5	57.5	8.4
Germany	1 448	4 132	11 104	2 152	7.7	21.9	59.0	11.4
Italy	379	3 997	2 871	1 645	4.3	45.0	32.3	18.5
Netherlands	1 068	834	3 367	715	17.8	13.9	56.3	11.9
Spain	257	2 245	1 968	293	5.4	47.1	41.3	6.2
United Kingdom	1 221	3 280	3 187	751	14.5	38.9	37.8	8.9
EFTA	527	2 564	5 595	940	5.5	26.6	58.1	9.8
EASTERN EUROPE	2 443	3 763	4 175	786	21.9	33.7	37.4	7.0
former Czechoslovakia	628	1 675	2 351	489	12.2	32.6	45.7	9.5
Poland	1 642	1 341	318	107	48.2	39.3	9.3	3.1
Romania	10	304	362	110	1.3	38.7	46.1	14.0
former USSR	...	...	...	...	...	...	...	...
Canada	269	1 588	2 664	442	5.4	32.0	53.7	8.9
United States	384	976	1 984	656	9.6	24.4	49.6	16.4
Japan	325	2 944	12 212	2 981	1.8	15.9	66.1	16.1
Korea, Republic of	470	1 893	7 516	1 817	4.0	16.2	64.3	15.5
Brazil	4 640	2 113	4 648	183	40.1	18.2	40.1	1.6
World Total Available	14 959	37 603	78 246	15 067	10.3	25.8	53.6	10.3

per cent and 65.8 per cent in 1992, respectively. Italy and Spain had a larger share of long products than flat products in both 1980 and 1992, even though both countries rapidly increased their shares of flat products. Poland and Brazil had high shares of ingots and semis in 1992 with figures of 48.2 per cent and 40.1 per cent, respectively. Interestingly, an increase in exports of ingots and semis by Poland occurred suddenly after the dissolution of the former CMEA system.

Figure 3-16 Structure of exports by product group in selected countries (share)

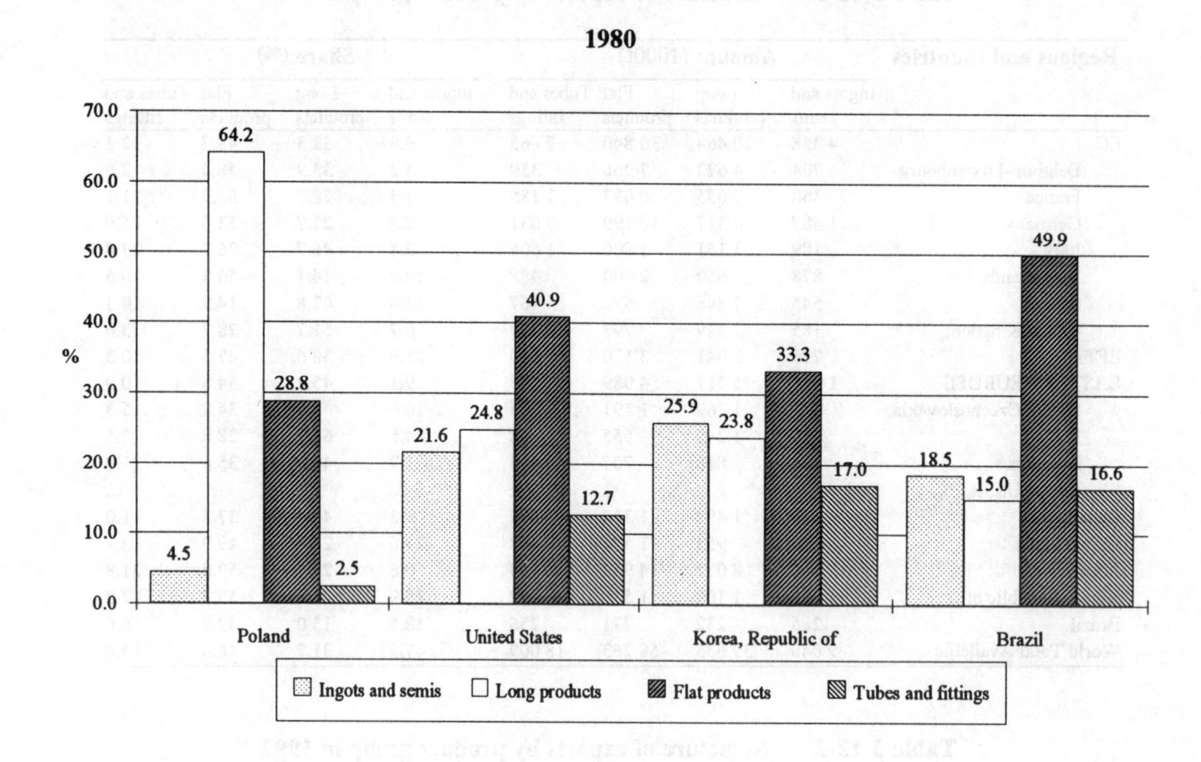

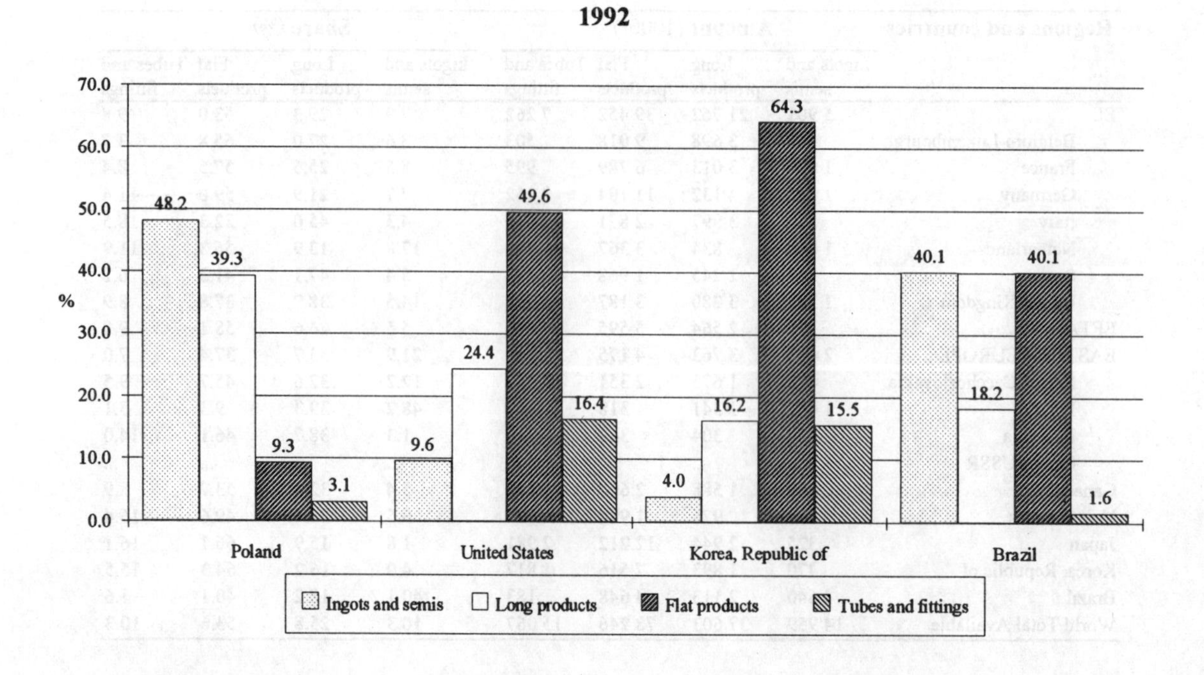

3.3.2. Evolution in product structure of steel imports

Because exports and imports are both sides of trade, the general trend in structural changes on a worldwide basis is the same. Among the import structures of major importers, the following points should be mentioned. Belgium-Luxembourg had a high share of imports of ingots and semis throughout the period. Because Belgium-Luxembourg is a big exporter among the member States of the EC, this suggests that these countries work as a processing centre in the EC, getting supplies of ingots and semis, carrying out hot-rolling and other processing, and exporting finished products to other member States.

Contrary to its high share of exports of long products, Italy has a higher share of imports of flat products. This suggests that production in Italy is still rather concentrated in long products and it tends to import flat products in exchange for long product exports.

The United States also had a relatively high share of ingots and semis in 1992. Japan increased its share of flat product imports to 73.6 per cent in 1992. Japan also seems to be accelerating the trend of exporting higher-end flat products and importing low-end flat products.

Table 3-13-1 Structure of imports by product group in 1980

Regions and countries	Amount (1000t)				Share (%)			
	Ingots and semis	Long products	Flat products	Tubes and fittings	Ingots and semis	Long products	Flat products	Tubes and fittings
EC	5 041	12 427	22 731	2 816	11.7	28.9	52.8	6.5
Belgium-Luxembourg	628	991	1 100	243	21.2	33.5	37.1	8.2
France	980	2 260	4 373		12.9	29.7	57.4	...
Germany	1 943	4 378	4 387	748	17.0	38.2	38.3	6.5
Italy	1 063	1 186	4 578	289	14.9	16.7	64.3	4.1
Netherlands	182	1 537	1 426	732	4.7	39.6	36.8	18.9
United Kingdom	351	996	2 896	464	7.5	21.2	61.5	9.9
EFTA	386	2 140	3 491	840	5.6	31.2	50.9	12.3
United States	141	3 792	6 275	6 448	0.8	22.8	37.7	38.7
Japan	449	46	585	27	40.6	4.2	52.8	2.4
Total Available	6 017	18 405	33 082	10 131	8.9	27.2	48.9	15.0

Table 3-13-2 Structure of imports by product group in 1992

Regions and countries	Amount (1000t)				Share (%)			
	Ingots and semis	Long products	Flat products	Tubes and fittings	Ingots and semis	Long products	Flat products	Tubes and fittings
EC	5 518	17 368	33 687	5 274	8.9	28.1	54.5	8.5
Belgium-Luxembourg	1 020	1 350	1 998	330	21.7	28.7	42.5	7.0
France	357	2 972	5 927	808	3.5	29.5	58.9	...
Germany	1 205	6 152	8 389	1 646	6.9	35.4	48.2	9.5
Italy	1 298	1 659	6 845	668	12.4	15.8	65.4	6.4
Netherlands	42	1 893	2 246	719	0.9	38.6	45.8	14.7
United Kingdom	222	1 152	3 255	584	4.3	22.1	62.4	11.2
EFTA	768	2 398	3 352	1 049	10.1	31.7	44.3	13.9
United States	2 129	3 178	8 795	1 512	13.6	20.4	56.3	9.7
Japan	615	712	520	296	28.7	33.2	24.3	13.8
Total Available	9 030	23 656	46 354	8 131	10.4	27.1	53.2	9.3

3.3.3. Evolution in product structure of steel trade balance

Because of the limited availability of import data, the trade balance by products can only be calculated for most of the developed economies. As seen in figure 3-17, there are different structure patterns even among the EC member countries. Germany and the Netherlands have deficits in long products with surplus flat products. On the other hand, Italy has a huge deficit in flat products with a smaller surplus in long products. The United Kingdom has a large surplus in long products with a small deficit in flat products. However, the United Kingdom successfully decreased its deficit in flat products from 2.1 million tonnes in 1980 to only 68 thousand tonnes in 1992.

EFTA, which had a deficit of flat products of 0.3 million tonnes in 1980, recorded a surplus for those products of 2.2 million tonnes in 1992. The United States, a constant net importer, has deficits in all products, but the major source is flat products. Japan has a large surplus in flat products as well as in long products and tubes and fittings, and a small deficit in ingots and semis.

Table 3-14-1 Structure of trade balance by product group in 1980

Regions and countries	Amount (1000t)				Share (%)			
	Ingots and semis	Long products	Flat products	Tubes and fittings	Ingots and semis	Long products	Flat products	Tubes and fittings
EC	- 683	8 037	8 109	4 849	-3.4	39.6	39.9	23.9
Belgium-Luxembourg	76	3 632	6 866	116	0.7	34.0	64.2	1.1
France	- 620	815	1 714	1 185	-20.0	26.3	55.4	31.9
Germany	- 456	- 61	5 812	2 283	-6.0	-0.8	76.7	30.1
Italy	- 874	1 965	- 2 778	1 317	-236.2	531.1	-750.8	355.9
Netherlands	696	- 887	1 174	- 244	94.2	-120.0	158.9	-33.0
United Kingdom	- 166	433	- 2 103	- 105	-8.6	22.3	-108.3	-5.4
EFTA	390	- 99	- 321	- 150	216.7	-55.0	-178.3	-83.3
United States	689	-2 839	-4 704	-2 960	7.0	-28.9	-47.9	-30.2
Japan	- 260	7 986	14 372	6 426	-0.9	28.0	50.4	22.5
Total Available	136	13 085	17 456	8 165	0.4	33.7	44.9	21.0

Table 3-14-2 Structure of trade balance by product group in 1992

Regions and countries	Amount (1000t)				Share (%)			
	Ingots and semis	Long products	Flat products	Tubes and fittings	Ingots and semis	Long products	Flat products	Tubes and fittings
EC	383	4 394	5 765	1 988	3.1	35.1	46.0	15.9
Belgium-Luxembourg	- 533	2 348	7 020	173	-5.9	26.1	77.9	1.9
France	643	41	862	187	37.1	2.4	49.7	17.2
Germany	243	- 2 020	2 715	506	16.8	-139.9	188.0	35.0
Italy	- 919	2 338	- 3 974	977	-58.2	148.2	-251.8	61.9
Netherlands	1 026	- 1 059	1 121	- 4	94.6	-97.7	103.4	-0.4
United Kingdom	999	2 128	- 68	167	31.0	66.0	-2.1	5.2
EFTA	- 241	166	2 243	- 109	11.7	-8.1	-108.9	5.3
United States	-1 745	-2 202	-6 811	- 856	-15.0	-19.0	-58.6	-7.4
Japan	- 290	2 232	7 692	2 685	-2.4	18.1	62.4	21.8
Total Available	- 1 893	4 590	8 889	3 708	-12.4	30.0	58.1	24.2

These observations may lead to the following conclusions. Even advanced steel producing countries have some imbalances between the consumption and production of specific products. Most of them are trying to decrease the gap between demand and supply of those products. At the same time, however, with an increase in ties inside regional economic zones or increases in interdependency with other countries, dependence on some products produced by trading partners remains or has even been increased. Advanced steel producers are developing much higher-end products to keep their competitive edge and their positions as leading exporters of steel products. World trade in steel has been increasing and its structure is becoming a much more complex and interrelated one.

Figure 3-17 Structure of trade balance by steel product group in the EC

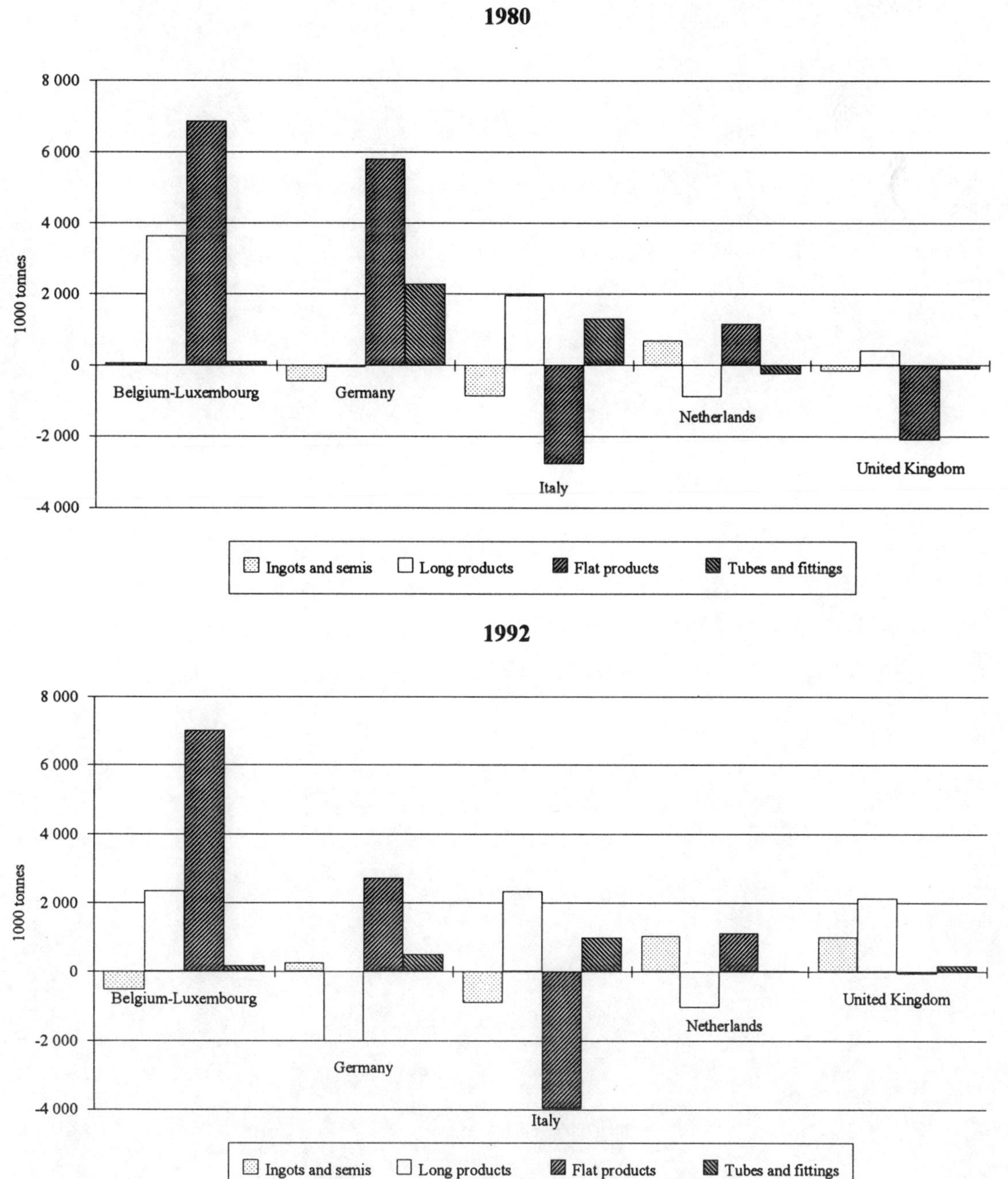

CHAPTER 4

CONCLUSIONS

Background

Along with the rapid and dramatic changes in the political and economic structure of the world, the structure of consumption and trade in steel also changed in the 1980s. The major factors which affected the structure of steel consumption and trade were rapid economic growth in the Far East, strengthened ties inside economic regions, particularly in the EC, fluctuations in the dollar and the dissolution of the CMEA and the former USSR and difficulties in those regions.

The European Community, the United States and Japan are the sole economic superpowers since the dissolution of the former USSR. They represent three pillars of the world economy, with 66 per cent of the total GDP, 45 per cent of world finished steel consumption and only 14 per cent of the world's population. Because they have not only the most advanced steel producing base, but also a wide range of efficient steel-using sectors, the evolution of steel production and consumption in these regions is an indication of future activity in other parts of the world. At the same time, different developments in steel have occurred in the rest of the world.

Balance between consumption, production and net exports

Comparison of the balance between consumption, production and net exports shows that the structure of world crude steel production has shifted from an oligopolistic situation, intense concentration in a small number of large producing countries, to greater diversification among various economic and geographic regions and countries. As a result of this, with a few exceptions, many countries, particularly in the Far East region, are improving their net export position or becoming self-dependent in steel at least in terms of volume. Large net exporters such as the EC and Japan in 1980, saw dramatically decreased net export volumes by 1992.

At the same time, maximizing the use of continuous casting and other process technologies has reduced the requirements for crude steel in finished steel production. The continuously cast steel ratio increased from 28.1 per cent in 1980 to 65.0 per cent in 1992. This technological development brought about relatively modest growth in world crude steel production compared to steady growth in finished steel consumption.

Developments in steel consumption

World finished steel consumption experienced a gradual increase during the period, with an eruption in the Far East and a rapid decline in the former CMEA region. The annual average total volume was 546 million tonnes in 1980-1983 and 657 million tonnes in 1990-1992. The share of the Far East, excluding Japan, increased from 12.9 per cent in 1980-1983 to 18.5 per cent in 1990-1992 and 25.1 per cent in 1992. The share of the former CMEA region, eastern Europe and the former USSR, decreased from 28.7 per cent in 1980-1983 to 15.5 per cent in 1992. Two major steel-consumers among the developed market economies, the EC and Japan, kept their shares in steel consumption steady at the level of 13 per cent and 17 per cent during the period, respectively. However, the other big consumer in the west, the United States, slightly reduced its share from 15.0 per cent in 1980-1983 to 13.7 per cent in 1990-1992.

World finished steel consumption per capita dropped from 130.7 kilograms in 1980 to 115.2 kilograms in 1992. However, this declining trend was a recent one triggered by a sharp drop in the former CMEA region and not applicable all over the world. Even highly developed market economies such as Japan, Italy, Belgium-Luxembourg, Denmark and Austria experienced increases as well as the rapidly growing countries such as China and the Republic of Korea.

Four groups of countries are identifiable when charting the relationship between GDP per capita and steel consumption per capita.

- First is the developing market economies which have less than US$8000 GDP and less than 250 kilograms of steel consumption per capita. In this group, there is a linear increasing trend in steel consumption accompanied by an increase in GDP. In 1989, this group made a general up and rightward shift reflecting higher GDP and higher steel consumption, but with much diversity in the positions of individual countries.

- The second group is a collection of highly developed countries with more than US$8000 GDP per capita. All these countries have higher steel consumption at the level of between 200 and 500 kilograms per capita. Even though there was a clear rightward shift to higher GDP for this group from 1980 to 1989, steel consumption per capita stayed at the same level.

- The third group consists of large steel consuming economies such as Japan and Italy. Even though they are highly developed market economies, they were isolated from the second group. However, the distance between these two countries and the second group has become closer. The Republic of Korea is rapidly becoming the third member of this group, increasing steel consumption per capita at an even higher pace than its relatively high rate of GDP growth.

- The fourth group is the former CMEA countries which had relatively high steel consumption compared to their GDP. There was a diversity of positions among these countries, reflecting both their past economic structures and the consequences of centralized economic planning. However, with the structural changes necessitated by their transition to market economies and their difficult economic situation in recent years, steel consumption has decreased rapidly and they are now closer to the group of developing market economies.

World steel intensity experienced a sharp drop at the beginning of the 1980s. This may have been caused by world-wide efforts to reduce steel consumption in the economy as a whole in an effort to reduce energy consumption after the second oil shock. However, the importance of steel in the economy as a whole, i.e. steel intensity, remained almost the same throughout the period after that. The Republic of Korea, Egypt, Greece and Portugal are countries showing an increasing trend in steel intensity, while most of the developed economies experienced slight declines. The former CMEA countries had a higher steel intensity than market economy countries under their old system. But this has come down to close to world average levels in recent years. There was a wide range of values in steel intensity among economies with a smaller GDP. This is a reflection of the different economic structures in the initial stages of development.

After a decline at the beginning of the 1980s, the share of gross fixed capital formation in world GDP increased constantly in the rest of the 1980s. This could be one of the factors which brought a stop to the decline in steel intensity. Two major world steel consumers, the EC and Japan, saw the same trend in which both the share of gross fixed capital formation in GDP and the steel intensity of GDP increased in the late 1980s. The Republic of Korea experienced a sharp increase in the steel intensity of GDP with an increased share of gross fixed capital formation in GDP, while eastern Europe saw the opposite trend. The difference in the share of gross fixed capital formation in GDP and the intensity of net indirect steel exports of GDP clearly explains the different magnitude of steel intensity of GDP among countries.

Distribution of steel consumption among major steel-using sectors

Seventy to 90 per cent of steel consumed is used in manufacturing and construction in all countries. The shares were 80 to 90 per cent in France, the United Kingdom and Japan, while they were lower at 75 to 80 per cent in the EC. Manufacturing, particularly of metal products, machinery and transport equipment, together with construction, are the predominant steel consuming sectors at any stage of economic development in any country.

The general perception is that manufacturing becomes the major steel consumer in more advanced stages of development, while steel consumption in construction is more important in the initial stages of economic development. However, this proved to be erroneous. Japan, an already mature, industrialized country, and the Republic of Korea, a newly developed country, have an equal distribution of steel consumption between manufacturing and construction.

The share of the construction sector in steel consumption is high and rising in many mature, highly developed countries or regions such as France, the United Kingdom, the EC and Japan. This could be connected with the increasing use of steel in high-rise buildings and new infrastructure projects. Taking into account the predominant use of long products in construction, the fall in the share of long products in the total product mix, which up to now has been one of the characteristics of technical progress in the steel industry, should be carefully examined in future in some regions and countries.

This development should create a very favourable environment for the further development of mini-mills, a sector mainly specializing in the production of long products. This observation is particularly important for the former CMEA countries which generally have an outmoded infrastructure and do not have mini-mills, but do have abundant resources of steel scrap.

Steel intensity by industrial sectors

The steel intensity of industrial sectors varied from 20 to 140 grams per United States dollar in the lowest steel consuming sectors to 50 to 700 grams per United States dollar in the most intensive steel consuming sectors.

The absence of a clear decreasing trend in the steel intensity of GDP even in highly developed economies was confirmed by a similar trend in the steel intensities of the most steel-consuming sectors, i.e. manufacturing plus construction. Further analysis confirmed that the major reason for this was that the decline of steel intensity in manufacturing was compensated by an increase in steel intensity in construction. This observation also implies that technological progress leading to reductions in steel consumption has temporarily reached its limit at least in the construction sector and even in advanced countries.

Construction is by far the most intensive steel consumer among industrial sectors in many countries. This is the case in Japan, the Republic of Korea and the EC. The steel intensity of construction is 350 grams per 1980 United States dollar in Japan and 700 grams per 1980 United States dollar in the Republic of Korea, 5 and 10 times more than the steel intensity in manufacturing, respectively. In the United Kingdom, Finland and Brazil, manufacturing has recently become the most intensive steel consumer. However, it is very important to bear in mind that no country shows a declining trend in steel intensity in construction. Rather it is increasing in many countries. On the other hand, steel intensity in manufacturing is declining or stagnant in many countries, while it is rising in some cases.

Steel intensity of manufacturing sectors

Fabricated metal production is the most intensive steel consumer in manufacturing in most cases. Electrical machinery is the least intensive steel consumer. The second most intensive steel consuming sector in manufacturing is transport equipment in Japan and the former Yugoslavia, but non-electrical machinery in the United Kingdom and Finland.

From the results of earlier research, one would expect that savings in steel consumption due to technological progress and the higher product quality of steel as well as shifts towards high-tech sectors in manufacturing would have resulted in a general decrease in steel intensities in all manufacturing sectors. This is the case in Japan but it is not obvious in other developed countries. The steel intensity of all four manufacturing sectors has stayed unchanged in the former Yugoslavia.

World trade in steel

Even though there was a boom in the middle of the 1980s due to increased imports by the United States and China and a successive decline after that, in general world steel trade increased during the 1980s. The annual average volume of exports increased from 140.0 million tonnes in 1980-1983 to 170.4 million tonnes in 1990-1992. The ratio of trade to world steel consumption, a measure of the importance of trade in consumption, increased from 24.6 per cent in 1980 to 28.5 per cent in 1992. Given that many countries are becoming more self-dependent in steel in terms of volume, this increase in trade suggests that the exchange of different products is increasing.

The Republic of Korea and Brazil increased their exports by 7.2 million tonnes and 10.1 million tonnes from 1980 to 1992, respectively. Their share of total world exports increased to 6.7 per cent and 6.6 per cent, respectively. 79 per cent of exports by Brazil went outside Latin America in 1992, while 32 per cent of exports by the Republic of Korea went outside the Far East. Small steel producing regions such as Africa, the Middle East and Oceania also increased their exports. Japan, a traditional big exporter, reached its export peak of 31.7 million tonnes in 1984, but experienced a sharp drop to 17.0 million tonnes in 1990-1992. Japan's world export share declined from 21.0 per cent in 1980-1983 to 10.0 per cent in 1990-1992. Because of a rapid decline in extra-regional exports, two thirds of Japan's exports were to the Far East region by 1992.

The European Community had a higher world share of exports outside the region, 21.5 per cent in 1980, but it declined to 16.6. per cent in 1992. The EC's intra-regional trade is so big that it accounts for almost one fourth of world trade in steel. This amount increased from 33.3 million tonnes in 1980 to 45.4 million tonnes in 1992. This accelerated trade in steel inside the region could be considered one of the important, positive aspects of efforts to unify the market in the region. Among the EC member States, Belgium-Luxembourg had the highest dependency rate on exports inside the region, while France and Germany had relatively lower dependency rates.

Regarding steel imports, the Far East saw sharp increases, accounting for 67.5 per cent of the total increase in world steel imports of 33.2 million tonnes from the early 1980s to the early 1990s. China, Japan and the Republic of Korea were the major importers. There was also a large increase of 22.4 million tonnes in the imports of the EC from 1980-1983 to 1990-1992. Excluding intra-regional trade, the EC had a share of 9.4 per cent of total world imports in 1992. The United States experienced a surge in imports in the middle of the 1980s, but its share of total world imports declined from 12.3 per cent in 1980-1983 to 10.5 per cent in 1990-1992.

As a result of the evolution in exports and imports in steel, the Far East, excluding Japan, rapidly increased its deficit in its steel trade balance. Large net exporters in 1980, the EC and Japan, decreased their volumes quite rapidly. Other America became a net exporter, owing to the increase in exports by Brazil.

Both export and import ratios to total world steel consumption increased on a worldwide basis. The United States did not experience significant changes in either its export or import ratios with the exception of the middle of the 1980s. But the European Community increased both export and import ratios mainly due to the expansion of intra-regional trade. It absorbed more imports from outside the region than before, while its export dependence outside the region declined. The former CMEA countries experienced a sharp increase in export ratios with the opposite trend for import ratios.

Regarding the product structure of trade in steel, there was a clear increasing trend in flat products and semi-finished products with a decreasing trend in long products and tubes and fittings. Many developing countries who increased their steel production capacity still needed to import high-end flat products from developed countries. For advanced countries, it is sometimes less costly to import commercial grade hot-rolled coils and even cold-rolled coils from developing countries. Strategic alliances between steel producers are increasingly sourcing hot-rolled coil or semi-finished products in other countries, and this is accelerating trade in steel. Low oil prices have been cooling trade in steel tubes and pipes as have the effects of the sluggish economy in the former USSR, a traditional large importer of tubes and pipes.

All countries have some specialties in steel production and the general trend is to overcome demand and supply gaps in specific product categories. Given the more global, interrelated world economy, the exchange of different steel products is becoming a major factor in accelerating steel trade. World trade in steel has been increasing and its structure is becoming a much more complex and interrelated.

ANNEX TABLES

Data for Europe, the United States, Canada and Japan are from the ECE/ITD database or other information made available to the Economic Commission for Europe (ECE). For other countries, data are from the International Iron and Steel Institute (IISI). Data for the European Community include the unified Germany from 1991 and eastern European data include the former German Democratic Republic (GDR) up to 1990.

Annex table 1 Crude steel production (1 of 2)

(1000 tonnes)

Regions and countries	1980	1981	1982	1983	1984	1985	1986
AFRICA-TOTAL	**10 535**	**10 525**	**9 958**	**9 140**	**9 381**	**11 211**	**11 488**
South Africa	9 067	9 005	8 280	7 180	7 732	8 507	8 895
Others	1 468	1 520	1 678	1 960	1 649	2 704	2 593
MIDDLE EAST-TOTAL	**2 194**	**2 263**	**2 306**	**2 651**	**3 304**	**3 703**	**3 668**
Egypt	968	1 015	1 074	979	928	1 028	1 013
Iran	550	565	550	734	854	836	838
Others	676	683	682	938	1 522	1 839	1 817
FAR EAST-TOTAL	**177 969**	**169 503**	**171 607**	**172 964**	**187 210**	**193 232**	**194 073**
China	37 121	35 604	37 160	40 021	43 475	46 794	52 208
India	9 514	10 765	10 997	10 237	10 549	11 936	12 197
Indonesia	543	621	693	983	1 171	1 374	1 729
Japan	111 395	101 676	99 548	97 179	105 586	105 279	98 275
Korea, DPR	5 800	5 500	5 800	6 100	6 500	6 500	6 600
Korea, Republic of	8 558	10 753	11 758	11 915	13 034	13 539	14 555
Others	5 038	4 584	5 651	6 529	6 895	7 810	8 509
NORTH AMERICA-TOTAL	**117 600**	**123 595**	**78 009**	**89 594**	**97 736**	**94 705**	**88 113**
Canada	15 902	14 811	11 872	12 832	14 699	14 637	14 081
United States	101 698	108 784	66 137	76 762	83 037	80 068	74 032
Others	-	-	-	-	-	-	-
OTHER AMERICA-TOTAL	**29 204**	**27 339**	**27 164**	**29 105**	**33 579**	**36 200**	**37 848**
Argentina	2 687	2 526	2 913	2 943	2 647	2 942	3 235
Brazil	15 337	13 226	12 995	14 727	18 386	20 455	21 233
Mexico	7 156	7 663	7 056	6 978	7 560	7 399	7 225
Venezuela	1 975	2 030	2 226	2 367	2 777	3 055	3 401
Others	2 049	1 894	1 974	2 090	2 209	2 349	2 754
OCEANIA-TOTAL	**7 819**	**7 856**	**6 623**	**5 910**	**6 582**	**6 836**	**6 961**
Australia	7 589	7 635	6 371	5 676	6 302	6 609	6 674
New Zealand	230	221	252	234	280	227	287
Others	-	-	-	-	-	-	-
EUROPE-TOTAL	**370 602**	**365 057**	**347 962**	**354 160**	**371 292**	**372 821**	**371 476**
EC-TOTAL	**142 116**	**139 892**	**125 253**	**123 358**	**134 296**	**135 471**	**125 776**
Belgium-Luxembourg	16 940	16 074	13 406	13 448	15 288	14 628	13 419
Denmark	734	612	560	493	548	528	632
France	23 176	21 258	18 402	17 582	19 000	18 808	17 857
Germany	...	...	...	...	...	...	...
former FR Germany	43 838	41 610	35 880	35 729	39 389	40 497	37 134
Greece	870	909	933	755	915	972	1 009
Ireland	2	32	61	141	166	203	208
Italy	26 501	24 777	24 009	21 811	24 061	23 898	22 882
Netherlands	5 272	5 472	4 353	4 484	5 739	5 518	5 284
Portugal	663	555	502	674	690	665	717
Spain	12 842	13 021	13 450	13 262	13 379	14 032	11 906
United Kingdom	11 278	15 572	13 697	14 979	15 121	15 722	14 728
EFTA-TOTAL	**13 161**	**12 636**	**12 175**	**12 748**	**14 101**	**13 923**	**13 499**
Austria	4 624	4 656	4 258	4 411	4 870	4 661	4 292
Finland	2 509	2 428	2 414	2 416	2 632	2 518	2 586
Iceland	-	-	-	-	-	-	-
Norway	862	848	768	876	916	944	836
Sweden	4 237	3 770	3 900	4 210	4 705	4 813	4 710
Switzerland	929	934	835	835	978	987	1 075
EASTERN EUROPE-TOTAL	**61 224**	**57 610**	**56 358**	**57 575**	**60 083**	**59 492**	**61 212**
Albania	...	...	60	70	80	90	100
Bulgaria	2 567	2 483	2 586	2 825	2 878	2 945	2 898
former Czechoslovakia	14 925	15 271	14 992	15 024	14 831	15 036	15 112
former GDR	7 308	7 467	7 168	7 219	7 573	7 853	7 967
Hungary	3 764	3 645	3 702	3 617	3 751	3 647	3 715
Poland	19 485	15 719	14 795	16 227	16 533	16 126	17 144
Romania	13 175	13 025	13 055	12 593	14 437	13 795	14 276
former USSR-TOTAL	**147 931**	**148 517**	**147 153**	**152 511**	**154 238**	**154 653**	**160 537**
OTHER EUROPE-TOTAL	**6 170**	**6 402**	**7 023**	**7 968**	**8 574**	**9 282**	**10 452**
Turkey	2 536	2 425	3 183	3 834	4 330	4 802	5 887
former Yugoslavia	3 634	3 977	3 840	4 134	4 244	4 480	4 565
OTHERS	...	...	...	...	...	...	...
WORLD-TOTAL	**715 923**	**706 138**	**643 629**	**663 524**	**709 084**	**718 708**	**713 627**

Annex table 1 Crude steel production (2 of 2)

(1000 tonnes)

Regions and countries	1987	1988	1989	1990	1991	1992	1993
AFRICA-TOTAL	**11 597**	**11 330**	**11 445**	**11 080**	**12 584**	**12 302**	**11 667**
South Africa	8 991	8 837	9 337	8 619	9 358	9 061	8 613
Others	2 606	2 493	2 108	2 461	3 226	3 241	3 054
MIDDLE EAST-TOTAL	**4 344**	**5 343**	**5 726**	**6 261**	**7 296**	**8 068**	**9 669**
Egypt	1 433	2 025	2 114	2 247	2 556	2 524	2 811
Iran	839	978	1 081	1 425	2 203	2 937	3 672
Others	2 072	2 340	2 531	2 589	2 537	2 607	3 186
FAR EAST-TOTAL	**202 730**	**219 895**	**228 746**	**239 060**	**249 639**	**251 123**	**267 885**
China	56 280	59 430	61 590	66 349	71 000	80 935	89 453
India	13 121	14 309	14 608	14 963	17 100	18 117	18 531
Indonesia	2 059	2 054	2 383	2 892	3 089	2 949	3 000
Japan	98 513	105 681	107 908	110 339	109 649	98 131	99 623
Korea, DPR	6 730	6 830	6 930	7 000	7 000	7 000	7 000
Korea, Republic of	16 782	19 118	21 873	23 125	26 001	28 054	33 016
Others	9 245	12 473	13 454	14 392	15 800	15 937	17 262
NORTH AMERICA-TOTAL	**94 996**	**105 757**	**103 764**	**101 910**	**92 563**	**98 162**	**103 089**
Canada	14 736	14 728	15 332	12 184	12 825	13 840	14 296
United States	80 260	91 029	88 432	89 726	79 738	84 322	88 793
Others	-	-	-	-	-	-	-
OTHER AMERICA-TOTAL	**40 101**	**42 710**	**42 759**	**38 511**	**39 549**	**41 368**	**43 246**
Argentina	3 633	3 652	3 908	3 657	2 972	2 661	2 851
Brazil	22 228	24 657	25 055	20 567	22 617	23 934	25 149
Mexico	7 642	7 779	7 851	8 734	7 964	8 435	8 998
Venezuela	3 699	3 646	3 196	2 998	3 304	3 443	3 349
Others	2 899	2 976	2 749	2 555	2 692	2 895	2 899
OCEANIA-TOTAL	**6 509**	**6 947**	**7 417**	**7 395**	**6 990**	**7 113**	**8 128**
Australia	6 100	6 387	6 735	6 676	6 184	6 355	7 278
New Zealand	409	560	682	719	806	758	850
Others	-	-	-	-	-	-	-
EUROPE-TOTAL	**375 621**	**388 151**	**384 945**	**366 049**	**329 153**	**304 718**	**285 017**
EC-TOTAL	**126 552**	**137 321**	**139 492**	**136 677**	**137 220**	**132 050**	**131 836**
Belgium-Luxembourg	13 085	14 880	14 675	14 920	14 656	13 344	13 412
Denmark	606	650	624	610	632	591	603
France	17 689	18 598	18 692	19 016	18 434	17 961	17 109
Germany	...	...	...	...	42 169	39 711	37 625
former FR Germany	36 248	41 023	41 073	38 434	*38 778*	*36 728*	...
Greece	908	959	956	999	980	924	980
Ireland	220	271	324	325	293	257	326
Italy	22 859	23 760	25 213	25 467	25 112	24 842	25 705
Netherlands	5 082	5 518	5 681	5 412	5 171	5 439	6 000
Portugal	750	832	762	717	547	749	745
Spain	11 691	11 880	12 752	12 936	12 932	12 182	12 862
United Kingdom	17 414	18 950	18 740	17 841	16 294	16 050	16 469
EFTA-TOTAL	**13 282**	**14 035**	**13 926**	**12 945**	**12 722**	**12 884**	**13 501**
Austria	4 301	4 560	4 718	4 291	4 187	3 953	4 149
Finland	2 669	2 798	2 921	2 860	2 890	3 077	3 256
Iceland	-	-	-	-	-	-	-
Norway	851	910	678	376	438	446	505
Sweden	4 595	4 779	4 693	4 455	4 252	4 358	4 591
Switzerland	866	988	916	963	955	1 050	1 000
EASTERN EUROPE-TOTAL	**62 532**	**61 262**	**59 129**	**48 962**	**33 149**	**29 340**	**29 688**
Albania	100	100	112	79	16	-	-
Bulgaria	3 044	2 880	2 899	2 180	1 615	1 552	1 941
former Czechoslovakia	15 416	15 380	15 465	14 877	12 071	11 044	*10 659*
former GDR	8 243	8 133	7 829	5 566	*3 391*	*2 983*	...
Hungary	3 622	3 582	3 315	2 866	1 914	1 533	1 736
Poland	17 145	16 873	15 094	13 633	10 403	9 835	9 906
Romania	14 962	14 314	14 415	9 761	7 130	5 376	5 446
former USSR-TOTAL	**161 874**	**163 037**	**160 096**	**154 414**	**132 839**	***118 514***	***97 735***
OTHER EUROPE-TOTAL	**11 381**	**12 496**	**12 302**	**13 051**	**13 223**	**11 930**	**12 257**
Turkey	7 044	8 009	7 854	9 443	9 398	10 343	11 519
former Yugoslavia	4 337	4 487	4 448	3 608	3 825	*1 587*	*738*
OTHERS	...	...	...	...	...	...	...
WORLD-TOTAL	**735 898**	**780 133**	**784 802**	**770 266**	**737 774**	**722 854**	**728 701**

Annex table 2 Apparent crude steel consumption (1 of 2)

(1000 tonnes)

Regions and countries	1980	1981	1982	1983	1984	1985	1986
AFRICA-TOTAL	**15 669**	**16 255**	**13 153**	**12 838**	**12 878**	**13 717**	**12 298**
South Africa	7 197	7 264	5 792	5 322	5 700	5 019	5 077
Others	8 472	8 991	7 361	7 516	7 178	8 698	7 221
MIDDLE EAST-TOTAL	**15 339**	**14 980**	**19 224**	**20 958**	**19 093**	**20 361**	**17 571**
Egypt	2 032	2 133	2 400	2 522	2 323	3 288	5 594
Iran	3 556	2 821	4 272	5 984	4 739	6 254	4 467
Others	9 751	10 027	12 552	12 451	12 030	10 819	7 510
FAR EAST-TOTAL	**165 195**	**156 850**	**158 281**	**165 652**	**185 580**	**197 712**	**200 906**
China	43 275	39 213	40 949	52 349	60 834	72 540	74 965
India	10 900	14 000	13 900	11 600	13 900	14 400	15 290
Indonesia	2 971	3 101	3 157	2 846	2 627	2 368	2 817
Japan	79 007	71 136	69 504	65 614	74 367	73 377	69 941
Korea, DPR	5 800	5 500	5 800	6 630	7 030	7 030	7 130
Korea, Republic of	6 100	7 480	7 630	8 916	10 669	11 313	12 191
Others	17 142	16 420	17 341	17 697	16 153	16 684	18 572
NORTH AMERICA-TOTAL	**128 530**	**143 202**	**93 408**	**107 601**	**126 459**	**122 710**	**110 335**
Canada	12 939	13 472	9 133	11 132	13 282	13 641	12 512
United States	115 591	129 730	84 275	96 469	113 177	109 069	97 823
Others	...	...	...	...	...	...	...
OTHER AMERICA-TOTAL	**36 546**	**33 891**	**29 204**	**23 606**	**28 239**	**29 023**	**31 426**
Argentina	3 574	2 582	2 698	2 915	3 070	2 180	2 510
Brazil	14 309	12 045	10 609	8 637	10 673	11 994	14 529
Mexico	10 569	11 369	8 182	6 296	7 331	7 694	6 393
Venezuela	2 797	2 575	2 966	1 687	2 097	1 769	2 627
Others	5 297	5 319	4 748	4 070	5 070	5 387	5 368
OCEANIA-TOTAL	**6 679**	**7 922**	**6 514**	**5 840**	**6 675**	**6 659**	**6 557**
Australia	6 074	7 195	5 716	5 121	5 897	5 989	5 927
New Zealand	605	727	798	720	778	670	630
Others	...	...	...	...	...	...	...
EUROPE-TOTAL	**348 118**	**328 049**	**327 644**	**328 444**	**336 344**	**331 643**	**341 395**
EC-TOTAL	**115 776**	**103 279**	**102 581**	**96 588**	**101 749**	**100 098**	**103 329**
Belgium-Luxembourg	2 967	3 210	4 617	3 396	4 248	3 641	3 491
Denmark	1 759	1 770	2 270	1 334	1 568	1 804	1 793
France	19 425	16 695	16 324	14 583	14 709	13 951	13 603
Germany	...	...	...	...	...	...	...
former FR Germany	33 783	30 989	26 847	29 835	31 101	30 788	30 677
Greece	1 644	1 394	1 577	1 921	1 319	1 512	1 611
Ireland	385	492	468	388	416	395	360
Italy	26 943	20 834	21 762	19 097	22 060	21 870	22 607
Netherlands	4 288	3 505	3 544	3 209	3 981	4 132	4 006
Portugal	1 294	1 430	1 639	1 260	1 192	1 114	1 183
Spain	9 505	8 060	9 335	7 504	6 825	6 541	9 178
United Kingdom	13 783	14 900	14 197	14 062	14 330	14 350	14 820
EFTA-TOTAL	**13 769**	**11 988**	**11 931**	**11 082**	**11 669**	**10 978**	**11 978**
Austria	2 873	2 535	2 421	2 354	2 384	2 244	2 486
Finland	2 134	1 872	2 088	1 811	1 933	1 762	1 948
Iceland	68	58	55	51	53	50	53
Norway	1 821	1 525	1 730	1 338	1 399	1 411	1 559
Sweden	4 139	3 430	3 523	3 493	3 711	3 283	3 534
Switzerland	2 734	2 568	2 114	2 035	2 189	2 228	2 399
EASTERN EUROPE-TOTAL	**58 933**	**52 337**	**54 169**	**53 463**	**52 949**	**53 028**	**54 100**
Albania	189	106	220	229	239	249	259
Bulgaria	2 920	3 025	3 137	3 315	2 973	3 073	2 930
former Czechoslovakia	10 778	11 218	11 099	11 058	10 809	10 973	11 061
former GDR	10 054	7 437	9 854	9 522	8 673	9 153	9 056
Hungary	3 614	3 668	3 745	3 579	3 326	3 448	3 509
Poland	19 147	15 319	14 357	14 732	15 184	15 076	15 932
Romania	12 230	11 565	11 756	11 029	11 745	11 056	11 352
former USSR-TOTAL	**150 419**	**150 936**	**150 463**	**157 768**	**159 570**	**157 255**	**161 582**
OTHER EUROPE-TOTAL	**9 221**	**9 509**	**8 500**	**9 543**	**10 407**	**10 283**	**10 406**
Turkey	3 454	3 402	3 264	4 257	5 267	5 208	4 731
former Yugoslavia	5 766	6 106	5 236	5 286	5 139	5 074	5 675
OTHERS	**197**	**178**	**215**	**201**	**222**	**225**	**303**
WORLD-TOTAL	**716 272**	**701 326**	**647 643**	**665 140**	**715 491**	**722 049**	**720 790**

Annex table 2 Apparent crude steel consumption (2 of 2)

(1000 tonnes)

Regions and countries	1987	1988	1989	1990	1991	1992	1993
AFRICA-TOTAL	**12 324**	**12 868**	**12 756**	**12 702**	**12 556**	**12 614**	**...**
South Africa	5 233	5 748	6 029	5 525	5 013	4 375	...
Others	7 091	7 120	6 727	7 177	7 543	8 239	...
MIDDLE EAST-TOTAL	**15 768**	**16 974**	**16 990**	**18 139**	**20 020**	**21 715**	**...**
Egypt	3 753	4 443	4 648	4 611	4 928	4 995	...
Iran	5 796	4 684	5 384	6 859	8 288	9 575	...
Others	6 219	7 847	6 957	6 670	6 804	7 145	...
FAR EAST-TOTAL	**211 693**	**231 233**	**244 065**	**256 074**	**269 675**	**270 532**	**...**
China	71 414	70 417	71 320	68 419	71 043	86 520	...
India	17 640	19 040	20 036	21 700	20 300	19 090	...
Indonesia	2 563	2 673	3 068	4 533	4 432	4 357	...
Japan	75 751	86 871	93 278	99 032	99 151	84 040	80 589
Korea, DPR	7 260	7 360	7 460	7 530	7 530	7 530	...
Korea, Republic of	15 048	15 825	18 267	21 478	26 068	23 238	...
Others	22 017	29 047	30 636	33 382	41 152	45 756	...
NORTH AMERICA-TOTAL	**119 362**	**127 913**	**119 556**	**114 117**	**103 945**	**110 166**	**118 250**
Canada	13 472	15 022	14 214	10 903	10 620	10 299	13 493
United States	105 890	112 891	105 342	103 214	93 325	99 867	104 757
Others	...	...	...	...	...	...	...
OTHER AMERICA-TOTAL	**33 049**	**30 714**	**32 050**	**26 812**	**28 494**	**30 724**	**...**
Argentina	3 293	3 002	1 856	1 578	2 139	3 130	...
Brazil	14 972	11 959	14 807	10 252	9 962	10 468	...
Mexico	6 430	6 871	7 285	8 131	9 082	9 937	...
Venezuela	2 971	3 149	2 228	1 601	2 732	2 127	...
Others	5 383	5 732	5 875	5 250	4 579	5 062	...
OCEANIA-TOTAL	**6 708**	**7 465**	**7 063**	**6 056**	**5 475**	**4 869**	**...**
Australia	5 867	6 447	6 393	5 408	4 954	4 351	...
New Zealand	840	1 018	670	647	521	518	...
Others	...	...	...	...	...	...	...
EUROPE-TOTAL	**347 678**	**366 495**	**369 559**	**340 454**	**295 636**	**270 169**	**...**
EC-TOTAL	**102 033**	**120 354**	**126 520**	**124 023**	**122 341**	**119 318**	**...**
Belgium-Luxembourg	3 011	5 218	3 976	3 551	3 864	3 809	...
Denmark	1 565	1 677	1 796	1 806	1 797	2 025	...
France	13 740	15 332	16 871	18 085	16 646	16 096	13 897
Germany	...	...	...	...	38 950	38 443	31 222
former FR Germany	29 008	35 235	35 523	35 201	*35 572*	...	...
Greece	1 900	1 493	1 981	2 418	2 705	1 800	...
Ireland	382	482	517	490	450	432	...
Italy	23 514	26 334	28 027	27 694	27 185	26 883	23 193
Netherlands	3 390	4 315	4 714	4 951	4 004	4 232	4 452
Portugal	1 574	2 268	3 409	1 880	1 865	2 000	...
Spain	8 969	10 508	12 306	11 919	11 345	10 684	9 732
United Kingdom	14 980	17 490	17 400	16 028	13 530	12 914	...
EFTA-TOTAL	**11 096**	**12 524**	**13 056**	**11 872**	**9 959**	**10 067**	**...**
Austria	2 179	2 630	2 898	2 639	2 941	2 773	2 573
Finland	1 992	1 975	2 273	1 988	1 443	1 378	1 557
Iceland	62	59	49	43	51	43	42
Norway	1 241	1 257	1 183	944	601	1 119	1 148
Sweden	3 551	3 821	3 945	3 689	2 774	2 725	2 944
Switzerland	2 071	2 782	2 708	2 569	2 149	2 029	...
EASTERN EUROPE-TOTAL	**55 004**	**53 332**	**51 330**	**37 586**	**20 553**	**16 104**	**...**
Albania	259	259	271	238	149	106	...
Bulgaria	3 422	2 836	2 873	1 658	1 188	878	889
former Czechoslovakia	10 482	10 731	10 863	10 299	6 157	4 914	*3 910*
former GDR	9 158	8 862	10 252	5 390	...	...	...
Hungary	3 371	3 135	2 837	2 369	1 359	1 012	1 210
Poland	15 689	15 448	13 355	9 524	5 789	5 821	6 464
Romania	12 623	12 061	10 879	8 108	5 911	3 373	2 992
former USSR-TOTAL	**163 032**	**164 679**	**166 556**	**157 199**	**131 865**	***114 908***	**...**
OTHER EUROPE-TOTAL	**16 513**	**15 607**	**12 096**	**9 774**	**10 919**	**9 772**	**...**
Turkey	12 414	11 793	7 265	6 506	7 162	7 771	10 521
former Yugoslavia	4 099	3 814	4 831	3 268	3 757	*2 001*	...
OTHERS	**265**	**363**	**323**	**336**	**260**	**306**	**...**
WORLD-TOTAL	**746 847**	**794 026**	**802 362**	**774 690**	**736 061**	**721 094**	**...**

Annex table 3 Net steel exports - crude steel equivalent (1 of 2)

(1000 tonnes)

Regions and countries	1980	1981	1982	1983	1984	1985	1986
AFRICA-TOTAL	**- 5 134**	**- 5 730**	**- 3 195**	**- 3 698**	**- 3 497**	**- 2 506**	**- 810**
South Africa	1 870	1 741	2 488	1 858	2 032	3 488	3 818
Others	- 7 004	- 7 471	- 5 683	- 5 556	- 5 529	- 5 994	- 4 628
MIDDLE EAST-TOTAL	**- 13 145**	**- 12 717**	**- 16 918**	**- 18 307**	**- 15 789**	**- 16 658**	**- 13 903**
Egypt	- 1 064	- 1 118	- 1 326	- 1 543	- 1 395	- 2 260	- 4 581
Iran	- 3 006	- 2 256	- 3 722	- 5 250	- 3 885	- 5 418	- 3 629
Others	- 9 075	- 9 344	- 11 870	- 11 513	- 10 508	- 8 980	- 5 693
FAR EAST-TOTAL	**12 774**	**12 653**	**13 326**	**7 312**	**1 630**	**- 4 480**	**- 6 833**
China	- 6 154	- 3 609	- 3 789	- 12 328	- 17 359	- 25 746	- 22 757
India	- 1 386	- 3 235	- 2 903	- 1 363	- 3 351	- 2 464	- 3 093
Indonesia	- 2 428	- 2 480	- 2 464	- 1 863	- 1 456	- 994	- 1 088
Japan	32 388	30 540	30 044	31 565	31 219	31 902	28 334
Korea, DPR	...	...	...	- 530	- 530	- 530	- 530
Korea, Republic of	2 458	3 273	4 128	2 999	2 365	2 226	2 364
Others	- 12 104	- 11 836	- 11 690	- 11 168	- 9 258	- 8 874	- 10 063
NORTH AMERICA-TOTAL	**- 10 930**	**- 19 607**	**- 15 399**	**- 18 007**	**- 28 723**	**- 28 005**	**- 22 222**
Canada	2 963	1 339	2 739	1 700	1 417	996	1 569
United States	- 13 893	- 20 946	- 18 138	- 19 707	- 30 140	- 29 001	- 23 791
Others	...	...	...	...	...	...	...
OTHER AMERICA-TOTAL	**- 7 342**	**- 6 552**	**- 2 040**	**5 499**	**5 340**	**7 177**	**6 422**
Argentina	- 887	- 56	215	28	- 423	762	725
Brazil	1 028	1 181	2 386	6 090	7 713	8 461	6 704
Mexico	- 3 413	- 3 706	- 1 126	682	229	- 295	832
Venezuela	- 822	- 545	- 740	680	680	1 286	774
Others	- 3 248	- 3 425	- 2 774	- 1 980	- 2 861	- 3 038	- 2 614
OCEANIA-TOTAL	**1 140**	**- 66**	**109**	**70**	**- 93**	**177**	**404**
Australia	1 515	440	655	555	405	620	747
New Zealand	- 375	- 506	- 546	- 486	- 498	- 443	- 343
Others	...	...	...	...	...	...	...
EUROPE-TOTAL	**22 484**	**37 008**	**20 318**	**25 716**	**34 948**	**41 178**	**30 081**
EC-TOTAL	**26 340**	**36 613**	**22 672**	**26 770**	**32 547**	**35 373**	**22 447**
Belgium-Luxembourg	13 973	12 864	8 789	10 052	11 040	10 987	9 928
Denmark	- 1 025	- 1 158	- 1 710	- 841	- 1 020	- 1 276	- 1 161
France	3 751	4 563	2 078	2 999	4 291	4 857	4 254
Germany	...	...	...	...	...	...	...
former FR Germany	10 055	10 621	9 033	5 894	8 288	9 709	6 457
Greece	- 774	- 485	- 644	- 1 166	- 404	- 540	- 602
Ireland	- 383	- 460	- 407	- 247	- 250	- 192	- 152
Italy	- 442	3 943	2 247	2 714	2 001	2 028	275
Netherlands	984	1 967	809	1 275	1 758	1 386	1 278
Portugal	- 631	- 875	- 1 137	- 586	- 502	- 449	- 466
Spain	3 337	4 961	4 115	5 758	6 554	7 491	2 728
United Kingdom	- 2 505	672	- 500	917	791	1 372	- 92
EFTA-TOTAL	**- 608**	**648**	**244**	**1 666**	**2 432**	**2 945**	**1 521**
Austria	1 751	2 121	1 837	2 057	2 486	2 417	1 806
Finland	375	556	326	605	699	756	638
Iceland	- 68	- 58	- 55	- 51	- 53	- 50	- 53
Norway	- 959	- 677	- 962	- 462	- 483	- 467	- 723
Sweden	98	340	377	717	994	1 530	1 176
Switzerland	- 1 805	- 1 634	- 1 279	- 1 200	- 1 211	- 1 241	- 1 324
EASTERN EUROPE-TOTAL	**2 291**	**5 273**	**2 189**	**4 112**	**7 134**	**6 464**	**7 112**
Albania	...	...	- 160	- 159	- 159	- 159	- 159
Bulgaria	- 353	- 542	- 551	- 490	- 95	- 128	- 32
former Czechoslovakia	4 147	4 053	3 893	3 966	4 022	4 063	4 051
former GDR	- 2 746	30	- 2 686	- 2 303	- 1 100	- 1 300	- 1 089
Hungary	150	- 23	- 43	38	425	199	206
Poland	338	400	438	1 495	1 349	1 050	1 212
Romania	945	1 460	1 299	1 564	2 692	2 739	2 924
former USSR-TOTAL	**- 2 488**	**- 2 419**	**- 3 310**	**- 5 257**	**- 5 332**	**- 2 602**	**- 1 045**
OTHER EUROPE-TOTAL	**- 3 051**	**- 3 107**	**- 1 477**	**- 1 575**	**- 1 833**	**- 1 001**	**46**
Turkey	- 918	- 977	- 81	- 423	- 937	- 406	1 156
former Yugoslavia	- 2 132	- 2 129	- 1 396	- 1 152	- 895	- 594	- 1 110
OTHERS	**- 197**	**- 178**	**- 215**	**- 201**	**- 222**	**- 225**	**- 303**
WORLD-TOTAL	**- 349**	**4 812**	**- 4 014**	**- 1 616**	**- 6 407**	**- 3 341**	**- 7 163**

Annex table 3 Net steel exports - crude steel equivalent (2 of 2)

(1000 tonnes)

Regions and countries	1987	1988	1989	1990	1991	1992	1993
AFRICA-TOTAL	**- 727**	**- 1 538**	**- 1 311**	**- 1 622**	**28**	**- 312**	**...**
South Africa	3 758	3 089	3 308	3 094	4 345	4 686	...
Others	- 4 485	- 4 627	- 4 619	- 4 716	- 4 317	- 4 998	...
MIDDLE EAST-TOTAL	**- 11 424**	**- 11 631**	**- 11 264**	**- 11 878**	**- 12 724**	**- 13 647**	**...**
Egypt	- 2 320	- 2 418	- 2 534	- 2 364	- 2 372	- 2 471	...
Iran	- 4 957	- 3 706	- 4 303	- 5 434	- 6 085	- 6 638	...
Others	- 4 147	- 5 507	- 4 426	- 4 081	- 4 267	- 4 538	...
FAR EAST-TOTAL	**- 8 963**	**- 11 338**	**- 15 319**	**- 17 014**	**- 20 036**	**- 19 409**	**...**
China	- 15 134	- 10 987	- 9 730	- 2 070	- 43	- 5 585	...
India	- 4 519	- 4 731	- 5 428	- 6 737	- 3 200	- 973	...
Indonesia	- 504	- 619	- 685	- 1 641	- 1 343	- 1 408	...
Japan	22 762	18 810	14 630	11 307	10 498	14 091	19 034
Korea, DPR	- 530	- 530	- 530	- 530	- 530	- 530	...
Korea, Republic of	1 734	3 293	3 606	1 647	- 67	4 816	...
Others	- 12 772	- 16 574	- 17 182	- 18 990	- 25 352	- 29 819	...
NORTH AMERICA-TOTAL	**- 24 366**	**- 22 156**	**- 15 792**	**- 12 207**	**- 11 382**	**- 12 004**	**- 15 161**
Canada	1 264	- 294	1 118	1 281	2 205	3 541	803
United States	- 25 630	- 21 862	- 16 910	- 13 488	- 13 587	- 15 545	- 15 964
Others	...	...	...	...	...	...	...
OTHER AMERICA-TOTAL	**7 052**	**11 996**	**10 709**	**11 699**	**11 055**	**10 644**	**...**
Argentina	340	650	2 052	2 079	833	- 469	...
Brazil	7 256	12 698	10 248	10 315	12 655	13 466	...
Mexico	1 212	908	566	603	- 1 118	- 1 502	...
Venezuela	728	497	968	1 397	572	1 316	...
Others	- 2 484	- 2 756	- 3 126	- 2 695	- 1 887	- 2 167	...
OCEANIA-TOTAL	**- 199**	**- 518**	**354**	**1 339**	**1 515**	**2 244**	**...**
Australia	233	- 60	342	1 268	1 230	2 004	...
New Zealand	- 431	- 458	12	72	285	240	...
Others	...	...	...	...	...	...	...
EUROPE-TOTAL	**27 943**	**21 656**	**15 386**	**25 595**	**33 517**	**34 549**	**...**
EC-TOTAL	**24 519**	**16 967**	**12 972**	**12 654**	**14 879**	**12 732**	**...**
Belgium-Luxembourg	10 074	9 662	10 699	11 369	10 792	9 535	...
Denmark	- 959	- 1 027	- 1 172	- 1 196	- 1 165	- 1 434	...
France	3 949	3 266	1 821	931	1 788	1 865	3 212
Germany	...	...	...	...	3 219	1 268	6 403
former FR Germany	7 240	5 788	5 550	3 233	*3 206*	...	...
Greece	- 992	- 534	- 1 025	- 1 419	- 1 725	- 876	...
Ireland	- 162	- 211	- 193	- 165	- 157	- 175	...
Italy	- 655	- 2 574	- 2 814	- 2 227	- 2 073	- 2 041	2 512
Netherlands	1 692	1 203	967	461	1 167	1 207	1 548
Portugal	- 824	- 1 436	- 2 647	- 1 163	- 1 318	- 1 251	...
Spain	2 722	1 372	446	1 017	1 587	1 498	3 130
United Kingdom	2 434	1 460	1 340	1 813	2 764	3 136	...
EFTA-TOTAL	**2 186**	**1 511**	**870**	**1 073**	**2 763**	**2 817**	**...**
Austria	2 122	1 930	1 820	1 652	1 246	1 180	1 576
Finland	677	823	648	872	1 447	1 699	1 699
Iceland	- 62	- 59	49	- 43	- 51	- 43	- 42
Norway	- 390	- 347	- 505	- 568	- 163	- 673	- 643
Sweden	1 044	958	748	766	1 478	1 633	1 647
Switzerland	- 1 205	- 1 794	- 1 792	- 1 606	- 1 194	- 979	...
EASTERN EUROPE-TOTAL	**7 528**	**7 930**	**7 799**	**11 376**	**12 596**	**13 236**	**...**
Albania	- 159	- 159	- 159	- 159	- 133	- 106	...
Bulgaria	- 378	44	26	522	427	674	1 052
former Czechoslovakia	4 934	4 649	4 602	4 578	5 914	6 130	*6 749*
former GDR	- 915	- 729	- 2 423	176	...	...	...
Hungary	251	447	478	497	555	521	526
Poland	1 456	1 425	1 739	4 109	4 614	4 014	3 442
Romania	2 339	2 253	3 536	1 653	1 219	2 003	2 454
former USSR-TOTAL	**- 1 158**	**- 1 642**	**- 6 460**	**- 2 785**	**974**	***3 606***	**...**
OTHER EUROPE-TOTAL	**- 5 132**	**- 3 111**	**206**	**3 277**	**2 304**	**2 158**	**...**
Turkey	- 5 370	- 3 784	589	2 937	2 236	2 572	998
former Yugoslavia	238	673	- 383	340	68	- *414*	...
OTHERS	**- 265**	**- 363**	**- 323**	**- 336**	**- 260**	**- 306**	**...**
WORLD-TOTAL	**- 10 949**	**- 13 893**	**- 17 560**	**- 4 424**	**1 713**	**1 760**	**...**

Annex table 4 Share in total production of continuously cast steel (1 of 2)

(%)

Regions and countries	1980	1981	1982	1983	1984	1985	1986
AFRICA-TOTAL	**51.9**	**55.2**	**61.2**	**51.7**	**54.7**	**53.5**	**53.0**
South Africa	51.9	55.2	61.2	60.1	61.5	64.7	63.9
Others	...	...	...	20.8	22.7	18.3	15.4
MIDDLE EAST-TOTAL	**20.7**	**20.5**	**21.2**	**55.2**	**64.1**	**67.5**	**67.1**
Egypt	-	-	-	63.9	74.9	73.8	73.6
Iran	-	-	-	-	-	-	-
Others	67.3	67.8	71.6	89.3	93.4	94.6	94.5
FAR EAST-TOTAL	**41.2**	**47.8**	**52.9**	**59.6**	**62.1**	**62.1**	**61.5**
China	-	-	-	9.0	10.6	10.8	11.9
India	-	-	-	-	-	4.1	6.7
Indonesia	-	-	-	72.2	80.5	82.7	89.5
Japan	59.5	70.7	78.7	86.3	89.1	91.1	92.7
Korea, DPR	...	...	...	...	...	...	...
Korea, Republic of	32.4	44.3	51.1	56.6	60.6	63.3	71.1
Others	38.3	40.3	59.5	69.5	67.1	61.1	61.9
NORTH AMERICA-TOTAL	**21.0**	**22.5**	**28.4**	**32.9**	**38.9**	**44.3**	**52.2**
Canada	25.6	32.2	32.8	37.4	38.4	43.6	45.8
United States	20.3	21.1	27.6	32.1	39.0	44.4	53.4
Others	...	...	...	...	...	...	...
OTHER AMERICA-TOTAL	**32.4**	**35.8**	**40.8**	**49.1**	**46.7**	**48.5**	**50.1**
Argentina	53.3	49.2	51.8	48.6	47.4	62.5	64.9
Brazil	33.3	36.4	41.1	44.2	41.3	43.7	46.1
Mexico	29.3	31.9	37.9	55.2	54.0	51.0	47.3
Venezuela	40.5	62.2	70.0	76.8	72.8	72.5	76.4
Others	0.7	0.5	0.2	32.9	33.3	32.9	39.1
OCEANIA-TOTAL	**10.3**	**13.0**	**17.5**	**27.6**	**30.1**	**29.5**	**30.0**
Australia	10.3	13.0	17.5	24.6	27.0	27.1	27.0
New Zealand	...	...	...	100.0	100.0	100.0	100.0
Others	...	...	...	...	...	...	...
EUROPE-TOTAL	**23.6**	**26.4**	**29.5**	**32.3**	**36.1**	**38.9**	**40.8**
EC-TOTAL	**40.0**	**44.2**	**51.6**	**58.9**	**64.0**	**69.8**	**76.0**
Belgium-Luxembourg	18.7	25.2	29.7	34.9	43.4	51.5	62.0
Denmark	73.3	95.8	96.8	97.4	99.5	100.0	100.0
France	41.3	51.4	58.5	63.8	66.9	80.6	90.1
Germany	...	...	...	...	...	...	...
former FR Germany	46.0	53.6	61.9	71.8	76.9	79.5	84.6
Greece	...	...	...	100.0	100.0	100.0	100.0
Ireland	...	100.0	100.0	100.0	100.0	100.0	100.0
Italy	49.9	50.8	58.5	68.2	73.3	78.6	83.9
Netherlands	5.9	21.2	31.0	35.9	38.7	39.1	42.8
Portugal	42.4	38.2	47.0	43.2	40.7	44.1	44.6
Spain	48.1	35.1	41.5	45.2	49.8	57.6	60.9
United Kingdom	27.1	31.8	38.9	46.6	52.0	54.8	60.4
EFTA-TOTAL	**55.7**	**66.1**	**77.0**	**83.5**	**85.2**	**86.9**	**88.0**
Austria	51.2	62.4	77.3	87.6	89.0	93.4	94.6
Finland	90.1	91.8	93.4	93.5	94.4	93.5	93.2
Iceland	...	...	...	...	...	...	...
Norway	12.9	16.0	29.6	36.5	51.4	55.6	56.7
Sweden	49.0	65.4	76.0	80.0	79.6	80.6	82.0
Switzerland	...	...	...	100.0	100.0	100.0	100.0
EASTERN EUROPE-TOTAL	**9.8**	**10.9**	**11.9**	**13.3**	**18.5**	**19.5**	**21.1**
Albania	...	...	...	...	...	...	...
Bulgaria	...	...	...	7.0	10.0	9.3	13.9
former Czechoslovakia	1.5	1.5	2.5	5.1	7.3	7.7	8.3
former GDR	14.2	15.8	17.1	18.1	25.6	33.7	36.5
Hungary	36.1	35.4	33.6	39.3	46.8	46.6	52.1
Poland	4.0	3.8	4.3	4.0	10.1	10.3	10.6
Romania	18.1	20.7	22.4	26.0	30.2	29.9	31.9
former USSR-TOTAL	**10.7**	**12.2**	**12.6**	**12.4**	**12.7**	**13.6**	**15.0**
OTHER EUROPE-TOTAL	**24.5**	**31.9**	**51.2**	**57.4**	**62.2**	**60.3**	**67.3**
Turkey	7.2	13.4	55.7	63.6	72.0	66.6	78.4
former Yugoslavia	36.6	43.2	47.6	51.6	52.2	53.5	52.9
OTHERS	**...**	**...**	**...**	**...**	**...**	**...**	**...**
WORLD-TOTAL	**28.1**	**31.3**	**36.3**	**40.4**	**44.0**	**46.5**	**48.4**

Annex table 4 Share in total production of continuously cast steel (2 of 2)

(%)

Regions and countries	1987	1988	1989	1990	1991	1992	1993
AFRICA-TOTAL	**53.1**	**58.1**	**63.5**	**65.7**	**65.3**	**68.3**	**...**
South Africa	63.9	69.6	73.4	73.7	75.8	79.7	...
Others	15.7	17.1	19.8	37.7	34.7	36.5	...
MIDDLE EAST-TOTAL	**76.9**	**73.9**	**72.8**	**96.8**	**94.9**	**95.6**	**...**
Egypt	95.5	84.3	82.3	96.4	89.4	89.9	...
Iran	-	-	-	98.5	100.0	100.0	...
Others	95.2	95.7	96.0	96.1	96.1	96.2	...
FAR EAST-TOTAL	**62.1**	**63.6**	**65.1**	**66.9**	**68.5**	**67.5**	**...**
China	12.9	14.7	16.3	22.3	26.5	30.0	...
India	7.9	9.7	11.4	12.3	14.3	16.6	...
Indonesia	90.8	94.3	95.8	96.6	98.0	100.0	...
Japan	93.3	93.1	93.5	93.9	94.4	95.4	95.7
Korea, DPR	...	...	...	...	...	...	...
Korea, Republic of	83.5	88.3	94.1	96.1	96.4	96.8	...
Others	60.3	65.5	66.1	68.5	84.8	86.7	...
NORTH AMERICA-TOTAL	**57.7**	**67.6**	**66.4**	**68.6**	**76.9**	**80.1**	**86.5**
Canada	49.0	69.9	76.7	77.3	84.6	87.1	91.7
United States	59.3	67.2	64.6	67.4	75.7	78.9	85.7
Others	...	...	...	...	...	...	...
OTHER AMERICA-TOTAL	**51.8**	**53.8**	**57.6**	**61.3**	**62.0**	**63.2**	**...**
Argentina	65.3	67.9	73.3	75.7	81.5	81.5	...
Brazil	45.5	49.0	53.9	58.5	56.0	57.9	...
Mexico	54.2	55.9	58.1	60.8	64.4	64.5	...
Venezuela	79.6	78.8	78.3	79.0	92.1	96.4	...
Others	40.8	41.1	42.4	44.1	47.0	47.4	...
OCEANIA-TOTAL	**48.0**	**73.8**	**81.9**	**85.0**	**83.8**	**86.8**	**...**
Australia	44.5	71.5	80.0	83.4	81.6	85.2	...
New Zealand	100.0	100.0	100.0	100.0	100.0	100.0	...
Others	...	...	...	...	...	...	...
EUROPE-TOTAL	**43.3**	**46.1**	**48.2**	**50.4**	**54.2**	**57.8**	**61.7**
EC-TOTAL	**81.4**	**84.2**	**87.9**	**89.9**	**90.5**	**92.1**	**92.9**
Belgium-Luxembourg	73.6	74.7	74.2	78.0	79.0	81.9	81.1
Denmark	100.0	100.0	100.0	100.0	100.0	100.0	100.0
France	93.1	93.8	93.9	94.3	95.0	95.2	95.4
Germany	...	...	...	...	89.5	92.0	93.9
former FR Germany	88.0	88.6	89.8	91.3	*92.2*	*93.7*	
Greece	100.0	100.0	100.0	100.0	100.0	100.0	100.0
Ireland	100.0	100.0	100.0	100.0	100.0	100.0	100.0
Italy	89.9	92.9	94.0	94.8	95.1	96.1	96.5
Netherlands	65.0	75.6	87.1	93.5	95.0	95.5	96.9
Portugal	44.8	45.4	50.5	55.4	96.9	98.9	...
Spain	66.7	74.4	86.3	89.4	92.9	94.1	96.7
United Kingdom	64.9	70.5	80.2	84.7	86.4	87.0	86.9
EFTA-TOTAL	**88.8**	**88.5**	**89.3**	**93.2**	**94.6**	**94.7**	**94.5**
Austria	95.7	95.5	95.7	95.9	97.2	96.7	96.3
Finland	94.0	93.9	94.0	98.0	99.0	100.0	100.0
Iceland	...	...	...	...	...	...	...
Norway	53.9	53.4	60.3	94.4	94.1	100.0	100.0
Sweden	83.6	83.0	82.3	86.0	88.0	89.0	89.0
Switzerland	100.0	100.0	100.0	100.0	100.0	100.0	100.0
EASTERN EUROPE-TOTAL	**21.7**	**22.3**	**22.1**	**22.2**	**23.0**	**24.6**	**31.0**
Albania	...	...	...	...	...	...	...
Bulgaria	12.4	15.5	15.5	19.7	18.8	23.0	23.2
former Czechoslovakia	8.5	8.7	9.2	11.5	17.0	22.0	*34.4*
former GDR	37.6	39.6	41.0	41.1	...	...	...
Hungary	55.9	63.2	55.6	64.2	82.7	92.2	87.5
Poland	11.0	11.1	7.7	7.0	8.0	7.7	10.4
Romania	32.4	31.5	34.2	36.2	39.7	41.9	46.5
former USSR-TOTAL	**16.1**	**16.6**	**17.3**	**17.9**	**17.7**	***20.5***	***21.1***
OTHER EUROPE-TOTAL	**70.3**	**80.6**	**83.4**	**85.9**	**82.3**	**90.0**	**91.0**
Turkey	79.1	91.1	92.9	91.9	92.0	93.4	93.6
former Yugoslavia	56.1	61.9	64.7	70.3	60.7	*68.6*	*68.6*
OTHERS	**...**	**...**	**...**	**...**	**...**	**...**	**...**
WORLD-TOTAL	**51.1**	**54.9**	**56.7**	**59.3**	**63.0**	**65.0**	**...**

Annex table 5 Crude steel production per capita (1 of 2)

(kilograms)

Regions and countries	1980	1981	1982	1983	1984	1985	1986
AFRICA-TOTAL	**24.0**	**23.2**	**21.3**	**19.0**	**18.9**	**22.0**	**21.8**
South Africa	320.6	311.1	279.7	237.2	250.0	269.3	275.2
Others	3.6	3.6	3.8	4.3	3.5	5.7	5.2
MIDDLE EAST-TOTAL	**16.8**	**16.7**	**16.4**	**18.3**	**22.0**	**23.9**	**22.9**
Egypt	23.3	23.8	24.4	21.7	20.0	21.6	20.7
Iran	14.1	13.9	13.0	16.6	18.6	17.6	17.0
Others	13.5	13.1	12.6	16.8	26.3	30.8	29.3
FAR EAST-TOTAL	**75.2**	**70.4**	**70.0**	**69.3**	**73.7**	**74.7**	**73.7**
China	37.3	35.3	36.4	38.7	41.5	44.1	48.6
India	13.8	15.3	15.3	13.9	14.0	15.5	15.5
Indonesia	3.6	4.0	4.4	6.1	7.2	8.3	10.2
Japan	953.7	864.6	840.9	815.4	880.1	871.8	810.2
Korea, DPR	...	...	...	...	...	...	...
Korea, Republic of	224.5	277.8	299.2	298.7	322.1	329.8	350.2
Others	13.5	11.9	14.3	16.1	16.5	18.3	19.4
NORTH AMERICA-TOTAL	**461.4**	**479.9**	**299.8**	**340.9**	**368.2**	**353.2**	**325.9**
Canada	664.2	611.3	484.3	517.4	585.8	576.7	549.9
United States	440.3	466.3	280.6	322.5	345.4	329.9	302.4
Others	-	-	-	-	-	-	-
OTHER AMERICA-TOTAL	**84.2**	**77.0**	**74.7**	**78.3**	**88.4**	**93.2**	**95.3**
Argentina	95.2	88.1	100.2	99.8	88.5	97.0	105.3
Brazil	126.5	106.5	102.3	113.4	138.5	150.9	153.3
Mexico	101.6	106.1	95.4	92.1	97.4	93.2	89.0
Venezuela	131.5	131.1	139.6	144.3	164.7	176.4	191.0
Others	18.3	16.5	16.8	17.4	18.0	18.7	21.4
OCEANIA-TOTAL	**439.1**	**435.3**	**362.2**	**319.0**	**350.7**	**359.7**	**362.0**
Australia	516.4	512.1	421.3	370.2	405.4	419.4	418.3
New Zealand	73.9	70.4	79.6	73.3	87.0	69.9	87.7
Others	-	-	-	-	-	-	-
EUROPE-TOTAL	**468.5**	**458.6**	**434.3**	**439.3**	**457.7**	**456.7**	**452.7**
EC-TOTAL	**446.8**	**438.7**	**391.8**	**384.9**	**418.0**	**420.6**	**389.9**
Belgium-Luxembourg	1 658.2	1 571.9	1 309.6	1 312.3	1 490.2	1 424.5	1 305.9
Denmark	143.3	119.5	109.3	96.3	107.0	103.1	123.4
France	430.1	392.7	338.3	321.7	346.0	340.9	322.5
Germany	...	...	...	...	...	...	...
former FR Germany	712.0	677.0	584.9	583.4	644.3	663.6	609.5
Greece	90.2	93.7	95.6	76.9	92.6	97.8	101.3
Ireland	0.6	9.3	17.6	40.4	47.1	57.2	58.0
Italy	469.6	438.0	423.4	383.7	422.2	418.3	400.3
Netherlands	372.6	384.9	304.8	312.5	398.1	381.0	363.5
Portugal	67.9	56.4	50.6	67.4	68.5	65.5	70.4
Spain	342.1	344.9	354.3	347.4	348.5	363.5	307.3
United Kingdom	200.2	276.2	242.7	265.1	267.3	277.7	259.8
EFTA-TOTAL	**420.7**	**403.1**	**387.6**	**404.9**	**446.9**	**440.3**	**426.5**
Austria	612.5	617.5	565.5	586.5	648.4	621.3	572.3
Finland	524.9	505.3	499.9	497.7	539.6	513.6	525.9
Iceland	-	-	-	-	-	-	-
Norway	211.0	206.9	186.8	212.3	221.3	227.4	200.8
Sweden	509.8	453.2	468.4	505.2	564.0	576.4	564.2
Switzerland	146.9	147.0	130.8	130.2	151.8	152.5	165.9
EASTERN EUROPE-TOTAL	**559.6**	**524.4**	**510.9**	**519.8**	**540.2**	**532.7**	**546.2**
Albania	...	...	...	...	...	...	...
Bulgaria	289.7	279.6	290.5	316.7	321.9	328.7	323.1
former Czechoslovakia	974.8	994.9	974.4	974.1	959.3	970.1	972.9
former GDR	436.6	446.6	429.2	432.8	454.5	471.8	478.6
Hungary	351.4	340.7	346.4	338.9	351.8	342.5	349.5
Poland	547.7	437.8	408.4	443.9	448.3	433.5	457.8
Romania	593.4	583.9	582.5	559.3	638.2	607.0	625.2
former USSR-TOTAL	**557.1**	**554.3**	**544.3**	**559.2**	**560.6**	**557.2**	**574.1**
OTHER EUROPE-TOTAL	**92.5**	**94.0**	**101.2**	**112.6**	**118.9**	**126.3**	**140.0**
Turkey	57.1	53.2	68.0	79.9	88.1	95.4	114.5
former Yugoslavia	163.0	177.0	169.7	181.4	184.9	193.8	196.2
OTHERS	...	...	...	...	...	...	...
WORLD-TOTAL	**164.7**	**159.6**	**142.9**	**144.7**	**152.0**	**151.5**	**147.8**

Annex table 5 Crude steel production per capita (2 of 2)

(kilograms)

Regions and countries	1987	1988	1989	1990	1991	1992
AFRICA-TOTAL	**21.4**	**20.2**	**19.9**	**18.7**	**20.5**	**19.4**
South Africa	272.0	261.6	270.5	244.5	259.7	246.1
Others	5.1	4.7	3.9	4.4	5.6	5.4
MIDDLE EAST-TOTAL	**26.2**	**31.3**	**32.5**	**34.5**	**39.0**	**41.9**
Egypt	28.6	39.3	40.1	41.6	46.2	44.6
Iran	16.4	18.5	19.7	25.2	37.9	49.1
Others	32.3	35.2	36.8	36.5	34.5	34.2
FAR EAST-TOTAL	**75.7**	**80.7**	**82.5**	**84.8**	**87.0**	**86.0**
China	51.6	53.8	55.0	58.5	61.7	69.4
India	16.3	17.5	17.5	17.5	19.6	20.3
Indonesia	12.0	11.7	13.4	16.0	16.8	15.8
Japan	808.6	863.6	877.9	893.7	884.1	787.6
Korea, DPR	...	...	...	...	...	...
Korea, Republic of	398.9	449.1	507.8	530.6	590.6	630.8
Others	20.5	27.0	28.4	29.7	31.8	31.3
NORTH AMERICA-TOTAL	**348.4**	**384.7**	**374.3**	**364.7**	**328.8**	**346.2**
Canada	570.3	565.0	583.1	459.3	479.7	513.7
United States	325.2	365.8	352.5	354.7	313.0	328.6
Others	-	-	-	-	-	-
OTHER AMERICA-TOTAL	**98.8**	**103.1**	**101.0**	**89.2**	**89.7**	**92.0**
Argentina	116.7	115.8	122.4	113.1	90.9	80.4
Brazil	157.1	170.7	170.0	136.8	147.5	153.2
Mexico	92.0	91.6	90.5	98.6	88.0	91.3
Venezuela	202.3	194.3	166.0	151.9	163.3	166.1
Others	22.0	22.1	19.9	18.1	18.7	19.6
OCEANIA-TOTAL	**334.6**	**353.1**	**372.7**	**367.5**	**343.7**	**346.1**
Australia	377.6	390.6	407.0	398.7	365.2	371.1
New Zealand	123.9	168.4	203.4	212.8	236.7	221.0
Others	-	-	-	-	-	-
EUROPE-TOTAL	**455.4**	**468.2**	**462.0**	**437.2**	**391.2**	**360.4**
EC-TOTAL	**391.8**	**424.4**	**430.5**	**421.2**	**401.6**	**385.8**
Belgium-Luxembourg	1 272.5	1 445.8	1 424.9	1 447.7	1 421.0	1 292.6
Denmark	118.3	126.9	121.9	119.1	123.4	115.3
France	318.3	333.5	334.0	338.5	327.0	317.4
Germany	...	...	...	...	546.8	515.5
former FR Germany	595.9	675.5	677.4	634.9	641.3	608.0
Greece	91.0	95.9	95.4	99.4	97.4	91.7
Ireland	60.8	74.2	87.9	87.4	78.0	67.8
Italy	399.6	415.1	440.2	444.3	437.7	432.6
Netherlands	348.3	376.8	386.5	366.9	349.3	366.1
Portugal	73.5	81.3	74.3	69.7	53.0	72.4
Spain	300.6	304.3	325.4	328.9	327.6	307.4
United Kingdom	306.9	333.6	329.6	313.4	285.9	281.3
EFTA-TOTAL	**419.1**	**442.5**	**438.5**	**407.2**	**399.9**	**404.6**
Austria	573.6	608.3	629.6	572.7	559.0	528.0
Finland	541.3	565.7	588.9	575.0	579.7	615.9
Iceland	-	-	-	-	-	-
Norway	203.8	217.3	161.4	89.3	103.7	105.3
Sweden	550.6	572.8	562.6	534.2	510.1	523.0
Switzerland	133.4	152.0	140.7	147.7	146.3	160.7
EASTERN EUROPE-TOTAL	**556.2**	**543.0**	**522.4**	**431.1**	**340.8**	**300.6**
Albania	...	...	...	...	...	...
Bulgaria	339.0	320.4	322.1	242.0	179.1	172.1
former Czechoslovakia	990.4	985.9	989.3	949.6	768.5	701.2
former GDR	495.2	488.6	470.3	334.3	203.7	179.3
Hungary	341.4	338.2	313.6	271.6	181.5	145.5
Poland	454.9	444.8	395.3	354.8	269.4	253.5
Romania	652.1	620.9	622.3	419.4	304.9	228.9
former USSR-TOTAL	**574.6**	**574.5**	**560.0**	**536.2**	**458.1**	**405.9**
OTHER EUROPE-TOTAL	**150.0**	**162.1**	**157.2**	**164.2**	**163.9**	**145.6**
Turkey	134.3	149.7	143.9	169.8	165.7	178.9
former Yugoslavia	185.3	190.5	187.7	151.3	159.6	65.9
OTHERS	...	...	...	...	...	...
WORLD-TOTAL	**149.8**	**156.1**	**154.5**	**149.1**	**140.4**	**135.2**

Annex table 6 Apparent crude steel consumption per capita (1 of 2)

(kilograms)

Regions and countries	1980	1981	1982	1983	1984	1985	1986
AFRICA-TOTAL	**35.7**	**35.8**	**28.1**	**26.7**	**26.0**	**26.9**	**23.4**
South Africa	254.5	251.0	195.6	175.8	184.3	158.9	157.1
Others	20.6	21.2	16.8	16.7	15.4	18.2	14.6
MIDDLE EAST-TOTAL	**117.6**	**110.7**	**137.0**	**144.4**	**127.2**	**131.4**	**109.6**
Egypt	48.9	49.9	54.6	55.9	50.1	69.1	114.5
Iran	91.4	69.4	100.8	135.6	103.3	131.3	90.4
Others	194.7	192.8	232.7	222.8	208.1	181.1	121.2
FAR EAST-TOTAL	**69.8**	**65.1**	**64.5**	**66.3**	**73.0**	**76.5**	**76.3**
China	43.4	38.9	40.1	50.6	58.1	68.4	69.7
India	15.8	19.9	19.3	15.7	18.5	18.7	19.5
Indonesia	19.7	20.1	20.1	17.8	16.1	14.2	16.6
Japan	676.4	604.9	587.1	550.6	619.9	607.7	576.6
Korea, DPR	...	...	...	...	...	...	...
Korea, Republic of	160.0	193.2	194.2	223.6	263.6	275.6	293.3
Others	45.8	42.7	43.8	43.5	38.7	39.0	42.3
NORTH AMERICA-TOTAL	**504.2**	**556.0**	**359.0**	**409.4**	**476.4**	**457.7**	**408.1**
Canada	540.5	556.1	372.5	448.8	529.4	537.5	488.6
United States	500.5	556.0	357.6	405.3	470.8	449.3	399.6
Others	-	-	-	-	-	-	-
OTHER AMERICA-TOTAL	**105.3**	**95.4**	**80.4**	**63.5**	**74.3**	**74.8**	**79.2**
Argentina	126.6	90.1	92.8	98.8	102.6	71.9	81.7
Brazil	118.0	97.0	83.5	66.5	80.4	88.5	104.9
Mexico	150.1	157.4	110.6	83.1	94.5	96.9	78.7
Venezuela	186.2	166.3	186.1	102.9	124.4	102.1	147.6
Others	47.3	46.4	40.4	33.9	41.3	42.9	41.7
OCEANIA-TOTAL	**375.1**	**438.9**	**356.2**	**315.2**	**355.7**	**350.3**	**341.0**
Australia	413.3	482.6	378.0	334.0	379.3	380.0	371.4
New Zealand	194.5	231.4	252.1	225.4	241.6	206.2	192.5
Others	-	-	-	-	-	-	-
EUROPE-TOTAL	**440.1**	**412.1**	**409.0**	**407.4**	**414.6**	**406.2**	**416.1**
EC-TOTAL	**364.0**	**323.9**	**320.9**	**301.4**	**316.7**	**310.8**	**320.3**
Belgium-Luxembourg	290.5	313.9	451.0	331.4	414.1	354.5	339.7
Denmark	343.4	345.5	443.1	260.4	306.1	352.2	350.1
France	360.5	308.4	300.1	266.8	267.9	252.9	245.7
Germany	...	...	...	...	...	...	...
former FR Germany	548.7	504.2	437.6	487.2	508.8	504.5	503.5
Greece	170.5	143.7	161.6	195.7	133.6	152.2	161.8
Ireland	113.1	143.3	135.3	111.1	118.0	111.2	100.3
Italy	477.4	368.3	383.7	335.9	387.1	382.8	395.5
Netherlands	303.1	246.6	248.1	223.6	276.1	285.3	275.6
Portugal	132.5	145.3	165.2	126.0	118.3	109.7	116.1
Spain	253.2	213.5	245.9	196.6	177.8	169.4	236.9
United Kingdom	244.7	264.2	251.5	248.9	253.4	253.5	261.5
EFTA-TOTAL	**440.2**	**382.4**	**379.8**	**352.0**	**369.8**	**347.2**	**378.4**
Austria	380.5	336.2	321.5	312.9	317.4	299.2	331.4
Finland	446.4	389.6	432.3	373.1	396.2	359.4	396.1
Iceland	297.2	250.2	237.2	217.6	223.5	207.3	218.5
Norway	445.7	372.0	420.7	324.3	338.0	339.8	374.4
Sweden	498.0	412.3	423.1	419.1	444.9	393.2	423.3
Switzerland	432.2	404.1	331.1	317.3	339.8	344.3	370.2
EASTERN EUROPE-TOTAL	**538.7**	**476.4**	**491.0**	**482.7**	**476.1**	**474.8**	**482.8**
Albania	...	...	...	...	...	...	...
Bulgaria	329.5	340.5	352.5	371.6	332.5	343.0	326.6
former Czechoslovakia	704.0	730.9	721.3	716.9	699.1	708.0	712.1
former GDR	600.7	444.8	590.1	570.8	520.5	549.9	544.1
Hungary	337.4	342.8	350.5	335.3	312.0	323.8	330.1
Poland	538.2	426.7	396.3	403.0	411.7	405.2	425.5
Romania	550.9	518.5	524.6	489.8	519.2	486.5	497.2
former USSR-TOTAL	**566.5**	**563.3**	**556.6**	**578.5**	**580.0**	**566.6**	**577.8**
OTHER EUROPE-TOTAL	**138.2**	**139.7**	**122.4**	**134.8**	**144.3**	**140.0**	**139.4**
Turkey	77.7	74.6	69.7	88.7	107.1	103.5	92.1
former Yugoslavia	258.6	271.8	231.4	231.9	223.9	219.5	243.9
OTHERS	**197.9**	**178.2**	**215.6**	**199.9**	**221.0**	**222.8**	**297.5**
WORLD-TOTAL	**164.8**	**158.5**	**143.8**	**145.1**	**153.4**	**152.2**	**149.3**

Annex table 6 Apparent crude steel consumption per capita (2 of 2)

(kilograms)

Regions and countries	1987	1988	1989	1990	1991	1992
AFRICA-TOTAL	**22.7**	**23.0**	**22.1**	**21.4**	**20.5**	**19.9**
South Africa	158.3	170.1	174.7	156.7	139.1	118.8
Others	13.9	13.5	12.4	12.9	13.1	13.8
MIDDLE EAST-TOTAL	**95.2**	**99.3**	**96.4**	**99.9**	**107.0**	**112.8**
Egypt	74.8	86.3	88.1	85.3	89.0	88.2
Iran	113.2	88.4	98.3	121.2	142.5	160.2
Others	96.8	118.1	101.3	94.0	92.5	93.8
FAR EAST-TOTAL	**79.0**	**84.8**	**88.0**	**90.8**	**93.9**	**92.6**
China	65.5	63.7	63.7	60.3	61.7	74.2
India	22.0	23.2	24.0	25.4	23.3	21.4
Indonesia	14.9	15.3	17.3	25.1	24.2	23.4
Japan	621.7	709.9	758.9	802.2	799.4	674.5
Korea, DPR	...	...	...	...	...	...
Korea, Republic of	357.7	371.7	424.1	492.8	592.1	522.5
Others	48.9	62.9	64.7	68.9	82.9	90.0
NORTH AMERICA-TOTAL	**437.8**	**465.3**	**431.3**	**408.3**	**369.2**	**388.5**
Canada	521.4	576.3	540.5	411.0	397.3	382.3
United States	429.0	453.6	419.9	408.1	366.3	389.1
Others	-	-	-	-	-	-
OTHER AMERICA-TOTAL	**81.5**	**74.1**	**75.7**	**62.1**	**64.6**	**68.3**
Argentina	105.8	95.2	58.1	48.8	65.4	94.6
Brazil	105.8	82.8	100.5	68.2	65.0	67.0
Mexico	77.4	80.9	84.0	91.8	100.4	107.6
Venezuela	162.5	167.8	115.7	81.1	135.0	102.6
Others	40.9	42.5	42.6	37.3	31.8	34.3
OCEANIA-TOTAL	**344.8**	**379.4**	**354.9**	**300.9**	**269.2**	**236.9**
Australia	363.2	394.3	386.3	323.0	292.5	254.1
New Zealand	254.7	306.1	199.8	191.6	153.1	151.0
Others	-	-	-	-	-	-
EUROPE-TOTAL	**421.6**	**442.1**	**443.6**	**406.6**	**351.3**	**319.5**
EC-TOTAL	**315.9**	**372.0**	**390.5**	**382.2**	**358.0**	**348.6**
Belgium-Luxembourg	292.8	507.0	386.1	344.6	374.6	369.0
Denmark	305.6	327.5	350.8	352.7	350.8	395.2
France	247.2	274.9	301.4	322.0	295.3	284.5
Germany	...	...	...	...	505.1	499.0
former FR Germany	476.9	580.2	585.8	581.5	588.2	...
Greece	190.4	149.3	197.6	240.7	268.8	178.6
Ireland	105.7	132.1	140.1	131.8	119.8	114.0
Italy	411.0	460.0	489.3	483.1	473.8	468.1
Netherlands	232.3	294.6	320.7	335.6	270.5	284.9
Portugal	154.2	221.7	332.3	182.8	180.8	193.4
Spain	230.6	269.2	314.0	303.0	287.4	269.6
United Kingdom	264.0	307.9	306.0	281.6	237.4	226.3
EFTA-TOTAL	**350.2**	**394.8**	**411.1**	**373.5**	**313.0**	**316.2**
Austria	290.6	350.8	386.7	352.2	392.7	370.4
Finland	404.0	399.4	458.3	399.7	289.5	275.8
Iceland	252.5	235.9	196.8	170.0	198.8	168.0
Norway	297.2	300.1	281.7	224.1	142.3	264.2
Sweden	425.5	458.0	473.0	442.4	332.8	327.0
Switzerland	319.1	428.0	416.0	394.0	329.3	310.6
EASTERN EUROPE-TOTAL	**489.2**	**472.7**	**453.5**	**330.9**	**211.3**	**165.0**
Albania	...	...	...	...	...	...
Bulgaria	381.1	315.4	319.2	184.0	131.8	97.3
former Czechoslovakia	673.4	687.9	694.9	657.4	392.0	312.0
former GDR	550.1	532.4	615.8	323.7	...	...
Hungary	317.7	296.0	268.3	224.5	128.9	96.1
Poland	416.2	407.2	349.8	247.9	149.9	150.0
Romania	550.2	523.2	469.7	348.4	252.8	143.6
former USSR-TOTAL	**578.7**	**580.2**	**582.6**	**545.8**	**454.7**	**393.5**
OTHER EUROPE-TOTAL	**217.7**	**202.5**	**154.5**	**123.0**	**135.3**	**119.3**
Turkey	236.7	220.4	133.1	117.0	126.3	134.4
former Yugoslavia	175.1	161.9	203.8	137.0	156.7	83.0
OTHERS	**258.2**	**350.8**	**309.1**	**319.2**	**244.6**	**285.5**
WORLD-TOTAL	**152.1**	**158.9**	**157.9**	**150.0**	**140.0**	**134.9**

Annex table 7 Crude steel production as a percentage of apparent crude steel consumption (1 of 2)

(%)

Regions and countries	1980	1981	1982	1983	1984	1985	1986
AFRICA-TOTAL	**67.2**	**64.8**	**75.7**	**71.2**	**72.8**	**81.7**	**93.4**
South Africa	126.0	124.0	143.0	134.9	135.6	169.5	175.2
Others	17.3	16.9	22.8	26.1	23.0	31.1	35.9
MIDDLE EAST-TOTAL	**14.3**	**15.1**	**12.0**	**12.6**	**17.3**	**18.2**	**20.9**
Egypt	47.6	47.6	44.8	38.8	39.9	31.3	18.1
Iran	15.5	20.0	12.9	12.3	18.0	13.4	18.8
Others	6.9	6.8	5.4	7.5	12.7	17.0	24.2
FAR EAST-TOTAL	**107.7**	**108.1**	**108.4**	**104.4**	**100.9**	**97.7**	**96.6**
China	85.8	90.8	90.7	76.5	71.5	64.5	69.6
India	87.3	76.9	79.1	88.3	75.9	82.9	79.8
Indonesia	18.3	20.0	22.0	34.5	44.6	58.0	61.4
Japan	141.0	142.9	143.2	148.1	142.0	143.5	140.5
Korea, DPR	100.0	100.0	100.0	92.0	92.5	92.5	92.6
Korea, Republic of	140.3	143.8	154.1	133.6	122.2	119.7	119.4
Others	29.4	27.9	32.6	36.9	42.7	46.8	45.8
NORTH AMERICA-TOTAL	**91.5**	**86.3**	**83.5**	**83.3**	**77.3**	**77.2**	**79.9**
Canada	122.9	109.9	130.0	115.3	110.7	107.3	112.5
United States	88.0	83.9	78.5	79.6	73.4	73.4	75.7
Others	...	...	...	...	...	...	...
OTHER AMERICA-TOTAL	**79.9**	**80.7**	**93.0**	**123.3**	**118.9**	**124.7**	**120.4**
Argentina	75.2	97.8	108.0	100.9	86.2	134.9	128.9
Brazil	107.2	109.8	122.5	170.5	172.3	170.5	146.1
Mexico	67.7	67.4	86.2	110.8	103.1	96.2	113.0
Venezuela	70.6	78.8	75.0	140.3	132.4	172.7	129.5
Others	38.7	35.6	41.6	51.3	43.6	43.6	51.3
OCEANIA-TOTAL	**117.1**	**99.2**	**101.7**	**101.2**	**98.6**	**102.7**	**106.2**
Australia	125.0	106.1	111.5	110.8	106.9	110.4	112.6
New Zealand	38.0	30.4	31.6	32.5	36.0	33.9	45.6
Others	...	...	...	...	...	...	...
EUROPE-TOTAL	**106.5**	**111.3**	**106.2**	**107.8**	**110.4**	**112.4**	**108.8**
EC-TOTAL	**122.8**	**135.5**	**122.1**	**127.7**	**132.0**	**135.3**	**121.7**
Belgium-Luxembourg	570.9	500.7	290.3	396.0	359.9	401.8	384.4
Denmark	41.7	34.6	24.7	37.0	34.9	29.3	35.2
France	119.3	127.3	112.7	120.6	129.2	134.8	131.3
Germany	...	...	...	...	...	...	...
former FR Germany	129.8	134.3	133.6	119.8	126.6	131.5	121.0
Greece	52.9	65.2	59.2	39.3	69.4	64.3	62.6
Ireland	0.5	6.5	13.0	36.3	39.9	51.4	57.8
Italy	98.4	118.9	110.3	114.2	109.1	109.3	101.2
Netherlands	122.9	156.1	122.8	139.7	144.2	133.5	131.9
Portugal	51.3	38.8	30.6	53.5	57.9	59.7	60.6
Spain	135.1	161.6	144.1	176.7	196.0	214.5	129.7
United Kingdom	81.8	104.5	96.5	106.5	105.5	109.6	99.4
EFTA-TOTAL	**95.6**	**105.4**	**102.0**	**115.0**	**120.8**	**126.8**	**112.7**
Austria	161.0	183.7	175.9	187.4	204.3	207.7	172.7
Finland	117.6	129.7	115.6	133.4	136.2	142.9	132.8
Iceland	-	-	-	-	-	-	-
Norway	47.3	55.6	44.4	65.5	65.5	66.9	53.6
Sweden	102.4	109.9	110.7	120.5	126.8	146.6	133.3
Switzerland	34.0	36.4	39.5	41.0	44.7	44.3	44.8
EASTERN EUROPE-TOTAL	**103.9**	**110.1**	**104.0**	**107.7**	**113.5**	**112.2**	**113.1**
Albania	...	...	27.2	30.6	33.5	36.1	38.6
Bulgaria	87.9	82.1	82.4	85.2	96.8	95.8	98.9
former Czechoslovakia	138.5	136.1	135.1	135.9	137.2	137.0	136.6
former GDR	72.7	100.4	72.7	75.8	87.3	85.8	88.0
Hungary	104.2	99.4	98.8	101.1	112.8	105.8	105.9
Poland	101.8	102.6	103.0	110.1	108.9	107.0	107.6
Romania	107.7	112.6	111.1	114.2	122.9	124.8	125.8
former USSR-TOTAL	**98.3**	**98.4**	**97.8**	**96.7**	**96.7**	**98.3**	**99.4**
OTHER EUROPE-TOTAL	**66.9**	**67.3**	**82.6**	**83.5**	**82.4**	**90.3**	**100.4**
Turkey	73.4	71.3	97.5	90.1	82.2	92.2	124.4
former Yugoslavia	63.0	65.1	73.3	78.2	82.6	88.3	80.4
OTHERS	...	...	...	...	...	...	...
WORLD-TOTAL	**100.0**	**100.7**	**99.4**	**99.8**	**99.1**	**99.5**	**99.0**

Annex table 7 Crude steel production as a percentage of apparent crude steel consumption (2 of 2)

(%)

Regions and countries	1987	1988	1989	1990	1991	1992	1993
AFRICA-TOTAL	**94.1**	**88.0**	**89.7**	**87.2**	**100.2**	**97.5**	...
South Africa	171.8	153.7	154.9	156.0	186.7	207.1	...
Others	36.8	35.0	31.3	34.3	42.8	39.3	...
MIDDLE EAST-TOTAL	**27.5**	**31.5**	**33.7**	**34.5**	**36.4**	**37.2**	...
Egypt	38.2	45.6	45.5	48.7	51.9	50.5	...
Iran	14.5	20.9	20.1	20.8	26.6	30.7	...
Others	33.3	29.8	36.4	38.8	37.3	36.5	...
FAR EAST-TOTAL	**95.8**	**95.1**	**93.7**	**93.4**	**92.6**	**92.8**	...
China	78.8	84.4	86.4	97.0	99.9	93.5	...
India	74.4	75.2	72.9	69.0	84.2	94.9	...
Indonesia	80.3	76.8	77.7	63.8	69.7	67.7	...
Japan	130.0	121.7	115.7	111.4	110.6	116.8	123.6
Korea, DPR	92.7	92.8	92.9	93.0	93.0	93.0	...
Korea, Republic of	111.5	120.8	119.7	107.7	99.7	120.7	...
Others	42.0	42.9	43.9	43.1	38.4	34.8	...
NORTH AMERICA-TOTAL	**79.6**	**82.7**	**86.8**	**89.3**	**89.0**	**89.1**	**87.2**
Canada	109.4	98.0	107.9	111.7	120.8	134.4	106.0
United States	75.8	80.6	83.9	86.9	85.4	84.4	84.8
Others	...	...	...	...	...	...	
OTHER AMERICA-TOTAL	**121.3**	**139.1**	**133.4**	**143.6**	**138.8**	**134.6**	...
Argentina	110.3	121.6	210.6	231.7	138.9	85.0	...
Brazil	148.5	206.2	169.2	200.6	227.0	228.6	...
Mexico	118.9	113.2	107.8	107.4	87.7	84.9	...
Venezuela	124.5	115.8	143.5	187.3	120.9	161.9	...
Others	53.9	51.9	46.8	48.7	58.8	57.2	...
OCEANIA-TOTAL	**97.0**	**93.1**	**105.0**	**122.1**	**127.7**	**146.1**	...
Australia	104.0	99.1	105.4	123.4	124.8	146.0	...
New Zealand	48.7	55.0	101.8	111.1	154.7	146.3	...
Others	...	...	...	...	...	...	...
EUROPE-TOTAL	**108.0**	**105.9**	**104.2**	**107.5**	**111.3**	**112.8**	...
EC-TOTAL	**124.0**	**114.1**	**110.3**	**110.2**	**112.2**	**110.7**	...
Belgium-Luxembourg	434.6	285.1	369.0	420.2	379.3	350.3	...
Denmark	38.7	38.8	34.7	33.8	35.2	29.2	...
France	128.7	121.3	110.8	105.1	110.7	111.6	123.1
Germany	...	...	...	...	108.3	103.3	120.5
former FR Germany	125.0	116.4	115.6	109.2	109.0	...	...
Greece	47.8	64.2	48.3	41.3	36.2	51.3	...
Ireland	57.5	56.2	62.7	66.3	65.1	59.5	...
Italy	97.2	90.2	90.0	92.0	92.4	92.4	110.8
Netherlands	149.9	127.9	120.5	109.3	129.1	128.5	134.8
Portugal	47.7	36.7	22.4	38.1	29.3	37.5	...
Spain	130.3	113.1	103.6	108.5	114.0	114.0	132.2
United Kingdom	116.2	108.3	107.7	111.3	120.4	124.3	...
EFTA-TOTAL	**119.7**	**112.1**	**106.7**	**109.0**	**127.7**	**128.0**	...
Austria	197.4	173.4	162.8	162.6	142.4	142.6	161.3
Finland	134.0	141.6	128.5	143.9	200.3	223.3	209.1
Iceland	-	-	-	-	-	-	-
Norway	68.6	72.4	57.3	39.8	72.9	39.9	44.0
Sweden	129.4	125.1	119.0	120.8	153.3	159.9	155.9
Switzerland	41.8	35.5	33.8	37.5	44.4	51.7	...
EASTERN EUROPE-TOTAL	**113.7**	**114.9**	**115.2**	**130.3**	**161.3**	**182.2**	...
Albania	38.6	38.6	41.3	33.2	10.8	...	...
Bulgaria	88.9	101.6	100.9	131.5	135.9	176.8	218.3
former Czechoslovakia	147.1	143.3	142.4	144.5	196.1	224.8	272.6
former GDR	90.0	91.8	76.4	103.3	...	...	...
Hungary	107.4	114.2	116.9	121.0	140.8	151.5	143.5
Poland	109.3	109.2	113.0	143.1	179.7	169.0	153.2
Romania	118.5	118.7	132.5	120.4	120.6	159.4	182.0
former USSR-TOTAL	**99.3**	**99.0**	**96.1**	**98.2**	**100.7**	**103.1**	...
OTHER EUROPE-TOTAL	**68.9**	**80.1**	**101.7**	**133.5**	**121.1**	**122.1**	...
Turkey	56.7	67.9	108.1	145.1	131.2	133.1	109.5
former Yugoslavia	105.8	117.7	92.1	110.4	101.8	79.3	...
OTHERS	...	...	...	...	...	...	...
WORLD-TOTAL	**98.5**	**98.3**	**97.8**	**99.4**	**100.2**	**100.2**	...

Annex table 8 Apparent finished steel consumption (1 of 2)

(1000 tonnes)

Regions and countries	1980	1981	1982	1983	1984	1985	1986
AFRICA-TOTAL	**11 644**	**12 169**	**10 290**	**10 081**	**10 144**	**10 716**	**9 608**
South Africa	5 200	5 331	4 692	4 311	4 617	4 066	4 113
Others	6 444	6 838	5 598	5 770	5 527	6 650	5 495
MIDDLE EAST-TOTAL	**13 059**	**12 771**	**16 446**	**18 239**	**16 753**	**17 787**	**15 327**
Egypt	1 700	1 784	2 007	2 157	2 021	2 856	4 858
Iran	2 975	2 360	3 574	5 006	3 964	5 231	3 737
Others	8 384	8 628	10 864	11 075	10 767	9 700	6 732
FAR EAST-TOTAL	**138 631**	**129 596**	**132 958**	**141 004**	**155 505**	**167 033**	**170 618**
China	32 403	29 361	30 661	39 439	45 950	54 812	56 742
India	8 874	10 382	10 496	9 857	9 555	10 744	11 520
Indonesia	2 485	2 594	2 641	2 466	2 306	2 085	2 506
Japan	71 307	63 272	63 601	61 168	69 097	69 861	67 702
Korea, DPR	4 377	4 150	4 377	5 003	5 305	5 305	5 381
Korea, Republic of	5 115	6 314	6 455	7 801	9 408	10 020	10 934
Others	14 070	13 522	14 727	15 269	13 885	14 206	15 833
NORTH AMERICA-TOTAL	**96 689**	**107 164**	**76 708**	**84 379**	**100 190**	**98 346**	**92 151**
Canada	10 282	11 507	7 410	8 670	10 450	10 923	10 481
United States	86 407	95 657	69 298	75 709	89 740	87 423	81 670
Others	...	...	...	...	...	...	...
OTHER AMERICA-TOTAL	**29 471**	**27 751**	**24 129**	**19 685**	**23 477**	**24 154**	**26 254**
Argentina	3 006	2 158	2 264	2 433	2 557	1 861	2 150
Brazil	11 648	9 856	8 747	7 157	8 803	9 932	12 077
Mexico	8 225	9 234	6 712	5 311	6 172	6 447	5 325
Venezuela	2 304	2 196	2 561	1 472	1 818	1 533	2 291
Others	4 289	4 306	3 844	3 311	4 127	4 382	4 411
OCEANIA-TOTAL	**5 092**	**6 063**	**5 033**	**4 605**	**5 275**	**5 249**	**5 164**
Australia	4 585	5 455	4 365	3 955	4 572	4 644	4 594
New Zealand	506	608	668	651	703	605	569
Others	...	...	...	...	...	...	...
EUROPE-TOTAL	**273 302**	**257 688**	**258 666**	**259 586**	**267 145**	**265 078**	**273 728**
EC-TOTAL	**96 868**	**86 280**	**86 629**	**82 128**	**86 634**	**86 243**	**89 462**
Belgium-Luxembourg	2 270	2 482	3 595	2 772	3 516	3 053	2 977
Denmark	1 402	1 406	1 709	1 066	1 416	1 630	1 620
France	16 021	13 996	13 843	12 470	12 639	12 246	12 114
Germany	...	...	...	...	...	...	...
former FR Germany	28 920	27 437	23 801	26 186	27 148	26 979	26 984
Greece	1 376	1 166	1 319	1 736	1 192	1 366	1 456
Ireland	322	445	423	351	376	357	325
Italy	22 534	17 450	18 455	16 444	19 146	19 138	19 943
Netherlands	3 215	2 692	2 874	2 624	3 270	3 396	3 312
Portugal	1 069	1 174	1 364	1 043	982	923	981
Spain	7 926	6 581	7 703	6 230	5 707	5 539	7 813
United Kingdom	11 813	11 451	11 542	11 207	11 241	11 615	11 938
EFTA-TOTAL	**10 866**	**9 800**	**9 755**	**9 169**	**10 071**	**9 656**	**10 463**
Austria	2 408	2 163	2 114	2 088	2 119	2 008	2 229
Finland	1 900	1 672	1 868	1 621	1 732	1 577	1 743
Iceland	52	44	43	40	41	38	41
Norway	1 367	1 141	1 305	1 050	1 173	1 192	1 319
Sweden	2 991	2 775	2 692	2 688	3 027	2 826	2 964
Switzerland	2 148	2 005	1 733	1 683	1 978	2 014	2 168
EASTERN EUROPE-TOTAL	**44 610**	**39 683**	**41 149**	**40 714**	**40 635**	**41 073**	**42 006**
Albania	143	80	166	173	180	188	195
Bulgaria	2 187	2 265	2 349	2 490	2 243	2 316	2 225
former Czechoslovakia	8 026	8 354	8 278	8 281	8 122	8 251	8 325
former GDR	7 637	5 663	7 520	7 278	6 709	7 456	7 411
Hungary	2 955	2 996	3 050	2 942	2 768	2 868	2 946
Poland	14 314	11 450	10 738	11 014	11 459	11 382	12 035
Romania	9 347	8 876	9 047	8 537	9 153	8 612	8 870
former USSR-TOTAL	**113 643**	**114 292**	**114 014**	**119 503**	**120 928**	**119 340**	**122 893**
OTHER EUROPE-TOTAL	**7 315**	**7 634**	**7 118**	**8 072**	**8 877**	**8 765**	**8 904**
Turkey	2 595	2 581	2 755	3 639	4 562	4 485	4 135
former Yugoslavia	4 720	5 053	4 363	4 433	4 315	4 281	4 769
OTHERS	**146**	**132**	**160**	**149**	**165**	**167**	**225**
WORLD-TOTAL	**568 034**	**553 334**	**524 389**	**537 727**	**578 654**	**588 530**	**593 075**

Annex table 8 Apparent finished steel consumption (2 of 2)

(1000 tonnes)

Regions and countries	1987	1988	1989	1990	1991	1992	1993
AFRICA-TOTAL	**9 638**	**10 093**	**10 036**	**10 218**	**10 082**	**10 174**	**...**
South Africa	4 239	4 660	4 880	4 480	4 061	3 587	...
Others	5 399	5 433	5 156	5 738	6 021	6 587	...
MIDDLE EAST-TOTAL	**13 799**	**14 887**	**14 846**	**16 324**	**17 989**	**19 523**	**...**
Egypt	3 370	3 921	4 090	4 145	4 384	4 447	...
Iran	4 849	3 918	4 504	6 185	7 491	8 655	...
Others	5 580	7 048	6 251	5 994	6 114	6 421	...
FAR EAST-TOTAL	**180 984**	**196 793**	**209 959**	**220 923**	**234 498**	**233 691**	**...**
China	54 134	53 530	54 358	52 647	55 035	67 402	...
India	13 830	14 930	15 670	16 900	15 850	15 280	...
Indonesia	2 284	2 396	2 756	4 076	3 994	3 938	...
Japan	72 891	80 961	88 306	92 807	93 132	79 029	74 155
Korea, DPR	5 479	5 554	5 630	5 682	5 682	5 682	...
Korea, Republic of	13 642	14 519	16 946	20 054	24 454	21 820	...
Others	18 724	24 904	26 294	28 757	36 351	40 539	...
NORTH AMERICA-TOTAL	**98 435**	**105 833**	**99 554**	**98 281**	**87 155**	**95 601**	**104 713**
Canada	11 400	12 700	11 700	9 521	9 190	10 152	12 045
United States	87 035	93 133	87 854	88 760	77 965	85 449	92 668
Others	...	...	...	...	...	...	...
OTHER AMERICA-TOTAL	**27 713**	**25 857**	**27 053**	**22 740**	**24 321**	**26 254**	**26 706**
Argentina	2 823	2 584	1 611	1 375	1 880	2 751	2 781
Brazil	12 435	9 988	12 466	8 693	8 415	8 868	10 161
Mexico	5 415	5 802	6 174	6 921	7 774	8 507	7 528
Venezuela	2 604	2 756	1 948	1 402	2 440	1 912	1 883
Others	4 436	4 727	4 854	4 350	3 812	4 216	4 353
OCEANIA-TOTAL	**5 624**	**6 500**	**6 212**	**5 353**	**4 826**	**4 314**	**...**
Australia	4 864	5 580	5 606	4 767	4 355	3 846	...
New Zealand	760	920	605	585	471	468	...
Others	...	...	...	...	...	...	...
EUROPE-TOTAL	**280 926**	**294 603**	**298 063**	**276 660**	**244 406**	**224 983**	**...**
EC-TOTAL	**89 490**	**102 256**	**108 807**	**108 040**	**109 218**	**106 754**	**...**
Belgium-Luxembourg	2 614	4 539	3 456	3 106	3 384	3 351	...
Denmark	1 415	1 516	1 623	1 632	1 624	1 830	...
France	12 292	13 730	15 105	16 208	14 933	14 444	12 475
Germany	...	...	...	...	34 654	34 333	27 963
former FR Germany	25 810	27 390	28 677	28 854	*29 207*	...	...
Greece	1 717	1 350	1 790	2 186	2 445	1 627	...
Ireland	346	436	467	443	406	390	...
Italy	20 933	23 551	25 111	24 837	24 392	24 157	20 853
Netherlands	2 904	3 758	4 179	4 431	3 592	3 799	4 005
Portugal	1 305	1 884	2 847	1 583	1 678	1 805	...
Spain	7 704	9 136	10 892	10 603	10 129	9 572	8 753
United Kingdom	12 449	14 967	14 660	14 157	11 981	11 446	...
EFTA-TOTAL	**9 726**	**11 137**	**11 144**	**10 628**	**8 931**	**9 139**	**...**
Austria	1 957	2 361	2 602	2 371	2 647	2 494	2 312
Finland	1 784	1 769	2 036	1 791	1 302	1 245	1 407
Iceland	48	45	38	33	39	33	32
Norway	1 050	1 062	1 006	846	538	1 011	1 037
Sweden	3 015	3 384	3 013	3 265	2 462	2 422	2 618
Switzerland	1 872	2 515	2 448	2 322	1 943	1 934	...
EASTERN EUROPE-TOTAL	**43 152**	**41 870**	**40 336**	**29 246**	**15 948**	**12 500**	**...**
Albania	195	195	205	180	112	80	...
Bulgaria	2 592	2 158	2 187	1 270	909	677	685
former Czechoslovakia	7 891	8 081	8 187	7 790	4 698	3 783	3 034
former GDR	7 508	7 289	8 453	4 445	...	...	...
Hungary	2 847	2 678	2 394	2 027	1 196	904	1 073
Poland	11 858	11 678	10 042	7 154	4 355	4 377	4 881
Romania	10 261	9 789	8 869	6 380	4 678	2 679	2 395
former USSR-TOTAL	**124 220**	**125 571**	**127 132**	**120 115**	**100 716**	**87 911**	**...**
OTHER EUROPE-TOTAL	**14 337**	**13 770**	**10 645**	**8 632**	**9 593**	**8 679**	**...**
Turkey	10 879	10 518	6 508	5 809	6 396	6 955	9 419
former Yugoslavia	3 458	3 251	4 137	2 823	3 197	1 724	...
OTHERS	**197**	**270**	**240**	**250**	**193**	**227**	**...**
WORLD-TOTAL	**617 316**	**654 838**	**665 962**	**650 748**	**623 470**	**614 767**	**...**

Annex table 9 Apparent finished steel consumption per capita (1 of 2)

(kilograms)

Regions and countries	1980	1981	1982	1983	1984	1985	1986
AFRICA-TOTAL	**26.5**	**26.8**	**22.0**	**20.9**	**20.5**	**21.0**	**18.3**
South Africa	183.8	184.2	158.5	142.4	149.3	128.7	127.2
Others	15.7	16.1	12.8	12.8	11.9	13.9	11.1
MIDDLE EAST-TOTAL	**100.1**	**94.3**	**117.2**	**125.6**	**111.6**	**114.8**	**95.6**
Egypt	40.9	41.8	45.7	47.8	43.6	60.0	99.4
Iran	76.5	58.1	84.3	113.4	86.4	109.8	75.6
Others	167.5	165.9	201.4	198.2	186.3	162.4	108.6
FAR EAST-TOTAL	**58.6**	**53.8**	**54.2**	**56.5**	**61.2**	**64.6**	**64.8**
China	32.5	29.1	30.0	38.1	43.9	51.7	52.8
India	12.9	14.7	14.6	13.4	12.7	14.0	14.7
Indonesia	16.5	16.8	16.8	15.4	14.1	12.5	14.8
Japan	610.5	538.0	537.2	513.3	576.0	578.5	558.2
Korea, DPR	...	...	...	...	...	...	...
Korea, Republic of	134.2	163.1	164.3	195.6	232.5	244.1	263.1
Others	37.6	35.1	37.2	37.6	33.3	33.2	36.0
NORTH AMERICA-TOTAL	**379.3**	**416.1**	**294.8**	**321.0**	**377.4**	**366.8**	**340.8**
Canada	429.5	474.9	302.3	349.6	416.5	430.4	409.3
United States	374.1	410.0	294.1	318.1	373.3	360.2	333.7
Others	-	-	-	-	-	-	-
OTHER AMERICA-TOTAL	**85.0**	**78.1**	**66.4**	**53.0**	**61.8**	**62.2**	**66.1**
Argentina	106.4	75.3	77.9	82.5	85.5	61.3	70.0
Brazil	96.0	79.4	68.9	55.1	66.3	73.3	87.2
Mexico	116.8	127.9	90.7	70.1	79.6	81.2	65.6
Venezuela	153.3	141.9	160.6	89.8	107.8	88.5	128.7
Others	38.3	37.6	32.7	27.6	33.6	34.9	34.3
OCEANIA-TOTAL	**285.9**	**335.9**	**275.2**	**248.6**	**281.1**	**276.2**	**268.5**
Australia	312.0	365.9	288.7	257.9	294.1	294.7	287.9
New Zealand	162.7	193.6	210.9	203.7	218.3	186.4	174.0
Others	-	-	-	-	-	-	-
EUROPE-TOTAL	**345.5**	**323.7**	**322.9**	**322.0**	**329.3**	**324.7**	**333.6**
EC-TOTAL	**304.6**	**270.6**	**271.0**	**256.3**	**269.7**	**267.8**	**277.4**
Belgium-Luxembourg	222.2	242.7	351.2	270.5	342.7	297.3	289.7
Denmark	273.7	274.4	333.6	208.1	276.5	318.2	316.3
France	297.3	258.5	254.5	228.2	230.2	222.0	218.8
Germany	...	...	...	...	...	...	...
former FR Germany	469.7	446.4	388.0	427.6	444.1	442.1	442.9
Greece	142.7	120.2	135.2	176.8	120.7	137.6	146.2
Ireland	94.6	129.6	122.2	100.4	106.7	100.5	90.7
Italy	399.3	308.4	325.4	289.3	336.0	335.0	348.9
Netherlands	227.2	189.4	201.2	182.8	226.8	234.4	227.8
Portugal	109.4	119.2	137.5	104.2	97.5	90.9	96.3
Spain	211.1	174.3	202.9	163.2	148.7	143.5	201.6
United Kingdom	209.7	203.1	204.5	198.3	198.7	205.1	210.6
EFTA-TOTAL	**347.4**	**312.6**	**310.5**	**291.2**	**319.2**	**305.4**	**330.6**
Austria	319.0	286.9	280.8	277.6	282.2	267.7	297.1
Finland	397.6	347.9	386.9	333.9	355.1	321.7	354.4
Iceland	228.6	192.5	182.5	167.4	171.9	159.5	168.1
Norway	334.6	278.4	317.4	254.5	283.4	287.1	316.8
Sweden	359.9	333.6	323.3	322.5	362.9	338.4	355.1
Switzerland	339.6	315.5	271.5	262.4	307.0	311.2	334.5
EASTERN EUROPE-TOTAL	**407.8**	**361.2**	**373.0**	**367.6**	**365.4**	**367.8**	**374.9**
Albania	...	...	...	...	...	...	...
Bulgaria	246.7	255.0	263.9	279.1	251.0	258.5	248.0
former Czechoslovakia	524.2	544.3	538.0	536.9	525.3	532.3	535.9
former GDR	456.3	338.8	450.3	436.3	402.6	448.0	445.2
Hungary	275.9	280.0	285.4	275.6	259.7	269.4	277.1
Poland	402.3	318.9	296.4	301.3	310.7	305.9	321.4
Romania	421.0	397.9	403.7	379.2	404.7	379.0	388.5
former USSR-TOTAL	**428.0**	**426.6**	**421.7**	**438.2**	**439.5**	**430.0**	**439.5**
OTHER EUROPE-TOTAL	**109.6**	**112.1**	**102.5**	**114.1**	**123.1**	**119.3**	**119.3**
Turkey	58.4	56.6	58.9	75.8	92.8	89.1	80.5
former Yugoslavia	211.7	224.9	192.8	194.5	188.0	185.2	205.0
OTHERS	...	...	...	...	...	...	...
WORLD-TOTAL	**130.7**	**125.0**	**116.4**	**117.3**	**124.1**	**124.1**	**122.9**

Annex table 9 Apparent finished steel consumption per capita (2 of 2)

(kilograms)

Regions and countries	1987	1988	1989	1990	1991	1992
AFRICA-TOTAL	**17.7**	**18.0**	**17.4**	**17.2**	**16.4**	**16.1**
South Africa	128.2	137.9	141.4	127.1	112.7	97.4
Others	10.6	10.3	9.5	10.3	10.4	11.1
MIDDLE EAST-TOTAL	**83.3**	**87.1**	**84.2**	**89.9**	**96.2**	**101.4**
Egypt	67.2	76.2	77.5	76.7	79.2	78.5
Iran	94.7	73.9	82.2	109.3	128.8	144.8
Others	86.9	106.0	91.0	84.5	83.1	84.3
FAR EAST-TOTAL	**67.5**	**72.2**	**75.7**	**78.3**	**81.7**	**80.0**
China	49.7	48.4	48.5	46.4	47.8	57.8
India	17.2	18.2	18.7	19.8	18.2	17.2
Indonesia	13.3	13.7	15.5	22.6	21.8	21.1
Japan	598.3	661.6	718.4	751.7	750.9	634.3
Korea, DPR	...	...	...	...	...	...
Korea, Republic of	324.3	341.0	393.4	460.1	555.4	490.6
Others	41.6	53.9	55.6	59.3	73.2	79.7
NORTH AMERICA-TOTAL	**361.0**	**384.9**	**359.1**	**351.7**	**309.6**	**337.1**
Canada	441.2	487.2	444.9	358.9	343.8	376.8
United States	352.6	374.2	350.2	350.9	306.0	333.0
Others	-	-	-	-	-	-
OTHER AMERICA-TOTAL	**68.3**	**62.4**	**63.9**	**52.6**	**55.2**	**58.4**
Argentina	90.7	82.0	50.4	42.5	57.5	83.1
Brazil	87.9	69.1	84.6	57.8	54.9	56.8
Mexico	65.2	68.3	71.2	78.1	85.9	92.1
Venezuela	142.4	146.9	101.2	71.0	120.6	92.3
Others	33.7	35.1	35.2	30.9	26.4	28.6
OCEANIA-TOTAL	**289.1**	**330.3**	**312.1**	**266.0**	**237.3**	**209.9**
Australia	301.1	341.3	338.8	284.7	257.2	224.6
New Zealand	230.2	276.7	180.6	173.2	138.3	136.5
Others	-	-	-	-	-	-
EUROPE-TOTAL	**340.6**	**355.4**	**357.8**	**330.4**	**290.5**	**266.1**
EC-TOTAL	**277.0**	**316.1**	**335.8**	**332.9**	**319.6**	**311.9**
Belgium-Luxembourg	254.2	441.0	335.6	301.4	328.1	324.6
Denmark	276.3	296.0	317.0	318.8	317.1	357.2
France	221.2	246.2	269.9	288.5	264.9	255.3
Germany	...	...	...	...	449.4	445.6
former FR Germany	424.3	451.0	472.9	476.6	483.0	...
Greece	172.1	134.9	178.6	217.6	243.0	161.4
Ireland	95.5	119.4	126.7	119.1	108.1	102.9
Italy	365.9	411.4	438.4	433.3	425.1	420.6
Netherlands	199.0	256.6	284.3	300.4	242.7	255.7
Portugal	127.9	184.1	277.5	153.9	162.7	174.5
Spain	198.1	234.0	278.0	269.6	256.6	241.6
United Kingdom	219.4	263.5	257.8	248.7	210.2	200.6
EFTA-TOTAL	**306.9**	**351.1**	**350.9**	**334.3**	**280.7**	**287.0**
Austria	261.0	315.0	347.3	316.5	353.4	333.1
Finland	361.9	357.7	410.5	360.1	261.2	249.2
Iceland	194.2	181.5	151.4	130.4	152.9	128.4
Norway	251.4	253.6	239.5	200.9	127.4	238.7
Sweden	361.3	405.6	361.2	391.5	295.3	290.7
Switzerland	288.4	386.9	376.0	356.1	297.7	296.0
EASTERN EUROPE-TOTAL	**383.8**	**371.1**	**356.3**	**257.5**	**164.0**	**128.1**
Albania	...	...	...	...	...	...
Bulgaria	288.7	240.1	243.0	141.0	100.8	75.1
former Czechoslovakia	506.9	518.0	523.7	497.2	299.1	240.2
former GDR	451.1	437.9	507.7	267.0	...	...
Hungary	268.3	252.9	226.5	192.1	113.4	85.8
Poland	314.6	307.8	263.0	186.2	112.8	112.8
Romania	447.2	424.6	382.9	274.1	200.1	114.0
former USSR-TOTAL	**440.9**	**442.4**	**444.7**	**417.1**	**347.3**	**301.1**
OTHER EUROPE-TOTAL	**189.0**	**178.7**	**136.0**	**108.6**	**118.9**	**105.9**
Turkey	207.4	196.6	119.3	104.4	112.8	120.3
former Yugoslavia	147.7	138.0	174.5	118.4	133.4	71.5
OTHERS	**...**	**...**	**...**	**...**	**...**	**...**
WORLD-TOTAL	**125.7**	**131.1**	**131.1**	**126.0**	**118.6**	**115.0**

Annex table 10 Steel intensity of GDP - crude steel consumption (1 of 2)

(In grams per US$ at 1980 prices)

Regions and countries	1980	1981	1982	1983	1984	1985
AFRICA-TOTAL	**40.0**	**42.3**	**33.8**	**33.3**	**33.0**	**34.0**
South Africa	90.3	87.0	69.9	65.6	66.9	59.4
Others	27.2	29.9	24.0	24.7	23.6	27.2
MIDDLE EAST-TOTAL	**37.4**	**37.9**	**48.0**	**51.5**	**45.9**	**48.6**
Egypt	86.2	87.1	89.0	87.0	75.4	100.0
Iran	37.9	27.8	36.6	46.6	35.1	44.1
Others	33.3	37.3	48.8	49.9	48.2	44.4
FAR EAST-TOTAL	**83.7**	**75.6**	**73.2**	**72.6**	**76.5**	**76.8**
China	145.3	126.0	121.1	140.6	142.8	151.5
India	63.1	76.4	72.9	56.5	65.3	64.0
Indonesia	38.0	34.8	35.6	29.7	25.8	22.3
Japan	74.3	64.4	61.0	56.1	60.9	57.1
Korea, DPR	...	...	...	...	...	...
Korea, Republic of	96.7	110.9	105.7	110.0	120.3	118.7
Others	75.1	66.7	68.2	66.5	57.0	57.0
NORTH AMERICA-TOTAL	**43.3**	**47.1**	**31.6**	**35.0**	**38.4**	**35.9**
Canada	49.2	49.3	34.3	40.5	45.8	44.9
United States	42.7	46.9	31.3	34.4	37.7	35.0
Others	...	...	...	...	...	...
OTHER AMERICA-TOTAL	**49.1**	**45.1**	**39.6**	**32.5**	**37.4**	**37.2**
Argentina	49.5	38.0	42.9	44.9	45.9	34.1
Brazil	56.6	49.1	42.8	35.4	41.2	43.1
Mexico	54.3	53.7	38.9	31.2	35.1	35.9
Venezuela	40.1	37.9	46.4	25.5	32.1	26.6
Others	34.3	33.6	31.2	27.3	32.9	34.2
OCEANIA-TOTAL	**36.7**	**42.2**	**35.3**	**29.9**	**32.8**	**31.4**
Australia	38.0	43.7	35.4	30.0	33.1	32.1
New Zealand	27.2	31.6	34.5	29.5	30.3	26.6
Others	...	...	...	...	...	...
EUROPE-TOTAL 1/	**67.5**	**63.1**	**62.1**	**60.8**	**60.5**	**58.2**
EC-TOTAL 2/	**37.0**	**33.0**	**32.5**	**30.2**	**31.0**	**29.8**
Belgium-Luxembourg	24.2	26.6	37.6	27.6	33.7	28.5
Denmark	26.5	26.9	33.5	19.2	21.6	23.8
France	29.2	24.8	23.7	21.1	21.0	19.5
Germany	...	...	...	...	...	...
former FR Germany	41.4	38.0	33.1	36.2	36.7	35.7
Greece	40.8	34.6	39.0	47.3	31.6	35.1
Ireland	20.0	24.6	22.9	19.1	19.6	18.2
Italy	59.4	45.1	47.0	41.0	45.8	44.4
Netherlands	25.3	20.8	21.3	19.1	23.0	23.3
Portugal	51.5	56.4	63.3	48.3	46.2	41.8
Spain	44.7	38.0	43.5	34.3	30.7	28.7
United Kingdom	25.8	28.2	26.5	25.4	25.4	24.5
EFTA-TOTAL	**33.1**	**28.6**	**28.3**	**25.7**	**26.2**	**23.9**
Austria	37.3	33.0	31.1	29.6	29.6	27.3
Finland	41.3	35.6	38.4	32.3	33.5	29.6
Iceland	20.6	16.9	15.9	15.3	15.3	13.9
Norway	31.5	26.2	29.6	21.9	21.6	20.8
Sweden	33.1	27.5	27.9	27.2	27.8	24.0
Switzerland	26.8	24.9	20.7	19.7	20.9	20.4
EASTERN EUROPE-TOTAL	**120.5**	**108.5**	**111.5**	**105.6**	**100.2**	**97.5**
Albania	...	...	...	...	...	...
Bulgaria	86.2	85.1	86.6	88.0	76.4	77.1
former Czechoslovakia	106.6	111.0	108.9	106.2	101.6	101.0
former GDR	106.8	75.8	97.7	90.3	78.4	78.8
Hungary	75.0	73.1	73.0	68.8	62.1	64.3
Poland	140.7	125.0	123.1	119.6	116.7	110.2
Romania	165.2	156.1	152.7	135.1	135.7	127.8
former USSR-TOTAL	**151.1**	**146.0**	**140.1**	**140.6**	**136.7**	**131.7**
OTHER EUROPE-TOTAL	**71.5**	**71.6**	**62.7**	**69.6**	**73.3**	**71.3**
Turkey	60.7	57.3	52.3	65.8	77.0	72.5
former Yugoslavia	80.0	83.3	71.6	72.9	69.8	70.1
OTHERS	**59.6**	**51.9**	**60.7**	**54.6**	**57.0**	**55.3**
WORLD-TOTAL 1/	**60.4**	**58.2**	**53.4**	**53.2**	**54.7**	**53.3**

1/ Because of a lack of data for the former USSR in 1990 and 1991, and for some eastern European countries in 1991, figures for these regions are reliable only up to 1989.

2/ EC-TOTAL includes figures for former FR Germany in 1991.

Annex table 10 Steel intensity of GDP - crude steel consumption (2 of 2)

(In grams per US$ at 1980 prices)

Regions and countries		1986	1987	1988	1989	1990	1991
AFRICA-TOTAL		**29.9**	**29.6**	**30.0**	**29.0**	**28.1**	**27.2**
South Africa		59.9	60.4	64.0	65.8	60.6	55.5
Others		22.1	21.5	21.0	19.3	19.9	20.3
MIDDLE EAST-TOTAL		**44.8**	**38.9**	**43.1**	**41.5**	**44.0**	**...**
Egypt		168.3	111.5	127.0	129.0	124.8	130.4
Iran		34.2	45.0	38.9	43.5	50.3	58.1
Others		32.9	25.6	33.0	27.9	27.9	...
FAR EAST-TOTAL		**74.5**	**74.1**	**75.3**	**75.7**	**74.9**	**75.2**
China		144.7	124.2	110.6	107.2	97.4	94.9
India		65.0	71.5	70.7	69.6	71.5	65.2
Indonesia		25.2	22.0	21.9	23.3	31.6	29.4
Japan		53.1	55.3	59.7	61.3	61.5	59.0
Korea, DPR		...	...	...	...	...	...
Korea, Republic of		113.7	125.6	118.6	129.4	139.8	157.1
Others		59.8	66.3	81.6	80.7	82.7	96.1
NORTH AMERICA-TOTAL		**31.3**	**32.6**	**33.5**	**30.5**	**29.1**	**26.8**
Canada		39.9	41.1	43.9	40.6	31.4	31.0
United States		30.4	31.8	32.4	29.6	28.9	26.4
Others		...	...	...	...	...	...
OTHER AMERICA-TOTAL		**39.0**	**39.7**	**36.5**	**37.5**	**31.3**	**32.2**
Argentina		37.1	47.5	45.2	29.1	24.9	32.1
Brazil		49.0	48.6	38.5	45.7	32.9	31.6
Mexico		31.0	30.6	32.3	33.2	35.6	38.3
Venezuela		36.9	39.8	40.1	31.0	21.1	33.0
Others		32.5	31.3	32.9	33.3	29.3	24.8
OCEANIA-TOTAL		**30.4**	**30.0**	**32.8**	**30.0**	**25.5**	**23.3**
Australia		31.2	29.7	31.9	30.5	25.6	23.7
New Zealand		24.4	32.8	39.2	25.8	24.7	20.3
Others		...	...	...	...	...	...
EUROPE-TOTAL	1/	**58.2**	**57.6**	**58.5**	**57.4**	**35.6**	**32.1**
EC-TOTAL	2/	**29.9**	**28.8**	**32.7**	**33.3**	**31.7**	**30.1**
Belgium-Luxembourg		26.8	22.5	37.1	27.1	23.3	26.1
Denmark		22.9	19.9	21.2	22.5	22.1	21.8
France		18.6	18.3	19.7	20.9	21.8	19.9
Germany		...	...	...	...	...	...
former FR Germany		34.7	32.2	38.1	37.1	35.1	34.4
Greece		36.9	44.0	33.2	42.1	52.0	57.5
Ireland		16.7	17.0	20.5	20.6	18.3	16.3
Italy		44.6	45.1	48.5	50.1	48.5	47.0
Netherlands		22.1	18.6	23.1	24.2	24.5	19.4
Portugal		42.8	55.1	76.9	105.3	55.5	53.3
Spain		38.9	36.1	40.2	45.1	42.2	39.2
United Kingdom		24.3	23.5	26.3	25.6	23.4	20.2
EFTA-TOTAL		**25.5**	**23.0**	**25.2**	**25.5**	**22.7**	**19.1**
Austria		29.9	25.8	29.9	31.7	27.5	29.8
Finland		32.1	31.6	29.7	32.4	28.3	21.8
Iceland		13.8	14.8	14.1	12.0	10.5	12.2
Norway		22.1	17.1	17.3	15.9	12.3	7.7
Sweden		25.3	24.7	26.0	26.1	24.3	18.5
Switzerland		21.4	18.1	23.6	22.2	20.5	17.3
EASTERN EUROPE-TOTAL		**96.5**	**95.9**	**90.9**	**87.6**	**70.0**	**63.0**
Albania		...	...	...	...	...	...
Bulgaria		70.5	77.7	62.8	63.9	41.7	...
former Czechoslovakia		99.9	94.1	94.0	93.7	91.7	64.8
former GDR		75.0	73.4	68.9	78.5	46.1	...
Hungary		64.7	59.7	55.7	50.4	43.7	...
Poland		111.8	107.8	102.0	88.0	71.2	47.1
Romania		128.6	142.4	136.5	130.7	106.0	90.5
former USSR-TOTAL		**130.9**	**128.5**	**122.9**	**120.8**	**...**	**...**
OTHER EUROPE-TOTAL		**68.2**	**105.4**	**98.7**	**75.3**	**59.9**	**72.6**
Turkey		60.8	148.5	135.8	82.8	68.0	73.4
former Yugoslavia		76.0	56.1	53.5	66.1	48.4	71.0
OTHERS		**71.3**	**58.7**	**74.3**	**61.1**	**60.6**	**47.6**
WORLD-TOTAL	1/	**51.6**	**51.6**	**52.6**	**51.5**	**42.5**	**42.6**

Annex table 11 Steel intensity of GDP - finished steel consumption (1 of 2)

(In grams per US$ at 1980 prices)

Regions and countries		1980	1981	1982	1983	1984	1985
AFRICA-TOTAL		**29.7**	**31.7**	**26.4**	**26.2**	**26.0**	**26.5**
South Africa		65.2	63.8	56.6	53.2	54.2	48.1
Others		20.7	22.8	18.3	19.0	18.1	20.8
MIDDLE EAST-TOTAL		**31.8**	**32.3**	**41.0**	**44.8**	**40.3**	**42.5**
Egypt		72.1	72.9	74.5	74.4	65.6	86.9
Iran		31.7	23.3	30.6	39.0	29.4	36.9
Others		28.6	32.1	42.2	44.4	43.1	39.8
FAR EAST-TOTAL		**72.8**	**64.8**	**63.8**	**64.4**	**66.6**	**67.3**
China		108.8	94.3	90.6	105.9	107.8	114.5
India		51.4	56.6	55.0	48.0	44.9	47.7
Indonesia		31.8	29.1	29.8	25.7	22.6	19.7
Japan		67.0	57.3	55.9	52.3	56.6	54.4
Korea, DPR		...	...	...	...	...	...
Korea, Republic of		81.1	93.6	89.5	96.3	106.1	105.1
Others		61.7	55.0	57.9	57.4	49.0	48.5
NORTH AMERICA-TOTAL		**32.6**	**35.3**	**25.9**	**27.4**	**30.5**	**28.8**
Canada		39.1	42.1	27.8	31.6	36.0	36.0
United States		31.9	34.6	25.7	27.0	29.9	28.1
Others		...	...	...	...	...	...
OTHER AMERICA-TOTAL		**39.6**	**36.9**	**32.7**	**27.1**	**31.1**	**30.9**
Argentina		41.6	31.8	36.0	37.5	38.2	29.1
Brazil		46.1	40.2	35.3	29.3	34.0	35.7
Mexico		42.2	43.6	31.9	26.3	29.5	30.1
Venezuela		33.0	32.3	40.1	22.2	27.8	23.0
Others		27.8	27.2	25.2	22.2	26.8	27.8
OCEANIA-TOTAL		**28.0**	**32.3**	**27.3**	**23.6**	**25.9**	**24.8**
Australia		28.7	33.1	27.0	23.2	25.7	24.9
New Zealand		22.7	26.4	28.9	26.6	27.4	24.0
Others		...	...	...	...	...	...
EUROPE-TOTAL	1/	**53.0**	**49.6**	**49.0**	**48.1**	**48.0**	**46.5**
EC-TOTAL	2/	**31.0**	**27.5**	**27.5**	**25.7**	**26.4**	**25.7**
Belgium-Luxembourg		18.5	20.6	29.2	22.5	27.9	23.9
Denmark		21.1	21.4	25.2	15.3	19.5	21.5
France		24.1	20.8	20.1	18.0	18.0	17.1
Germany		...	...	...	...	...	...
former FR Germany		35.5	33.6	29.3	31.8	32.1	31.3
Greece		34.2	29.0	32.6	42.8	28.6	31.8
Ireland		16.7	22.3	20.7	17.2	17.7	16.5
Italy		49.7	37.8	39.9	35.3	39.7	38.9
Netherlands		19.0	16.0	17.3	15.6	18.9	19.1
Portugal		42.6	46.3	52.7	39.9	38.0	34.6
Spain		37.3	31.0	35.9	28.5	25.7	24.3
United Kingdom		22.1	21.6	21.6	20.2	19.9	19.8
EFTA-TOTAL		**26.1**	**23.4**	**23.1**	**21.3**	**22.6**	**21.0**
Austria		31.3	28.2	27.2	26.3	26.4	24.4
Finland		36.8	31.8	34.4	29.0	30.0	26.5
Iceland		15.9	13.0	12.2	11.8	11.8	10.7
Norway		23.7	19.6	22.3	17.2	18.1	17.5
Sweden		23.9	22.2	21.4	20.9	22.6	20.6
Switzerland		21.1	19.4	17.0	16.3	18.9	18.5
EASTERN EUROPE-TOTAL		**91.2**	**82.3**	**84.7**	**80.4**	**76.9**	**75.5**
Albania		...	...	...	...	...	...
Bulgaria		64.5	63.7	64.9	66.1	57.7	58.1
former Czechoslovakia		79.4	82.7	81.2	79.5	76.3	75.9
former GDR		81.1	57.8	74.6	69.0	60.6	64.2
Hungary		61.3	59.7	59.4	56.6	51.7	53.5
Poland		105.2	93.5	92.0	89.4	88.1	83.2
Romania		126.3	119.8	117.5	104.6	105.7	99.6
former USSR-TOTAL		**114.1**	**110.6**	**106.2**	**106.5**	**103.6**	**100.0**
OTHER EUROPE-TOTAL		**56.7**	**57.5**	**52.5**	**58.8**	**62.5**	**60.8**
Turkey		45.6	43.5	44.2	56.3	66.7	62.4
former Yugoslavia		65.5	68.9	59.7	61.1	58.6	59.1
OTHERS		**44.3**	**38.6**	**45.1**	**40.6**	**42.3**	**41.1**
WORLD-TOTAL	1/	**48.3**	**46.3**	**43.6**	**43.4**	**44.7**	**43.9**

1/ Because of a lack of data for the former USSR in 1990 and 1991, and for some eastern European countries in 1991, figures for these regions are reliable only up to 1989.

2/ EC-TOTAL includes figures for former FR Germany in 1991.

Annex table 11 Steel intensity of GDP - finished steel consumption (2 of 2)

(In grams per US$ at 1980 prices)

Regions and countries		1986	1987	1988	1989	1990	1991
AFRICA-TOTAL		**23.4**	**23.1**	**23.6**	**22.8**	**22.6**	**21.8**
South Africa		48.5	49.0	51.9	53.2	49.1	44.9
Others		16.8	16.4	16.0	14.8	15.9	16.2
MIDDLE EAST-TOTAL		**39.1**	**34.0**	**37.8**	**36.3**	**39.6**	**...**
Egypt		146.1	100.1	112.1	113.5	112.2	116.0
Iran		28.6	37.7	32.5	36.4	45.3	52.5
Others		29.5	23.0	29.6	25.1	25.1	...
FAR EAST-TOTAL		**65.6**	**65.6**	**66.2**	**67.1**	**66.6**	**67.3**
China		109.5	94.1	84.1	81.7	74.9	73.5
India		49.0	56.0	55.4	54.5	55.6	50.9
Indonesia		22.4	19.6	19.6	20.9	28.5	26.5
Japan		51.4	53.2	55.6	58.0	57.7	55.4
Korea, DPR		...	...	...	...	...	...
Korea, Republic of		102.0	113.9	108.8	120.1	130.5	147.4
Others		51.0	56.4	70.0	69.3	71.2	84.9
NORTH AMERICA-TOTAL		**26.1**	**26.9**	**27.7**	**25.4**	**25.3**	**22.5**
Canada		33.4	34.8	37.1	33.4	29.6	26.9
United States		25.4	26.1	26.8	24.7	24.8	22.1
Others		...	...	...	...	...	...
OTHER AMERICA-TOTAL		**32.6**	**33.3**	**30.7**	**31.6**	**26.5**	**27.4**
Argentina		31.8	40.7	38.9	25.3	21.7	28.3
Brazil		40.8	40.4	32.2	38.5	27.9	26.7
Mexico		25.8	25.8	27.2	28.1	30.3	32.8
Venezuela		32.2	34.9	35.1	27.1	18.5	29.5
Others		26.7	25.8	27.2	27.5	24.3	20.7
OCEANIA-TOTAL		**23.9**	**25.2**	**28.5**	**26.4**	**22.5**	**20.6**
Australia		24.2	24.6	27.6	26.8	22.5	20.8
New Zealand		22.0	29.6	35.5	23.3	22.3	18.4
Others		...	...	...	...	...	...
EUROPE-TOTAL	1/	**46.6**	**46.6**	**47.1**	**46.3**	**30.5**	**27.7**
EC-TOTAL	2/	**25.9**	**25.2**	**27.8**	**28.6**	**27.6**	**26.2**
Belgium-Luxembourg		22.8	19.5	32.2	23.5	20.4	22.8
Denmark		20.7	18.0	19.2	20.3	20.0	19.7
France		16.5	16.4	17.6	18.7	19.6	17.8
Germany		...	...	...	...	...	...
former FR Germany		30.5	28.7	29.6	30.0	28.8	28.3
Greece		33.4	39.7	30.0	38.1	47.0	52.0
Ireland		15.1	15.3	18.5	18.6	16.5	14.7
Italy		39.4	40.1	43.4	44.9	43.5	42.2
Netherlands		18.3	15.9	20.1	21.5	21.9	17.4
Portugal		35.5	45.7	63.9	87.9	46.8	48.0
Spain		33.1	31.0	35.0	39.9	37.5	35.0
United Kingdom		19.6	19.5	22.5	21.5	20.6	17.9
EFTA-TOTAL		**22.2**	**20.2**	**22.4**	**21.7**	**20.3**	**17.1**
Austria		26.8	23.1	26.8	28.5	24.7	26.8
Finland		28.7	28.3	26.6	29.0	25.5	19.7
Iceland		10.6	11.4	10.8	9.3	8.0	9.4
Norway		18.7	14.5	14.6	13.5	11.0	6.9
Sweden		21.2	21.0	23.0	20.0	21.5	16.4
Switzerland		19.3	16.3	21.3	20.1	18.6	15.6
EASTERN EUROPE-TOTAL		**74.9**	**75.2**	**71.4**	**68.8**	**54.5**	**48.5**
Albania		...	...	...	...	...	...
Bulgaria		53.6	58.9	47.8	48.7	32.0	...
former Czechoslovakia		75.2	70.8	70.8	70.7	69.4	49.4
former GDR		61.4	60.2	56.7	64.7	38.1	...
Hungary		54.3	50.4	47.6	42.5	37.4	...
Poland		84.4	81.5	77.1	66.1	53.5	35.4
Romania		100.5	115.7	110.8	106.5	83.4	71.7
former USSR-TOTAL		**99.6**	**97.9**	**93.7**	**92.2**	**...**	
OTHER EUROPE-TOTAL		**58.4**	**91.5**	**87.1**	**66.2**	**52.9**	**63.8**
Turkey		53.1	130.1	121.1	74.2	60.7	65.6
former Yugoslavia		63.9	47.3	45.6	56.6	41.8	60.5
OTHERS		**53.0**	**43.6**	**55.2**	**45.4**	**45.0**	**35.3**
WORLD-TOTAL	1/	**42.9**	**43.1**	**43.8**	**43.2**	**37.1**	**36.7**

Annex table 12 Net steel exports (1 of 2) 1/

(%)

Regions and countries	1980	1981	1982	1983	1984	1985	1986
AFRICA-TOTAL	**- 3 458**	**- 4 070**	**- 2 246**	**- 2 424**	**- 2 792**	**- 1 809**	**- 449**
South Africa	1 869	1 612	2 076	1 841	1 465	2 774	3 073
Others	- 5 327	- 5 682	- 4 322	- 4 265	- 4 257	- 4 583	- 3 522
MIDDLE EAST-TOTAL	**- 11 208**	**- 10 862**	**- 14 497**	**- 15 953**	**- 13 869**	**- 14 546**	**- 12 117**
Egypt	- 890	- 935	- 1 109	- 1 320	- 1 214	- 1 963	- 3 978
Iran	- 2 515	- 1 887	- 3 114	- 4 392	- 3 250	- 4 532	- 3 036
Others	- 7 803	- 8 040	- 10 274	- 10 241	- 9 405	- 8 051	- 5 103
FAR EAST-TOTAL	**12 525**	**12 177**	**13 518**	**8 933**	**6 431**	**1 169**	**- 1 284**
China	- 4 608	- 2 702	- 2 837	- 9 288	- 13 112	- 19 454	- 17 225
India	- 1 768	- 2 706	- 2 452	- 1 702	- 1 415	- 2 151	- 2 428
Indonesia	- 2 031	- 2 075	- 2 061	- 1 614	- 1 278	- 875	- 968
Japan	28 522	26 859	26 595	28 078	27 680	28 587	25 365
Korea, DPR	...	...	...	- 400	- 400	- 400	- 400
Korea, Republic of	2 345	2 548	4 201	3 495	2 914	3 018	2 951
Others	- 9 935	- 9 747	- 9 928	- 9 636	- 7 958	- 7 556	- 8 579
NORTH AMERICA-TOTAL	**- 7 238**	**- 13 959**	**- 10 912**	**- 12 295**	**- 21 156**	**- 19 876**	**- 16 262**
Canada	2 577	1 144	2 264	1 722	1 251	982	1 278
United States	- 9 814	- 15 103	- 13 176	14 016	- 22 407	- 20 858	- 17 540
Others	...	...	...	...	...	...	...
OTHER AMERICA-TOTAL	**- 5 872**	**- 5 329**	**- 1 662**	**4 627**	**4 464**	**6 054**	**5 414**
Argentina	- 746	- 47	180	23	- 352	650	621
Brazil	837	966	1 967	5 047	6 362	7 007	5 573
Mexico	- 2 656	- 3 010	- 924	575	193	- 247	693
Venezuela	- 677	- 465	- 639	593	590	1 115	675
Others	- 2 630	- 2 773	- 2 246	- 1 611	- 2 329	- 2 471	- 2 148
OCEANIA-TOTAL	**830**	**- 89**	**43**	**- 10**	**- 136**	**81**	**269**
Australia	1 144	334	500	429	314	481	579
New Zealand	- 314	- 423	- 457	- 439	- 450	- 400	- 310
Others	...	...	...	...	...	...	...
EUROPE-TOTAL	**17 510**	**29 061**	**16 173**	**20 889**	**29 227**	**34 743**	**26 473**
EC-TOTAL	**20 312**	**28 564**	**17 766**	**21 469**	**27 211**	**30 075**	**20 627**
Belgium-Luxembourg	10 690	9 944	6 843	8 205	9 138	9 213	8 465
Denmark	- 773	- 925	- 1 247	- 660	- 919	- 1 150	- 1 046
France	3 094	3 826	1 762	2 565	3 687	4 263	3 788
Germany	...	...	...	...	...	...	...
former FR Germany	7 580	7 863	6 714	4 297	7 154	8 483	5 795
Greece	- 648	- 406	- 539	- 1 054	- 365	- 488	- 544
Ireland	- 320	- 416	- 368	- 223	- 226	- 174	- 137
Italy	- 370	3 303	1 905	2 337	1 736	1 775	243
Netherlands	738	1 511	656	1 042	1 444	1 139	1 056
Portugal	- 521	- 718	- 946	- 485	- 414	- 372	- 386
Spain	2 783	4 051	3 395	4 780	5 482	6 343	2 322
United Kingdom	- 1 940	532	- 409	664	494	1 042	1 071
EFTA-TOTAL	**- 179**	**876**	**534**	**1 648**	**2 172**	**2 624**	**1 177**
Austria	1 468	1 810	1 604	1 825	2 211	2 163	1 620
Finland	334	496	292	542	627	677	571
Iceland	- 52	- 44	- 43	- 40	- 41	- 38	- 41
Norway	- 684	- 462	- 677	- 333	- 405	- 405	- 611
Sweden	131	297	331	586	870	1 343	1 034
Switzerland	- 1 376	- 1 222	- 974	- 933	- 1 090	- 1 115	- 1 396
EASTERN EUROPE-TOTAL	**1 692**	**3 956**	**1 612**	**3 082**	**5 449**	**4 871**	**5 386**
Albania	- 143	- 80	- 121	- 120	- 120	- 120	- 120
Bulgaria	- 265	- 406	- 413	- 368	- 72	- 97	- 24
former Czechoslovakia	3 088	3 018	2 904	2 970	3 022	3 055	3 049
former GDR	- 2 086	23	- 2 050	- 1 760	- 851	- 1 059	- 891
Hungary	123	- 19	- 35	31	354	166	173
Poland	252	299	327	1 118	1 018	793	915
Romania	722	1 120	1 000	1 211	2 098	2 133	2 285
former USSR-TOTAL	**- 1 880**	**- 1 832**	**- 2 508**	**- 3 982**	**- 4 041**	**- 1 975**	**- 795**
OTHER EUROPE-TOTAL	**- 2 435**	**- 2 503**	**- 1 232**	**- 1 327**	**- 1 564**	**- 851**	**77**
Turkey	- 690	- 741	- 68	- 361	- 812	- 350	1 010
former Yugoslavia	- 1 745	- 1 762	- 1 164	- 966	- 752	- 501	- 933
OTHERS	**- 146**	**- 132**	**-160**	**- 149**	**- 165**	**- 167**	**- 225**
WORLD-TOTAL	**2 944**	**6 796**	**257**	**3 618**	**2 004**	**5 649**	**1 819**

1/ Figures show trade balance in steel products, calculated as steel exports in Table 5-13 minus steel imports in Table 5-14.

Annex table 12 Net steel exports (2 of 2)

(1000 tonnes)

Regions and countries	1987	1988	1989	1990	1991	1992	1993
AFRICA-TOTAL	**- 893**	**- 1 577**	**- 1 876**	**- 920**	**25**	**- 107**	**...**
South Africa	2 522	1 954	1 664	2 800	3 414	3 828	...
Others	- 3 415	- 3 531	- 3 540	- 3 720	- 3 389	- 3 935	...
MIDDLE EAST-TOTAL	**- 9 951**	**- 10 180**	**- 9 807**	**- 10 692**	**- 11 444**	**- 12 278**	**...**
Egypt	- 2 083	- 2 134	- 2 230	- 2 125	- 2 110	- 2 200	...
Iran	- 4 147	- 3 100	- 3 600	- 4 900	- 5 500	- 6 000	...
Others	- 3 721	- 4 946	- 3 977	- 3 667	- 3 834	- 4 078	...
FAR EAST-TOTAL	**- 2 662**	**- 5 455**	**- 8 667**	**- 9 671**	**- 17 702**	**- 14 700**	**...**
China	- 11 472	- 8 352	- 7 416	- 1 593	- 33	- 4 351	...
India	- 1 961	- 1 802	- 1 498	- 252	- 609	- 195	...
Indonesia	- 449	- 555	- 615	- 1 475	- 1 210	- 1 273	...
Japan	20 203	16 254	12 469	8 714	7 761	12 319	16 557
Korea, DPR	- 400	- 400	- 400	- 400	- 400	- 400	...
Korea, Republic of	2 279	3 610	3 540	1 695	- 817	5 619	...
Others	- 10 862	- 14 210	- 14 747	- 16 359	- 22 394	- 26 419	...
NORTH AMERICA-TOTAL	**- 16 469**	**- 21 646**	**- 10 879**	**- 10 459**	**- 7 156**	**- 8 879**	**- 13 405**
Canada	1 024	- 3 821	1 084	1 128	2 160	2 734	717
United States	- 17 493	- 17 825	- 11 963	- 11 587	- 9 316	- 11 613	- 14 122
Others	...	...	...	...	...	...	...
OTHER AMERICA-TOTAL	**5 929**	**10 092**	**9 152**	**10 061**	**9 404**	**9 088**	**11 133**
Argentina	291	559	1 781	1 811	732	- 412	- 159
Brazil	6 026	10 604	8 627	8 747	10 689	11 408	11 900
Mexico	1 021	767	480	513	- 957	- 1 286	- 605
Venezuela	638	435	847	1 223	511	1 183	74
Others	- 2 047	- 2 273	- 2 583	- 2 233	- 1 571	- 1 805	- 77
OCEANIA-TOTAL	**- 197**	**- 466**	**311**	**1 182**	**1 339**	**1 988**	**...**
Australia	193	- 52	300	1 118	1 082	1 771	...
New Zealand	- 390	- 414	11	65	257	217	...
Others	...	...	...	...	...	...	...
EUROPE-TOTAL	**23 443**	**18 377**	**13 489**	**20 904**	**26 787**	**25 682**	**...**
EC-TOTAL	**21 320**	**15 056**	**11 468**	**11 217**	**13 640**	**12 599**	**...**
Belgium-Luxembourg	8 747	8 404	9 298	10 457	9 991	9 008	3 028
Denmark	- 864	- 951	- 1 060	- 1 071	- 1 050	- 1 286	...
France	3 533	2 924	1 631	927	1 725	1 732	2 883
Germany	...	...	...	...	3 111	1 445	5 735
former FR Germany	6 355	5 172	4 941	3 238	...	...	...
Greece	- 897	- 483	- 926	- 1 009	- 1 129	- 900	...
Ireland	- 147	- 191	- 174	- 149	- 142	- 158	...
Italy	- 583	- 2 302	- 2 521	- 2 601	- 1 882	- 1 508	2 259
Netherlands	1 449	1 048	857	413	1 047	1 084	1 393
Portugal	- 683	- 1 193	- 2 210	- 1 781	- 1 856	- 1 385	...
Spain	2 338	1 193	395	905	1 417	1 342	2 815
United Kingdom	2 070	1 435	1 238	1 888	2 405	3 227	...
EFTA-TOTAL	**1 739**	**1 271**	**789**	**503**	**2 023**	**2 059**	**...**
Austria	1 906	1 733	1 635	1 481	1 120	1 054	1 417
Finland	606	737	580	768	1 302	1 532	1 536
Iceland	- 48	- 45	- 38	- 33	- 39	- 33	...
Norway	- 341	- 359	- 430	- 508	- 146	- 606	- 581
Sweden	920	815	670	302	875	998	1 464
Switzerland	- 1 304	- 1 609	- 1 628	- 1 507	- 1 089	- 885	...
EASTERN EUROPE-TOTAL	**5 772**	**6 103**	**5 964**	**8 775**	**8 770**	**9 266**	**11 308**
Albania	- 120	- 120	- 120	- 120	- 100	- 80	...
Bulgaria	- 286	34	20	400	327	534	811
former Czechoslovakia	3 715	3 501	3 468	3 462	4 513	4 793	5 467
former GDR	- 750	- 600	- 1 998	145	...	...	...
Hungary	212	382	404	614	- 193	435	467
Poland	1 101	1 077	1 307	2 947	3 473	3 014	2 599
Romania	1 901	1 829	2 882	1 327	750	571	1 964
former USSR-TOTAL	**- 882**	**- 1 252**	**- 4 931**	**- 2 128**	**744**	**2 052**	**...**
OTHER EUROPE-TOTAL	**- 4 505**	**- 2 801**	**200**	**2 538**	**1 610**	**- 295**	**...**
Turkey	- 4 706	- 3 375	528	2 244	1 552	22	893
former Yugoslavia	201	574	- 328	294	58	- 317	...
OTHERS	**- 197**	**- 270**	**- 240**	**- 250**	**- 193**	**- 227**	**...**
WORLD-TOTAL	**- 996**	**- 11 125**	**- 8 517**	**156**	**1 061**	**567**	**...**

Annex table 13 Steel exports (1 of 2)

(1000 tonnes)

Regions and countries	1980	1981	1982	1983	1984	1985	1986
AFRICA-TOTAL	**2 529**	**2 098**	**2 632**	**2 405**	**2 035**	**3 270**	**3 556**
South Africa	1 993	1 773	2 244	2 020	1 708	2 899	3 164
Others	536	325	388	385	327	371	392
MIDDLE EAST-TOTAL	**553**	**721**	**723**	**608**	**665**	**1 041**	**1 059**
Egypt	17	31	14	33	11	17	38
Iran	...	...	...	...	...	...	...
Others	536	690	709	575	654	1 024	1 021
FAR EAST-TOTAL	**36 124**	**35 541**	**37 729**	**39 759**	**41 080**	**40 812**	**37 793**
China	398	617	1 101	492	203	181	197
India	49	0	11	25	154	15	27
Indonesia	24	16	2	0	50	111	225
Japan	29 631	28 416	28 606	30 821	31 684	31 490	28 673
Korea, DPR	...	...	...	...	...	...	...
Korea, Republic of	4 482	4 663	5 546	5 639	5 848	5 650	5 860
Others	1 540	1 829	2 463	2 782	3 141	3 365	2 811
NORTH AMERICA-TOTAL	**7 363**	**6 337**	**5 020**	**3 820**	**3 826**	**3 794**	**4 366**
Canada	3 522	3 607	3 260	2 667	2 864	2 925	3 435
United States	3 842	2 730	1 760	1 154	962	869	931
Others	0	0	0	0	0	0	0
OTHER AMERICA-TOTAL	**2 246**	**2 963**	**4 062**	**7 991**	**8 936**	**10 147**	**9 754**
Argentina	338	608	789	677	537	1 111	1 174
Brazil	1 502	1 864	2 388	5 132	6 464	7 109	6 139
Mexico	67	42	293	1 004	894	433	1 209
Venezuela	241	389	293	852	801	1 266	913
Others	98	60	299	326	240	228	319
OCEANIA-TOTAL	**1 792**	**1 179**	**1 487**	**1 149**	**1 164**	**1 331**	**1 329**
Australia	1 667	1 134	1 400	1 029	1 014	1 181	1 179
New Zealand	125	45	87	120	150	150	150
Others							
EUROPE-TOTAL	**89 157**	**94 226**	**82 280**	**87 552**	**98 246**	**107 651**	**100 535**
EC-TOTAL	**63 326**	**65 927**	**55 470**	**59 045**	**66 816**	**71 112**	**64 567**
Belgium-Luxembourg	13 652	12 621	9 434	11 368	12 562	12 611	11 764
Denmark	647	584	474	798	639	542	580
France	10 707	10 861	8 900	9 031	10 618	10 842	10 422
Germany	...	...	...	...	...	...	...
former FR Germany	19 034	19 171	16 918	15 776	18 280	20 045	17 954
Greece	521	357	291	488	716	708	721
Ireland	36	35	55	115	143	187	191
Italy	6 746	8 210	6 935	7 231	7 760	7 971	7 221
Netherlands	4 616	4 962	4 049	4 279	5 148	5 189	5 209
Portugal	69	24	32	167	227	322	286
Spain	4 533	5 178	4 868	5 706	6 594	7 792	4 910
United Kingdom	2 766	3 925	3 513	4 086	4 130	4 903	5 308
EFTA-TOTAL	**6 677**	**7 059**	**6 807**	**7 855**	**8 924**	**9 323**	**8 203**
Austria	2 382	2 701	2 454	2 705	3 194	3 191	2 820
Finland	953	1 116	948	1 162	1 243	1 375	1 252
Iceland	...	...	...	...	...	...	...
Norway	652	669	573	705	733	755	570
Sweden	2 125	2 009	2 240	2 400	2 825	3 092	2 870
Switzerland	565	563	592	884	929	911	691
EASTERN EUROPE-TOTAL	**11 742**	**13 746**	**11 245**	**13 721**	**14 862**	**14 584**	**14 983**
Albania	...	...	...	...	...	...	...
Bulgaria	1 050	1 056	952	895	1 042	920	958
former Czechoslovakia	3 426	3 362	3 256	3 483	3 759	3 880	3 959
former GDR	2 153	4 493	2 498	3 712	3 570	3 424	3 494
Hungary	1 198	1 091	1 004	1 123	1 419	1 295	1 281
Poland	1 930	1 660	1 634	2 333	2 293	2 133	2 232
Romania	1 984	2 083	1 901	2 175	2 779	2 932	3 060
former USSR-TOTAL	**7 184**	**7 089**	**7 575**	**5 320**	**5 473**	**8 738**	**9 305**
OTHER EUROPE-TOTAL	**228**	**405**	**1 182**	**1 611**	**2 170**	**3 894**	**3 477**
Turkey	15	127	817	1 047	1 389	2 720	2 633
former Yugoslavia	213	278	365	564	781	1 174	844
OTHERS	**...**	**...**	**...**	**...**	**...**	**...**	**...**
WORLD-TOTAL	**139 765**	**143 064**	**133 933**	**143 283**	**155 952**	**168 046**	**158 392**

Annex table 13 Steel exports (2 of 2)

(1000 tonnes)

Regions and countries	1987	1988	1989	1990	1991	1992	1993
AFRICA-TOTAL	**2 981**	**2 498**	**2 227**	**3 241**	**3 857**	**4 322**	**...**
South Africa	2 636	2 145	1 842	2 948	3 571	4 018	...
Others	345	353	385	293	286	304	...
MIDDLE EAST-TOTAL	**1 211**	**1 046**	**1 070**	**1 484**	**1 005**	**1 007**	**...**
Egypt	61	66	70	175	190	200	...
Iran	...	...	...	...	...	...	...
Others	1 150	980	1 000	1 309	815	807	...
FAR EAST-TOTAL	**34 394**	**34 203**	**31 613**	**29 914**	**32 209**	**39 512**	**...**
China	277	159	781	2 090	3 293	3 701	...
India	43	119	246	1 142	427	763	...
Indonesia	691	845	560	459	706	670	...
Japan	25 177	23 125	19 719	15 835	16 721	18 462	22 622
Korea, DPR	...	...	...	...	...	...	...
Korea, Republic of	6 040	6 992	7 346	7 258	7 674	11 697	...
Others	2 166	2 963	2 961	3 131	3 388	4 219	...
NORTH AMERICA-TOTAL	**4 455**	**647**	**8 011**	**8 267**	**9 799**	**8 963**	**8 178**
Canada	3 787	...	3 715	3 976	4 675	4 963	4 463
United States	668	647	4 296	4 291	5 124	4 000	3 715
Others	...	...	...	...	...	...	...
OTHER AMERICA-TOTAL	**10 287**	**14 665**	**13 663**	**14 688**	**15 592**	**16 407**	**16 894**
Argentina	1 053	1 548	2 200	1 992	1 315	810	908
Brazil	6 546	10 717	8 932	8 940	10 845	11 584	12 237
Mexico	1 359	1 299	1 129	1 670	1 429	1 707	1 879
Venezuela	1 041	824	1 162	1 651	1 482	1 730	1 376
Others	288	277	240	435	521	576	494
OCEANIA-TOTAL	**1 243**	**862**	**1 378**	**2 133**	**2 421**	**2 822**	**...**
Australia	1 103	790	1 066	1 757	1 935	2 360	...
New Zealand	140	72	312	377	486	462	...
Others				...	...	...	...
EUROPE-TOTAL	**104 130**	**105 969**	**103 262**	**105 368**	**105 793**	**102 393**	**...**
EC-TOTAL	**66 230**	**66 736**	**70 056**	**71 448**	**75 637**	**74 447**	**...**
Belgium-Luxembourg	12 325	13 530	13 999	15 081	14 875	13 706	11 561
Denmark	636	644	639	605	657	679	...
France	10 473	10 638	11 529	11 444	12 018	11 797	10 354
Germany	...	...	...	...	19 508	18 836	19 095
former FR Germany	18 123	18 182	18 947	18 094	...	...	...
Greece	597	545	842	636	1 003	816	...
Ireland	213	244	316	312	276	260	...
Italy	7 212	6 774	7 721	8 213	8 103	8 962	12 112
Netherlands	5 262	5 678	5 760	5 593	6 220	5 984	5 727
Portugal	158	85	147	175	136	205	...
Spain	4 769	3 889	3 560	4 179	4 863	4 763	5 638
United Kingdom	6 461	6 527	6 597	7 116	7 976	8 439	...
EFTA-TOTAL	**8 958**	**8 933**	**8 898**	**8 601**	**9 104**	**9 625**	**...**
Austria	3 064	3 076	3 106	3 193	2 915	2 909	3 012
Finland	1 412	1 425	1 469	1 598	1 929	2 172	2 279
Iceland	...	...	...	...	...	-	-
Norway	759	718	650	563	592	611	681
Sweden	2 859	2 932	2 806	2 340	2 619	2 820	3 330
Switzerland	864	783	867	907	1 049	1 114	...
EASTERN EUROPE-TOTAL	**15 924**	**16 486**	**14 669**	**14 336**	**11 151**	**11 176**	**14 271**
Albania	...	...	...	...	...	...	...
Bulgaria	791	834	694	474	400	841	1 467
former Czechoslovakia	4 400	4 008	3 776	3 715	4 651	5 153	6 358
former GDR	3 943	4 650	3 102	3 355	...	...	...
Hungary	1 402	1 546	1 350	1 349	1 457	989	1 001
Poland	2 289	2 348	2 363	3 560	3 668	3 408	3 228
Romania	3 100	3 100	3 383	1 884	975	786	2 217
former USSR-TOTAL	**9 118**	**9 248**	**5 269**	**4 932**	**5 354**	**4 052**	**...**
OTHER EUROPE-TOTAL	**3 901**	**4 567**	**4 371**	**6 051**	**4 547**	**3 092**	**...**
Turkey	2 177	2 805	3 343	4 400	3 950	2 726	6 301
former Yugoslavia	1 724	1 762	1 028	1 651	597	366	...
OTHERS	**...**	**...**	**...**	**...**	**...**	**...**	**...**
WORLD-TOTAL	**158 702**	**159 890**	**161 225**	**165 096**	**170 677**	**175 425**	**...**

Annex table 14 Steel imports (1 of 2)

(1000 tonnes)

Regions and countries	1980	1981	1982	1983	1984	1985	1986
AFRICA-TOTAL	**5 987**	**6 168**	**4 878**	**4 829**	**4 827**	**5 079**	**4 005**
South Africa	124	161	168	179	243	125	91
Others	5 863	6 007	4 710	4 650	4 584	4 954	3 914
MIDDLE EAST-TOTAL	**11 761**	**11 583**	**15 220**	**16 561**	**14 534**	**15 587**	**13 176**
Egypt	907	966	1 123	1 353	1 225	1 980	4 016
Iran	2 515	1 887	3 114	4 392	3 250	4 532	3 036
Others	8 339	8 730	10 983	10 816	10 059	9 075	6 124
FAR EAST-TOTAL	**23 599**	**23 364**	**24 211**	**30 826**	**34 649**	**39 643**	**39 077**
China	5 006	3 319	3 938	9 780	13 315	19 635	17 422
India	1 817	2 706	2 463	1 727	1 569	2 166	2 455
Indonesia	2 055	2 091	2 063	1 614	1 328	986	1 193
Japan	1 109	1 557	2 011	2 743	4 004	2 903	3 308
Korea, DPR	...	...	...	400	400	400	400
Korea, Republic of	2 137	2 115	1 345	2 144	2 934	2 632	2 909
Others	11 475	11 576	12 391	12 418	11 099	10 921	11 390
NORTH AMERICA-TOTAL	**14 601**	**20 296**	**15 932**	**16 115**	**24 982**	**23 670**	**20 628**
Canada	945	2 463	996	945	1 613	1 943	2 157
United States	13 656	17 833	14 936	15 170	23 369	21 727	18 471
Others	0	0	0	0	0	0	0
OTHER AMERICA-TOTAL	**8 118**	**8 292**	**5 724**	**3 364**	**4 472**	**4 093**	**4 340**
Argentina	1 084	655	609	654	889	461	553
Brazil	665	898	421	85	102	102	566
Mexico	2 723	3 052	1 217	429	701	680	516
Venezuela	918	854	932	259	211	151	238
Others	2 728	2 833	2 545	1 937	2 569	2 699	2 467
OCEANIA-TOTAL	**962**	**1 268**	**1 444**	**1 159**	**1 300**	**1 250**	**1 060**
Australia	523	800	900	600	700	700	600
New Zealand	439	468	544	559	600	550	460
Others							
EUROPE-TOTAL	**71 647**	**65 165**	**66 107**	**66 663**	**69 019**	**72 908**	**74 062**
EC-TOTAL	**43 014**	**37 363**	**37 703**	**37 576**	**39 606**	**41 038**	**43 939**
Belgium-Luxembourg	2 962	2 677	2 591	3 163	3 424	3 398	3 299
Denmark	1 420	1 509	1 721	1 457	1 558	1 692	1 626
France	7 613	7 035	7 138	6 466	6 931	6 579	6 634
Germany	...	...	...	...	...	...	...
former FR Germany	11 454	11 308	10 204	11 479	11 126	11 562	12 159
Greece	1 169	763	830	1 542	1 081	1 196	1 265
Ireland	356	451	423	338	369	361	328
Italy	7 116	4 907	5 030	4 894	6 024	6 196	6 978
Netherlands	3 878	3 451	3 393	3 237	3 704	4 050	4 153
Portugal	590	742	978	652	641	694	672
Spain	1 750	1 127	1 473	926	1 112	1 449	2 588
United Kingdom	4 706	3 393	3 922	3 422	3 636	3 861	4 237
EFTA-TOTAL	**6 856**	**6 183**	**6 274**	**6 208**	**6 752**	**6 699**	**7 026**
Austria	914	891	850	880	983	1 028	1 200
Finland	619	620	656	620	616	698	681
Iceland	52	44	43	40	41	38	41
Norway	1 336	1 131	1 250	1 037	1 138	1 160	1 181
Sweden	1 994	1 712	1 909	1 814	1 955	1 749	1 836
Switzerland	1 941	1 785	1 566	1 817	2 019	2 026	2 087
EASTERN EUROPE-TOTAL	**10 050**	**9 790**	**9 633**	**10 639**	**9 413**	**9 713**	**9 597**
Albania	143	80	121	120	120	120	120
Bulgaria	1 315	1 462	1 365	1 263	1 114	1 017	982
former Czechoslovakia	338	344	352	513	737	825	910
former GDR	4 239	4 470	4 548	5 472	4 421	4 483	4 385
Hungary	1 075	1 110	1 039	1 092	1 065	1 129	1 108
Poland	1 678	1 361	1 307	1 215	1 275	1 340	1 317
Romania	1 262	963	901	964	681	799	775
former USSR-TOTAL	**9 064**	**8 921**	**10 083**	**9 302**	**9 514**	**10 713**	**10 100**
OTHER EUROPE-TOTAL	**2 663**	**2 908**	**2 414**	**2 938**	**3 734**	**4 745**	**3 400**
Turkey	705	868	885	1 408	2 201	3 070	1 623
former Yugoslavia	1 958	2 040	1 529	1 530	1 533	1 675	1 777
OTHERS	**146**	**132**	**160**	**149**	**165**	**167**	**225**
WORLD-TOTAL	**136 821**	**136 268**	**133 676**	**139 666**	**153 948**	**162 397**	**156 573**

Annex table 14 Steel imports (2 of 2)

(1000 tonnes)

Regions and countries	1987	1988	1989	1990	1991	1992	1993
AFRICA-TOTAL	**3 874**	**4 075**	**4 103**	**4 161**	**3 832**	**4 429**	**...**
South Africa	114	191	178	148	157	190	...
Others	3 760	3 884	3 925	4 013	3 675	4 239	...
MIDDLE EAST-TOTAL	**11 162**	**11 226**	**10 877**	**12 176**	**12 449**	**13 285**	**...**
Egypt	2 144	2 200	2 300	2 300	2 300	2 400	...
Iran	4 147	3 100	3 600	4 900	5 500	6 000	...
Others	4 871	5 926	4 977	4 976	4 649	4 885	...
FAR EAST-TOTAL	**37 056**	**39 658**	**40 280**	**39 585**	**49 911**	**54 212**	**...**
China	11 749	8 511	8 197	3 683	3 326	8 052	...
India	2 004	1 921	1 744	1 394	1 036	958	...
Indonesia	1 140	1 400	1 175	1 934	1 916	1 943	...
Japan	4 974	6 871	7 250	7 121	8 960	6 143	6 065
Korea, DPR	400	400	400	400	400	400	...
Korea, Republic of	3 761	3 382	3 806	5 563	8 491	6 078	...
Others	13 028	17 173	17 708	19 490	25 782	30 638	...
NORTH AMERICA-TOTAL	**20 924**	**22 293**	**18 890**	**18 726**	**16 955**	**17 842**	**21 583**
Canada	2 763	3 821	2 631	2 848	2 515	2 229	3 746
United States	18 161	18 472	16 259	15 878	14 440	15 613	17 837
Others	0	...	...	...	...	...	...
OTHER AMERICA-TOTAL	**4 358**	**4 573**	**4 511**	**4 627**	**6 188**	**7 319**	**5 080**
Argentina	762	989	419	181	583	1 222	721
Brazil	520	113	305	193	156	176	194
Mexico	338	532	649	1 157	2 386	2 993	1 761
Venezuela	403	389	315	428	971	547	120
Others	2 335	2 550	2 823	2 668	2 092	2 381	2 284
OCEANIA-TOTAL	**1 440**	**1 328**	**1 067**	**951**	**1 082**	**834**	**...**
Australia	910	842	766	639	853	589	...
New Zealand	530	486	301	312	229	245	...
Others						...	...
EUROPE-TOTAL	**80 687**	**87 593**	**89 773**	**84 465**	**79 006**	**76 711**	**...**
EC-TOTAL	**44 911**	**51 680**	**58 588**	**60 232**	**61 997**	**61 848**	**...**
Belgium-Luxembourg	3 578	5 126	4 701	4 624	4 884	4 698	8 533
Denmark	1 500	1 595	1 699	1 676	1 707	1 965	...
France	6 940	7 714	9 898	10 517	10 293	10 065	7 471
Germany	...	...	...	...	16 397	17 391	13 360
former FR Germany	11 768	13 010	14 006	14 856	...	...	...
Greece	1 494	1 028	1 768	1 645	2 132	1 716	...
Ireland	360	435	490	461	418	418	...
Italy	7 795	9 076	10 242	10 814	9 985	10 470	9 853
Netherlands	3 813	4 630	4 903	5 180	5 173	4 900	4 334
Portugal	841	1 278	2 357	1 956	1 991	1 590	...
Spain	2 431	2 696	3 165	3 274	3 446	3 421	2 823
United Kingdom	4 391	5 092	5 359	5 228	5 571	5 213	...
EFTA-TOTAL	**7 219**	**7 662**	**8 109**	**8 099**	**7 081**	**7 567**	**...**
Austria	1 158	1 343	1 471	1 711	1 795	1 855	1 595
Finland	806	688	889	831	627	641	743
Iceland	48	45	38	33	39	33	...
Norway	1 100	1 077	1 080	1 072	738	1 217	1 262
Sweden	1 939	2 117	2 136	2 038	1 744	1 822	1 866
Switzerland	2 168	2 392	2 495	2 414	2 138	1 999	...
EASTERN EUROPE-TOTAL	**10 152**	**10 383**	**8 705**	**5 561**	**2 381**	**1 910**	**2 963**
Albania	120	120	120	120	100	80	...
Bulgaria	1 077	800	674	74	73	307	656
former Czechoslovakia	685	507	308	253	138	360	891
former GDR	4 693	5 250	5 100	3 210	...	...	...
Hungary	1 190	1 164	946	735	1 650	554	534
Poland	1 188	1 271	1 056	613	195	394	629
Romania	1 199	1 271	501	557	225	215	253
former USSR-TOTAL	**10 000**	**10 500**	**10 200**	**7 060**	**4 610**	**2 000**	**...**
OTHER EUROPE-TOTAL	**8 406**	**7 368**	**4 171**	**3 513**	**2 937**	**3 387**	**...**
Turkey	6 883	6 180	2 815	2 156	2 398	2 704	5 408
former Yugoslavia	1 523	1 188	1 356	1 357	539	683	...
OTHERS	**197**	**270**	**240**	**250**	**193**	**227**	**...**
WORLD-TOTAL	**159 698**	**171 016**	**169 741**	**164 941**	**169 616**	**174 858**	**...**

Annex table 15 Steel imports as a percentage of apparent finished steel consumption (1 of 2)

(%)

Regions and countries	1980	1981	1982	1983	1984	1985	1986
AFRICA-TOTAL	**51.4**	**50.7**	**47.4**	**47.9**	**47.6**	**47.4**	**41.7**
South Africa	2.4	3.0	3.6	4.2	5.3	3.1	2.2
Others	91.0	87.8	84.1	80.6	82.9	74.5	71.2
MIDDLE EAST-TOTAL	**90.1**	**90.7**	**92.5**	**90.8**	**86.8**	**87.6**	**86.0**
Egypt	53.4	54.1	55.9	62.7	60.6	69.3	82.7
Iran	84.5	80.0	87.1	87.7	82.0	86.6	81.2
Others	99.5	101.2	101.1	97.7	93.4	93.6	91.0
FAR EAST-TOTAL	**17.0**	**18.0**	**18.2**	**21.9**	**22.3**	**23.7**	**22.9**
China	15.4	11.3	12.8	24.8	29.0	35.8	30.7
India	20.5	26.1	23.5	17.5	16.4	20.2	21.3
Indonesia	82.7	80.6	78.1	65.5	57.6	47.3	47.6
Japan	1.6	2.5	3.2	4.5	5.8	4.2	4.9
Korea, DPR	...	...	...	8.0	7.5	7.5	7.4
Korea, Republic of	41.8	33.5	20.8	27.5	31.2	26.3	26.6
Others	81.6	85.6	84.1	81.3	79.9	76.9	71.9
NORTH AMERICA-TOTAL	**15.1**	**18.9**	**20.8**	**19.1**	**24.9**	**24.1**	**22.4**
Canada	9.2	21.4	13.4	10.9	15.4	17.8	20.6
United States	15.8	18.6	21.6	20.0	26.0	24.9	22.6
Others	...	...	...	...	...	...	...
OTHER AMERICA-TOTAL	**27.5**	**29.9**	**23.7**	**17.1**	**19.0**	**16.9**	**16.5**
Argentina	36.1	30.4	26.9	26.9	34.8	24.8	25.7
Brazil	5.7	9.1	4.8	1.2	1.2	1.0	4.7
Mexico	33.1	33.1	18.1	8.1	11.4	10.5	9.7
Venezuela	39.8	38.9	36.4	17.6	11.6	9.8	10.4
Others	63.6	65.8	66.2	58.5	62.2	61.6	55.9
OCEANIA-TOTAL	**18.9**	**20.9**	**28.7**	**25.2**	**24.6**	**23.8**	**20.5**
Australia	11.4	14.7	20.6	15.2	15.3	15.1	13.1
New Zealand	86.7	77.0	81.5	85.9	85.3	90.9	80.8
Others	...	...	...	...	...	...	...
EUROPE-TOTAL	**26.2**	**25.3**	**25.6**	**25.7**	**25.8**	**27.5**	**27.1**
EC-TOTAL	**44.4**	**43.3**	**43.5**	**45.8**	**45.7**	**47.6**	**49.1**
Belgium-Luxembourg	130.5	107.9	72.1	114.1	97.4	111.3	110.8
Denmark	101.3	107.3	100.7	136.7	110.0	103.8	100.4
France	47.5	50.3	51.6	51.9	54.8	53.7	54.8
Germany	...	...	...	...	...	...	...
former FR Germany	39.6	41.2	42.9	43.8	41.0	42.9	45.1
Greece	85.0	65.4	62.9	88.8	90.6	87.5	86.9
Ireland	110.7	101.5	100.1	96.4	98.1	101.0	100.9
Italy	31.6	28.1	27.3	29.8	31.5	32.4	35.0
Netherlands	120.6	128.2	118.1	123.4	113.3	119.3	125.4
Portugal	55.2	63.2	71.7	62.5	65.2	75.2	68.5
Spain	22.1	17.1	19.1	14.9	19.5	26.2	33.1
United Kingdom	39.8	29.6	34.0	30.5	32.3	33.2	35.5
EFTA-TOTAL	**63.1**	**63.1**	**64.3**	**67.7**	**67.0**	**69.4**	**67.1**
Austria	38.0	41.2	40.2	42.1	46.4	51.2	53.8
Finland	32.6	37.1	35.1	38.3	35.6	44.3	39.1
Iceland	100.0	100.0	100.0	100.0	100.0	100.0	100.0
Norway	97.7	99.1	95.8	98.8	97.0	97.3	89.5
Sweden	66.7	61.7	70.9	67.5	64.6	61.9	61.9
Switzerland	90.4	89.0	90.4	108.0	102.1	100.6	96.3
EASTERN EUROPE-TOTAL	**22.5**	**24.7**	**23.4**	**26.1**	**23.2**	**23.6**	**22.8**
Albania	100.0	100.0	72.8	69.4	66.5	63.9	61.4
Bulgaria	60.1	64.6	58.1	50.7	49.7	43.9	44.1
former Czechoslovakia	4.2	4.1	4.3	6.2	9.1	10.0	10.9
former GDR	55.5	78.9	60.5	75.2	65.9	60.1	59.2
Hungary	36.4	37.0	34.1	37.1	38.5	39.4	37.6
Poland	11.7	11.9	12.2	11.0	11.1	11.8	10.9
Romania	13.5	10.9	10.0	11.3	7.4	9.3	8.7
former USSR-TOTAL	**8.0**	**7.8**	**8.8**	**7.8**	**7.9**	**9.0**	**8.2**
OTHER EUROPE-TOTAL	**36.4**	**38.1**	**33.9**	**36.4**	**42.1**	**54.1**	**38.2**
Turkey	27.2	33.6	32.1	38.7	48.2	68.5	39.2
former Yugoslavia	41.5	40.4	35.0	34.5	35.5	39.1	37.3
OTHERS	**100.0**	**100.0**	**100.0**	**100.0**	**100.0**	**100.0**	**100.0**
WORLD-TOTAL	**24.1**	**24.6**	**25.5**	**26.0**	**26.6**	**27.6**	**26.4**

Annex table 15 Steel imports as a percentage of apparent finished steel consumption (2 of 2)

(%)

Regions and countries	1987	1988	1989	1990	1991	1992	1993
AFRICA-TOTAL	**40.2**	**40.4**	**40.9**	**40.7**	**38.0**	**43.5**	**...**
South Africa	2.7	4.1	3.6	3.3	3.9	5.3	...
Others	69.6	71.5	76.1	69.9	61.0	64.4	...
MIDDLE EAST-TOTAL	**80.9**	**75.4**	**73.3**	**74.6**	**69.2**	**68.0**	**...**
Egypt	63.6	56.1	56.2	55.5	52.5	54.0	...
Iran	85.5	79.1	79.9	79.2	73.4	69.3	...
Others	87.3	84.1	79.6	83.0	76.0	76.1	...
FAR EAST-TOTAL	**20.5**	**20.2**	**19.2**	**17.9**	**21.3**	**23.2**	**...**
China	21.7	15.9	15.1	7.0	6.0	11.9	...
India	14.5	12.9	11.1	8.2	6.5	6.3	...
Indonesia	49.9	58.4	42.6	47.4	48.0	49.3	...
Japan	6.8	8.5	8.2	7.7	9.6	7.8	8.2
Korea, DPR	7.3	7.2	7.1	7.0	7.0	7.0	...
Korea, Republic of	27.6	23.3	22.5	27.7	34.7	27.9	...
Others	69.6	69.0	67.3	67.8	70.9	75.6	...
NORTH AMERICA-TOTAL	**21.3**	**21.1**	**19.0**	**18.9**	**19.0**	**18.6**	**20.6**
Canada	24.2	30.1	22.5	27.7	27.0	23.0	31.1
United States	20.9	19.8	18.5	17.9	18.0	18.1	19.2
Others	...	...	...	...	...	...	...
OTHER AMERICA-TOTAL	**15.7**	**17.7**	**16.7**	**20.3**	**25.4**	**27.9**	**19.0**
Argentina	27.0	38.3	26.0	13.2	31.0	44.4	25.9
Brazil	4.2	1.1	2.4	2.2	1.9	2.0	1.9
Mexico	6.2	9.2	10.5	16.7	30.7	35.2	23.4
Venezuela	15.5	14.1	16.2	30.5	39.8	28.6	6.4
Others	52.6	53.9	58.2	61.3	54.9	56.5	52.5
OCEANIA-TOTAL	**25.6**	**20.4**	**17.2**	**17.8**	**22.4**	**19.3**	**...**
Australia	18.7	15.1	13.7	13.4	19.6	15.3	...
New Zealand	69.8	52.8	49.7	53.3	48.6	52.3	...
Others	...	...	...	...	...	...	...
EUROPE-TOTAL	**28.7**	**29.7**	**30.1**	**30.4**	**32.4**	**34.0**	**...**
EC-TOTAL	**50.2**	**50.5**	**53.8**	**55.5**	**58.5**	**59.7**	**...**
Belgium-Luxembourg	136.9	112.9	136.0	175.5	169.1	169.3	...
Denmark	106.0	105.2	104.7	101.7	105.2	107.4	...
France	56.5	56.2	65.5	65.3	69.5	70.0	59.9
Germany	...	...	...	...	51.9	54.5	47.8
former FR Germany	45.6	47.5	48.8	51.5	...	...	...
Greece	87.0	76.2	98.8	86.1	105.8	98.9	...
Ireland	104.0	99.8	105.0	104.2	102.8	107.2	...
Italy	37.2	38.5	40.8	42.5	40.9	43.9	47.2
Netherlands	131.3	123.2	117.3	116.9	144.0	129.0	108.2
Portugal	64.4	67.8	82.8	81.3	85.0	77.2	...
Spain	31.6	29.5	29.1	30.9	34.0	35.7	32.3
United Kingdom	35.3	34.0	36.6	37.4	46.2	45.8	...
EFTA-TOTAL	**74.2**	**68.8**	**72.8**	**77.2**	**72.1**	**78.2**	**...**
Austria	59.2	56.9	56.5	72.1	67.8	74.2	69.0
Finland	45.2	38.9	43.7	45.9	47.9	51.4	52.8
Iceland	100.0	100.0	100.0	101.1	100.0	100.0	...
Norway	104.8	101.4	107.4	99.6	55.5	86.9	121.7
Sweden	64.3	62.6	70.9	71.0	68.1	71.0	71.3
Switzerland	115.8	95.1	101.9	104.0	110.0	103.4	...
EASTERN EUROPE-TOTAL	**23.5**	**24.8**	**21.6**	**18.8**	**13.9**	**14.0**	**...**
Albania	61.4	61.4	58.7	66.8	89.2	100.0	...
Bulgaria	41.5	37.1	30.8	5.8	8.0	46.4	95.8
former Czechoslovakia	8.7	6.3	3.8	3.2	2.9	9.5	29.4
former GDR	62.5	72.0	60.3	72.2	...	...	...
Hungary	41.8	43.5	39.5	38.4	87.2	59.3	49.8
Poland	10.0	10.9	10.5	8.4	4.5	9.0	12.9
Romania	11.7	13.0	5.6	8.4	4.4	5.6	10.6
former USSR-TOTAL	**8.1**	**8.4**	**8.0**	**5.9**	**4.6**	**2.3**	**...**
OTHER EUROPE-TOTAL	**58.6**	**53.5**	**39.2**	**39.0**	**29.5**	**30.9**	**...**
Turkey	63.3	58.8	43.3	34.8	35.5	29.3	57.4
former Yugoslavia	44.0	36.5	32.8	48.1	16.9	39.6	...
OTHERS	**100.0**	**100.0**	**100.0**	**100.0**	**100.0**	**100.0**	**...**
WORLD-TOTAL	**25.9**	**26.1**	**25.5**	**25.3**	**27.1**	**28.4**	**...**

Annex table 16-1 Destination and relative importance of steel exports by major exporting regions and countries : EC

Destination	Amount (1000t)				Share (%)			
	1980	1985	1990	1992	1980	1985	1990	1992
AFRICA	**4 387**	**3 244**	**2 361**	**3 178**	**6.9**	**4.6**	**3.3**	**4.3**
South Africa	94	60	102	112	0.1	0.1	0.1	0.2
Others	4 292	3 184	2 260	3 066	6.8	4.5	3.2	4.1
MIDDLE EAST	**4 540**	**4 826**	**2 651**	**3 046**	**7.2**	**6.8**	**3.7**	**4.1**
Egypt	519	1 100	271	239	0.8	1.5	0.4	0.3
Iran	1 165	1 009	1 152	965	1.8	1.4	1.6	1.3
Others	2 857	2 717	1 228	1 841	4.5	3.8	1.7	2.5
FAR EAST	**2 485**	**6 792**	**3 950**	**5 243**	**3.9**	**9.6**	**5.5**	**7.0**
China	755	3 977	302	1 112	1.2	5.6	0.4	1.5
India	803	1 118	760	725	1.3	1.6	1.1	1.0
Indonesia	163	58	213	139	0.3	0.1	0.3	0.2
Japan	38	118	277	185	0.1	0.2	0.4	0.2
Korea, DPR	8	2	10	10	0.0	0.0	0.0	0.0
Korea, Republic of	94	119	326	255	0.1	0.2	0.5	0.3
Others	623	1 401	2 063	2 816	1.0	2.0	2.9	3.8
NORTH AMERICA	**4 148**	**7 166**	**5 963**	**4 826**	**6.6**	**10.1**	**8.3**	**6.5**
Canada	311	857	641	585	0.5	1.2	0.9	0.8
United States	3 832	6 304	5 318	4 238	6.1	8.9	7.4	5.7
Others	5	5	4	4	0.0	0.0	0.0	0.0
OTHER AMERICA	**2 510**	**1 299**	**1 204**	**1 262**	**4.0**	**1.8**	**1.7**	**1.7**
Argentina	483	127	42	111	0.8	0.2	0.1	0.1
Brazil	283	58	120	80	0.4	0.1	0.2	0.1
Mexico	719	398	409	57	1.1	0.6	0.6	0.1
Venezuela	467	49	218	213	0.7	0.1	0.3	0.3
Others	559	668	415	801	0.9	0.9	0.6	1.1
OCEANIA	**68**	**94**	**133**	**100**	**0.1**	**0.1**	**0.2**	**0.1**
Australia	28	40	77	56	0.0	0.1	0.1	0.1
New Zealand	14	34	23	13	0.0	0.0	0.0	0.0
Others	27	21	34	31	0.0	0.0	0.0	0.0
EUROPE-TOTAL	**45 110**	**46 182**	**52 239**	**52 857**	**71.2**	**64.9**	**73.1**	**71.0**
EC	**33 276**	**31 942**	**43 422**	**45 379**	**52.5**	**44.9**	**60.8**	**61.0**
Belgium-Luxembourg	2 715	3 083	4 536	4 688	4.3	4.3	6.3	6.3
Denmark	918	1 003	946	1 070	1.4	1.4	1.3	1.4
France	7 881	6 647	9 689	9 273	12.4	9.3	13.6	12.5
Germany	7 937	7 525	7 392	8 748	12.5	10.6	10.3	11.8
former FR Germany	7 859	7 477	7 375	...	12.4	10.5	10.3	...
Greece	789	977	1 369	1 572	1.2	1.4	1.9	2.1
Ireland	339	286	419	481	0.5	0.4	0.6	0.6
Italy	4 459	4 349	6 447	6 645	7.0	6.1	9.0	8.9
Netherlands	3 142	3 161	4 595	4 331	5.0	4.4	6.4	5.8
Portugal	607	523	1 021	1 224	1.0	0.7	1.4	1.6
Spain	1 318	1 504	2 939	3 038	2.1	2.1	4.1	4.1
United Kingdom	3 249	2 932	4 087	4 310	5.1	4.1	5.7	5.8
EFTA	**5 013**	**4 873**	**5 954**	**6 034**	**7.9**	**6.9**	**8.3**	**8.1**
Austria	655	805	1 364	1 368	1.0	1.1	1.9	1.8
Finland	287	396	389	275	0.5	0.6	0.5	0.4
Iceland	24	24	15	26	0.0	0.0	0.0	0.0
Norway	891	682	824	1 271	1.4	1.0	1.2	1.7
Sweden	1 422	1 323	1 412	1 245	2.2	1.9	2.0	1.7
Switzerland	1 734	1 643	1 951	1 851	2.7	2.3	2.7	2.5
EASTERN EUROPE	**1 195**	**716**	**400**	**498**	**1.9**	**1.0**	**0.6**	**0.7**
Albania	44	7	14	22	0.1	0.0	0.0	0.0
Bulgaria	298	183	54	16	0.5	0.3	0.1	0.0
former Czechoslovakia	43	161	37	121	0.1	0.2	0.1	0.2
former GDR	78	48	17	...	0.1	0.1	0.0	...
Hungary	91	95	82	93	0.1	0.1	0.1	0.1
Poland	330	173	171	220	0.5	0.2	0.2	0.3
Romania	311	50	25	25	0.5	0.1	0.0	0.0
former USSR	**4 638**	**6 537**	**1 324**	**8**	**7.3**	**9.2**	**1.9**	**0.0**
OTHER EUROPE	**989**	**2 113**	**1 138**	**938**	**1.6**	**3.0**	**1.6**	**1.3**
Turkey	375	1 698	783	938	0.6	2.4	1.1	1.3
former Yugoslavia	614	416	355	1	1.0	0.6	0.5	0.0
OTHERS	**79**	**1 510**	**2 947**	**3 934**	**0.1**	**2.1**	**4.1**	**5.3**
WORLD-TOTAL	**63 327**	**71 112**	**71 449**	**74 446**	**100.0**	**100.0**	**100.0**	**100.0**

Annex table 16-2 **Destination and relative importance of steel exports by major exporting regions and countries : *Belgium-Luxembourg***

Destination	Amount (1000t)				Share (%)			
	1980	1985	1990	1992	1980	1985	1990	1992
AFRICA	**680**	**268**	**235**	**231**	**5.0**	**2.1**	**1.6**	**1.7**
South Africa	7	3	12	9	0.1	0.0	0.1	0.1
Others	673	265	223	222	4.9	2.1	1.5	1.6
MIDDLE EAST	**469**	**290**	**372**	**318**	**3.4**	**2.3**	**2.5**	**2.3**
Egypt	16	29	20	9	0.1	0.2	0.1	0.1
Iran	80	87	183	106	0.6	0.7	1.2	0.8
Others	373	173	170	204	2.7	1.4	1.1	1.5
FAR EAST	**204**	**489**	**721**	**613**	**1.5**	**3.9**	**4.8**	**4.5**
China	78	174	33	68	0.6	1.4	0.2	0.5
India	40	73	55	84	0.3	0.6	0.4	0.6
Indonesia	22	1	97	13	0.2	0.0	0.6	0.1
Japan	...	1	117	14	...	0.0	0.8	0.1
Korea, DPR	...	...	...	0	...	...	...	0.0
Korea, Republic of	1	65	53	18	0.0	0.5	0.3	0.1
Others	63	177	366	416	0.5	1.4	2.4	3.0
NORTH AMERICA	**825**	**773**	**788**	**520**	**6.0**	**6.1**	**5.2**	**3.8**
Canada	48	106	73	86	0.4	0.8	0.5	0.6
United States	777	667	714	435	5.7	5.3	4.7	3.2
Others	...	0	1	...	...	0.0	0.0	...
OTHER AMERICA	**202**	**116**	**90**	**135**	**1.5**	**0.9**	**0.6**	**1.0**
Argentina	53	5	2	10	0.4	0.0	0.0	0.1
Brazil	17	1	2	9	0.1	0.0	0.0	0.1
Mexico	17	17	29	...	0.1	0.1	0.2	...
Venezuela	33	4	9	13	0.2	0.0	0.1	0.1
Others	82	89	47	103	0.6	0.7	0.3	0.8
OCEANIA	**6**	**7**	**11**	**10**	**0.0**	**0.1**	**0.1**	**0.1**
Australia	2	3	6	3	0.0	0.0	0.0	0.0
New Zealand	...	0	0	1	...	0.0	0.0	0.0
Others	4	4	5	6	0.0	0.0	0.0	0.0
EUROPE-TOTAL	**11 251**	**9 287**	**12 997**	**11 873**	**82.4**	**73.6**	**86.2**	**86.6**
EC	**9 887**	**7 842**	**11 840**	**10 877**	**72.4**	**62.2**	**78.5**	**79.4**
Belgium-Luxembourg	.	.	.	.	.	.	.	.
Denmark	125	138	145	119	0.9	1.1	1.0	0.9
France	3 567	2 796	4 199	3 931	26.1	22.2	27.8	28.7
Germany	3 183	2 448	3 717	3 405	23.3	19.4	24.6	24.8
former FR Germany	3 135	2 447	3 710	...	23.0	19.4	24.6	...
Greece	71	84	71	32	0.5	0.7	0.5	0.2
Ireland	39	18	21	14	0.3	0.1	0.1	0.1
Italy	1 002	943	1 360	1 135	7.3	7.5	9.0	8.3
Netherlands	1 058	847	1 259	1 259	7.7	6.7	8.3	9.2
Portugal	96	82	171	129	0.7	0.6	1.1	0.9
Spain	197	54	250	269	1.4	0.4	1.7	2.0
United Kingdom	597	435	654	585	4.4	3.4	4.3	4.3
EFTA	**691**	**628**	**785**	**852**	**5.1**	**5.0**	**5.2**	**6.2**
Austria	46	66	138	143	0.3	0.5	0.9	1.0
Finland	48	36	72	31	0.4	0.3	0.5	0.2
Iceland	6	5	5	4	0.0	0.0	0.0	0.0
Norway	118	82	137	125	0.9	0.7	0.9	0.9
Sweden	198	193	203	178	1.5	1.5	1.3	1.3
Switzerland	275	246	230	372	2.0	1.9	1.5	2.7
EASTERN EUROPE	**147**	**18**	**30**	**11**	**1.1**	**0.1**	**0.2**	**0.1**
Albania	...	1	3	0	...	0.0	0.0	0.0
Bulgaria	61	6	7	0	0.4	0.0	0.0	0.0
former Czechoslovakia	1	0	1	3	0.0	0.0	0.0	0.0
former GDR	48	1	7	...	0.4	0.0	0.0	...
Hungary	2	1	3	2	0.0	0.0	0.0	0.0
Poland	11	8	8	4	0.1	0.1	0.1	0.0
Romania	24	0	0	1	0.2	0.0	0.0	0.0
former USSR	**444**	**632**	**194**	**...**	**3.3**	**5.0**	**1.3**	**...**
OTHER EUROPE	**82**	**167**	**148**	**133**	**0.6**	**1.3**	**1.0**	**1.0**
Turkey	65	154	136	133	0.5	1.2	0.9	1.0
former Yugoslavia	17	13	12	...	0.1	0.1	0.1	...
OTHERS	**15**	**1 382**	**- 134**	**5**	**0.1**	**11.0**	**-0.9**	**0.0**
WORLD-TOTAL	**13 652**	**12 611**	**15 081**	**13 706**	**100.0**	**100.0**	**100.0**	**100.0**

Annex table 16-3 Destination and relative importance of steel exports by major exporting regions and countries : *France*

Destination	Amount (1000t)				Share (%)			
	1980	1985	1990	1992	1980	1985	1990	1992
AFRICA	**819**	**681**	**482**	**359**	**7.6**	**6.3**	**4.2**	**3.0**
South Africa	6	8	9	9	0.1	0.1	0.1	0.1
Others	813	673	473	350	7.6	6.2	4.1	3.0
MIDDLE EAST	**467**	**486**	**348**	**327**	**4.4**	**4.5**	**3.0**	**2.8**
Egypt	34	92	48	24	0.3	0.8	0.4	0.2
Iran	25	48	92	64	0.2	0.4	0.8	0.5
Others	409	346	208	239	3.8	3.2	1.8	2.0
FAR EAST	**405**	**910**	**427**	**670**	**3.8**	**8.4**	**3.7**	**5.7**
China	148	523	16	212	1.4	4.8	0.1	1.8
India	144	222	102	99	1.3	2.0	0.9	0.8
Indonesia	4	14	19	22	0.0	0.1	0.2	0.2
Japan	6	3	14	5	0.1	0.0	0.1	0.0
Korea, DPR	...	...	...	0	...	...	...	0.0
Korea, Republic of	54	6	55	30	0.5	0.1	0.5	0.3
Others	49	143	222	301	0.5	1.3	1.9	2.6
NORTH AMERICA	**883**	**1 500**	**1 261**	**910**	**8.3**	**13.8**	**11.0**	**7.7**
Canada	60	199	151	83	0.6	1.8	1.3	0.7
United States	823	1 301	1 110	826	7.7	12.0	9.7	7.0
Others	0	0	0	0	0.0	0.0	0.0	0.0
OTHER AMERICA	**658**	**358**	**256**	**168**	**6.1**	**3.3**	**2.2**	**1.4**
Argentina	37	49	5	8	0.3	0.5	0.0	0.1
Brazil	148	13	32	20	1.4	0.1	0.3	0.2
Mexico	205	95	50	31	1.9	0.9	0.4	0.3
Venezuela	87	9	74	18	0.8	0.1	0.6	0.2
Others	181	191	95	90	1.7	1.8	0.8	0.8
OCEANIA	**19**	**19**	**36**	**36**	**0.2**	**0.2**	**0.3**	**0.3**
Australia	2	5	12	14	0.0	0.0	0.1	0.1
New Zealand	0	0	1	1	0.0	0.0	0.0	0.0
Others	17	14	23	21	0.2	0.1	0.2	0.2
EUROPE-TOTAL	**7 447**	**6 882**	**8 696**	**6 770**	**69.6**	**63.5**	**76.0**	**57.4**
EC	**5 912**	**5 390**	**7 751**	**6 008**	**55.2**	**49.7**	**67.7**	**50.9**
Belgium-Luxembourg	869	877	1 511	1 584	8.1	8.1	13.2	13.4
Denmark	118	112	88	95	1.1	1.0	0.8	0.8
France	.	.	.	.	.	.	.	.
Germany	1 654	1 527	2 141	...	15.4	14.1	18.7	...
former FR Germany	1 633	1 508	2 134	...	15.3	13.9	18.6	...
Greece	295	216	311	229	2.8	2.0	2.7	1.9
Ireland	45	25	19	13	0.4	0.2	0.2	0.1
Italy	1 670	1 502	1 860	1 989	15.6	13.9	16.2	16.9
Netherlands	325	292	498	379	3.0	2.7	4.4	3.2
Portugal	76	53	103	117	0.7	0.5	0.9	1.0
Spain	401	433	664	795	3.7	4.0	5.8	6.7
United Kingdom	479	373	564	808	4.5	3.4	4.9	6.8
EFTA	**653**	**617**	**659**	**626**	**6.1**	**5.7**	**5.8**	**5.3**
Austria	25	61	63	66	0.2	0.6	0.6	0.6
Finland	38	41	42	43	0.4	0.4	0.4	0.4
Iceland	2	1	0	1	0.0	0.0	0.0	0.0
Norway	90	84	127	175	0.8	0.8	1.1	1.5
Sweden	174	192	175	156	1.6	1.8	1.5	1.3
Switzerland	325	238	251	185	3.0	2.2	2.2	1.6
EASTERN EUROPE	**168**	**73**	**26**	**23**	**1.6**	**0.7**	**0.2**	**0.2**
Albania	0	0	0	...	0.0	0.0	0.0	...
Bulgaria	74	22	1	0	0.7	0.2	0.0	0.0
former Czechoslovakia	1	1	1	7	0.0	0.0	0.0	0.1
former GDR	21	19	7	...	0.2	0.2	0.1	...
Hungary	4	4	4	3	0.0	0.0	0.0	0.0
Poland	39	17	11	9	0.4	0.2	0.1	0.1
Romania	29	9	1	3	0.3	0.1	0.0	0.0
former USSR	**631**	**573**	**121**	**...**	**5.9**	**5.3**	**1.1**	**...**
OTHER EUROPE	**83**	**229**	**139**	**114**	**0.8**	**2.1**	**1.2**	**1.0**
Turkey	46	174	99	114	0.4	1.6	0.9	1.0
former Yugoslavia	36	55	41	...	0.3	0.5	0.4	...
OTHERS	**8**	**5**	**- 62**	**2 557**	**0.1**	**0.0**	**-0.5**	**21.7**
WORLD-TOTAL	**10 707**	**10 842**	**11 444**	**11 797**	**100.0**	**100.0**	**100.0**	**100.0**

Annex table 16-4 **Destination and relative importance of steel exports by major exporting regions and countries : *Germany***

Destination	Amount (1000t)				Share (%)			
	1980	1985	1990	1992	1980	1985	1990	1992
AFRICA	**721**	**722**	**469**	**505**	**3.8**	**3.6**	**2.6**	**2.7**
South Africa	42	30	46	55	0.2	0.2	0.3	0.3
Others	679	692	423	451	3.6	3.5	2.3	2.4
MIDDLE EAST	**1 180**	**1 060**	**796**	**852**	**6.2**	**5.3**	**4.4**	**4.5**
Egypt	82	99	68	66	0.4	0.5	0.4	0.4
Iran	504	341	373	374	2.6	1.7	2.1	2.0
Others	594	620	355	412	3.1	3.1	2.0	2.2
FAR EAST	**924**	**1 971**	**1 140**	**1 579**	**4.9**	**9.8**	**6.3**	**8.4**
China	255	1 123	150	482	1.3	5.6	0.8	2.6
India	392	508	433	316	2.1	2.5	2.4	1.7
Indonesia	93	12	30	49	0.5	0.1	0.2	0.3
Japan	4	4	39	60	0.0	0.0	0.2	0.3
Korea, DPR	0	2	0	0	0.0	0.0	0.0	0.0
Korea, Republic of	4	27	26	48	0.0	0.1	0.1	0.3
Others	176	295	462	623	0.9	1.5	2.6	3.3
NORTH AMERICA	**1 184**	**2 299**	**1 614**	**1 501**	**6.2**	**11.5**	**8.9**	**8.0**
Canada	77	204	102	164	0.4	1.0	0.6	0.9
United States	1 107	2 094	1 512	1 337	5.8	10.4	8.4	7.1
Others	0	0	0	...	0.0	0.0	0.0	...
OTHER AMERICA	**670**	**415**	**416**	**326**	**3.5**	**2.1**	**2.3**	**1.7**
Argentina	172	17	9	37	0.9	0.1	0.1	0.2
Brazil	73	37	47	25	0.4	0.2	0.3	0.1
Mexico	237	222	164	...	1.2	1.1	0.9	...
Venezuela	95	20	100	26	0.5	0.1	0.6	0.1
Others	94	119	95	238	0.5	0.6	0.5	1.3
OCEANIA	**10**	**13**	**24**	**13**	**0.1**	**0.1**	**0.1**	**0.1**
Australia	8	10	22	11	0.0	0.0	0.1	0.1
New Zealand	1	2	1	1	0.0	0.0	0.0	0.0
Others	2	1	0	1	0.0	0.0	0.0	0.0
EUROPE-TOTAL	**14 337**	**13 555**	**13 725**	**13 357**	**75.3**	**67.6**	**75.9**	**70.9**
EC	**8 684**	**8 211**	**10 275**	**10 847**	**45.6**	**41.0**	**56.8**	**57.6**
Belgium-Luxembourg	811	1 020	1 236	1 380	4.3	5.1	6.8	7.3
Denmark	535	472	389	419	2.8	2.4	2.1	2.2
France	2 521	1 955	2 297	2 421	13.2	9.8	12.7	12.9
Germany	.	.	.	.	.	.	.	.
former FR Germany	.	.	.	.	.	.	.	.
Greece	168	144	188	178	0.9	0.7	1.0	0.9
Ireland	59	21	16	17	0.3	0.1	0.1	0.1
Italy	1 182	1 181	1 633	1 898	6.2	5.9	9.0	10.1
Netherlands	1 528	1 621	2 102	2 001	8.0	8.1	11.6	10.6
Portugal	230	189	232	248	1.2	0.9	1.3	1.3
Spain	525	548	898	811	2.8	2.7	5.0	4.3
United Kingdom	1 125	1 060	1 285	1 473	5.9	5.3	7.1	7.8
EFTA	**2 183**	**1 971**	**2 230**	**2 083**	**11.5**	**9.8**	**12.3**	**11.1**
Austria	449	486	630	727	2.4	2.4	3.5	3.9
Finland	131	169	131	93	0.7	0.8	0.7	0.5
Iceland	6	5	3	5	0.0	0.0	0.0	0.0
Norway	303	219	216	182	1.6	1.1	1.2	1.0
Sweden	590	469	488	406	3.1	2.3	2.7	2.2
Switzerland	704	623	762	670	3.7	3.1	4.2	3.6
EASTERN EUROPE	**549**	**294**	**228**	**280**	**2.9**	**1.5**	**1.3**	**1.5**
Albania	5	4	5	1	0.0	0.0	0.0	0.0
Bulgaria	94	54	30	10	0.5	0.3	0.2	0.1
former Czechoslovakia	29	82	33	99	0.2	0.4	0.2	0.5
former GDR	.	.	.	.	.	.	.	.
Hungary	44	38	40	58	0.2	0.2	0.2	0.3
Poland	214	99	109	106	1.1	0.5	0.6	0.6
Romania	164	17	11	7	0.9	0.1	0.1	0.0
former USSR	**2 552**	**2 804**	**630**	**...**	**13.4**	**14.0**	**3.5**	**...**
OTHER EUROPE	**369**	**275**	**362**	**147**	**1.9**	**1.4**	**2.0**	**0.8**
Turkey	127	164	163	147	0.7	0.8	0.9	0.8
former Yugoslavia	242	110	199	...	1.3	0.5	1.1	...
OTHERS	**8**	**9**	**- 90**	**702**	**0.0**	**0.0**	**-0.5**	**3.7**
WORLD-TOTAL	**19 034**	**20 045**	**18 094**	**18 836**	**100.0**	**100.0**	**100.0**	**100.0**

Annex table 16-5 Destination and relative importance of steel exports by major exporting regions and countries : *Italy*

Destination	Amount (1000t)				Share (%)			
	1980	1985	1990	1992	1980	1985	1990	1992
AFRICA	**1 160**	**460**	**633**	**1 048**	**17.2**	**5.8**	**7.7**	**11.7**
South Africa	14	3	10	12	0.2	0.0	0.1	0.1
Others	1 147	457	623	1 036	17.0	5.7	7.6	11.6
MIDDLE EAST	**796**	**719**	**560**	**712**	**11.8**	**9.0**	**6.8**	**7.9**
Egypt	81	130	70	45	1.2	1.6	0.9	0.5
Iran	47	89	343	211	0.7	1.1	4.2	2.4
Others	668	500	147	457	9.9	6.3	1.8	5.1
FAR EAST	**215**	**488**	**414**	**617**	**3.2**	**6.1**	**5.0**	**6.9**
China	99	429	31	194	1.5	5.4	0.4	2.2
India	32	27	22	46	0.5	0.3	0.3	0.5
Indonesia	3	1	10	4	0.0	0.0	0.1	0.0
Japan	0	2	2	3	0.0	0.0	0.0	0.0
Korea, DPR	7	0	0	0	0.1	0.0	0.0	0.0
Korea, Republic of	1	3	4	23	0.0	0.0	0.0	0.3
Others	74	26	346	347	1.1	0.3	4.2	3.9
NORTH AMERICA	**192**	**599**	**493**	**278**	**2.8**	**7.5**	**6.0**	**3.1**
Canada	5	47	117	47	0.1	0.6	1.4	0.5
United States	188	552	376	231	2.8	6.9	4.6	2.6
Others	...	0	...	0	...	0.0	...	0.0
OTHER AMERICA	**247**	**79**	**72**	**157**	**3.7**	**1.0**	**0.9**	**1.7**
Argentina	73	51	7	8	1.1	0.6	0.1	0.1
Brazil	18	2	36	18	0.3	0.0	0.4	0.2
Mexico	81	19	7	...	1.2	0.2	0.1	...
Venezuela	36	3	5	86	0.5	0.0	0.1	1.0
Others	40	5	16	45	0.6	0.1	0.2	0.5
OCEANIA	**5**	**6**	**6**	**6**	**0.1**	**0.1**	**0.1**	**0.1**
Australia	2	6	5	4	0.0	0.1	0.1	0.0
New Zealand	0	0	1	1	0.0	0.0	0.0	0.0
Others	3	0	1	0	0.1	0.0	0.0	0.0
EUROPE-TOTAL	**4 104**	**5 611**	**4 291**	**5 945**	**60.8**	**70.4**	**52.3**	**66.3**
EC	**2 877**	**3 229**	**3 028**	**4 860**	**42.7**	**40.5**	**36.9**	**54.2**
Belgium-Luxembourg	58	143	190	173	0.9	1.8	2.3	1.9
Denmark	9	18	29	61	0.1	0.2	0.3	0.7
France	1 145	1 183	1 576	1 395	17.0	14.8	19.2	15.6
Germany	1 128	1 266	...	1 910	16.7	15.9	...	21.3
former FR Germany	1 120	1 264	...	...	16.6	15.9	...	...
Greece	147	86	239	396	2.2	1.1	2.9	4.4
Ireland	11	9	6	8	0.2	0.1	0.1	0.1
Italy	.	.	.	.	.	.	.	.
Netherlands	70	140	252	153	1.0	1.8	3.1	1.7
Portugal	26	12	44	71	0.4	0.2	0.5	0.8
Spain	41	107	415	387	0.6	1.3	5.1	4.3
United Kingdom	250	268	277	306	3.7	3.4	3.4	3.4
EFTA	**344**	**431**	**780**	**725**	**5.1**	**5.4**	**9.5**	**8.1**
Austria	102	153	338	359	1.5	1.9	4.1	4.0
Finland	3	3	9	15	0.0	0.0	0.1	0.2
Iceland	2	1	0	0	0.0	0.0	0.0	0.0
Norway	5	9	6	15	0.1	0.1	0.1	0.2
Sweden	17	13	29	35	0.3	0.2	0.4	0.4
Switzerland	216	252	398	300	3.2	3.2	4.8	3.3
EASTERN EUROPE	**146**	**154**	**63**	**92**	**2.2**	**1.9**	**0.8**	**1.0**
Albania	5	1	1	21	0.1	0.0	0.0	0.2
Bulgaria	39	18	9	1	0.6	0.2	0.1	0.0
former Czechoslovakia	11	76	1	9	0.2	1.0	0.0	0.1
former GDR	9	2	...	...	0.1	0.0	...	...
Hungary	16	27	25	20	0.2	0.3	0.3	0.2
Poland	36	19	23	28	0.5	0.2	0.3	0.3
Romania	30	10	4	14	0.4	0.1	0.1	0.2
former USSR	**579**	**1 354**	**148**	**...**	**8.6**	**17.0**	**1.8**	**...**
OTHER EUROPE	**158**	**443**	**272**	**268**	**2.3**	**5.6**	**3.3**	**3.0**
Turkey	53	263	195	268	0.8	3.3	2.4	3.0
former Yugoslavia	105	180	77	...	1.6	2.3	0.9	...
OTHERS	**26**	**8**	**1 744**	**200**	**0.4**	**0.1**	**21.2**	**2.2**
WORLD-TOTAL	**6 746**	**7 971**	**8 213**	**8 962**	**100.0**	**100.0**	**100.0**	**100.0**

Annex table 16-6 Destination and relative importance of steel exports by major exporting regions and countries : *Netherlands*

Destination	Amount (1000t)				Share (%)			
	1980	1985	1990	1992	1980	1985	1990	1992
AFRICA	**149**	**139**	**99**	**95**	**3.2**	**2.7**	**1.8**	**1.6**
South Africa	2	2	2	2	0.1	0.0	0.0	0.0
Others	147	137	97	93	3.2	2.6	1.7	1.6
MIDDLE EAST	**99**	**158**	**87**	**146**	**2.1**	**3.0**	**1.6**	**2.4**
Egypt	3	13	12	48	0.1	0.2	0.2	0.8
Iran	0	32	27	7	0.0	0.6	0.5	0.1
Others	96	114	48	91	2.1	2.2	0.9	1.5
FAR EAST	**145**	**349**	**102**	**265**	**3.1**	**6.7**	**1.8**	**4.4**
China	71	145	1	42	1.5	2.8	0.0	0.7
India	8	29	6	11	0.2	0.6	0.1	0.2
Indonesia	36	5	4	9	0.8	0.1	0.1	0.1
Japan	0	0	40	0	0.0	0.0	0.7	0.0
Korea, DPR	0	...	...	...	0.0	...	...	...
Korea, Republic of	0	11	1	1	0.0	0.2	0.0	0.0
Others	31	158	50	202	0.7	3.1	0.9	3.4
NORTH AMERICA	**249**	**450**	**478**	**656**	**5.4**	**8.7**	**8.5**	**11.0**
Canada	1	14	41	96	0.0	0.3	0.7	1.6
United States	248	436	437	560	5.4	8.4	7.8	9.4
Others	0	...	...	...	0.0	...	...	...
OTHER AMERICA	**216**	**74**	**68**	**95**	**4.7**	**1.4**	**1.2**	**1.6**
Argentina	96	0	2	2	2.1	0.0	0.0	0.0
Brazil	0	0	1	0	0.0	0.0	0.0	0.0
Mexico	6	0	7	6	0.1	0.0	0.1	0.1
Venezuela	80	2	3	10	1.7	0.0	0.1	0.2
Others	33	71	55	76	0.7	1.4	1.0	1.3
OCEANIA	**1**	**2**	**5**	**2**	**0.0**	**0.0**	**0.1**	**0.0**
Australia	1	1	2	1	0.0	0.0	0.0	0.0
New Zealand	0	1	0	0	0.0	0.0	0.0	0.0
Others	0	0	3	1	0.0	0.0	0.1	0.0
EUROPE-TOTAL	**3 755**	**4 015**	**4 755**	**4 601**	**81.4**	**77.4**	**85.0**	**76.9**
EC	**3 180**	**3 414**	**4 211**	**3 943**	**68.9**	**65.8**	**75.3**	**65.9**
Belgium-Luxembourg	817	706	1 021	928	17.7	13.6	18.3	15.5
Denmark	52	72	80	64	1.1	1.4	1.4	1.1
France	216	219	355	316	4.7	4.2	6.3	5.3
Germany	1 075	1 062	1 352	1 355	23.3	20.5	24.2	22.7
former FR Germany	1 075	1 061	1 349	...	23.3	20.5	24.1	...
Greece	30	262	162	274	0.6	5.1	2.9	4.6
Ireland	11	15	9	11	0.2	0.3	0.2	0.2
Italy	253	352	384	313	5.5	6.8	6.9	5.2
Netherlands	.	.	.	.	.	.	.	.
Portugal	66	40	48	70	1.4	0.8	0.9	1.2
Spain	113	230	107	84	2.4	4.4	1.9	1.4
United Kingdom	548	456	695	528	11.9	8.8	12.4	8.8
EFTA	**411**	**394**	**485**	**569**	**8.9**	**7.6**	**8.7**	**9.5**
Austria	16	10	148	30	0.3	0.2	2.7	0.5
Finland	8	31	20	10	0.2	0.6	0.4	0.2
Iceland	4	4	4	4	0.1	0.1	0.1	0.1
Norway	183	105	102	339	4.0	2.0	1.8	5.7
Sweden	115	103	106	102	2.5	2.0	1.9	1.7
Switzerland	84	141	105	85	1.8	2.7	1.9	1.4
EASTERN EUROPE	**18**	**7**	**7**	**5**	**0.4**	**0.1**	**0.1**	**0.1**
Albania	...	0	...	...	...	0.0	...	...
Bulgaria	1	0	0	0	0.0	0.0	0.0	0.0
former Czechoslovakia	0	...	0	1	0.0	...	0.0	0.0
former GDR	0	0	3	...	0.0	0.0	0.0	...
Hungary	4	1	2	1	0.1	0.0	0.0	0.0
Poland	9	5	1	3	0.2	0.1	0.0	0.1
Romania	3	...	0	0	0.1	...	0.0	0.0
former USSR	**117**	**28**	**3**	**1**	**2.5**	**0.5**	**0.1**	**0.0**
OTHER EUROPE	**30**	**172**	**49**	**82**	**0.6**	**3.3**	**0.9**	**1.4**
Turkey	20	157	45	82	0.4	3.0	0.8	1.4
former Yugoslavia	10	15	4	1	0.2	0.3	0.1	0.0
Others	**1**	**2**	**...**	**123**	**0.0**	**0.0**	**...**	**2.1**
WORLD-TOTAL	**4 616**	**5 189**	**5 593**	**5 983**	**100.0**	**100.0**	**100.0**	**100.0**

Annex table 16-7 **Destination and relative importance of steel exports by major exporting regions and countries : *Spain***

Destination	Amount (1000t)				Share (%)			
	1980	1985	1990	1992	1980	1985	1990	1992
AFRICA	**596**	**715**	**272**	**725**	**13.1**	**9.2**	**6.5**	**15.2**
South Africa	8	2	3	6	0.2	0.0	0.1	0.1
Others	589	713	269	719	13.0	9.1	6.4	15.1
MIDDLE EAST	**1 109**	**1 703**	**172**	**187**	**24.5**	**21.9**	**4.1**	**3.9**
Egypt	167	657	27	14	3.7	8.4	0.7	0.3
Iran	477	369	56	81	10.5	4.7	1.3	1.7
Others	465	677	89	92	10.2	8.7	2.1	1.9
FAR EAST	**251**	**1 686**	**381**	**507**	**5.5**	**21.6**	**9.1**	**10.6**
China	72	1 194	48	42	1.6	15.3	1.1	0.9
India	61	85	9	23	1.4	1.1	0.2	0.5
Indonesia	5	21	16	19	0.1	0.3	0.4	0.4
Japan	27	106	62	74	0.6	1.4	1.5	1.6
Korea, DPR	...	0	10	10	...	0.0	0.2	0.2
Korea, Republic of	...	2	84	59	...	0.0	2.0	1.2
Others	85	277	152	280	1.9	3.6	3.6	5.9
NORTH AMERICA	**474**	**557**	**349**	**200**	**10.4**	**7.2**	**8.4**	**4.2**
Canada	46	99	28	10	1.0	1.3	0.7	0.2
United States	427	458	321	190	9.4	5.9	7.7	4.0
Others	...	0	...	0	...	...	...	0.0
OTHER AMERICA	**398**	**188**	**207**	**255**	**8.8**	**2.4**	**4.9**	**5.4**
Argentina	36	4	14	32	0.8	0.0	0.3	0.7
Brazil	11	1	1	1	0.2	0.0	0.0	0.0
Mexico	148	35	123	...	3.3	0.5	2.9	...
Venezuela	132	6	20	38	2.9	0.1	0.5	0.8
Others	71	142	49	185	1.6	1.8	1.2	3.9
OCEANIA	**2**	**2**	**4**	**5**	**0.0**	**0.0**	**0.1**	**0.1**
Australia	1	1	3	3	0.0	0.0	0.1	0.1
New Zealand	0	0	0	1	0.0	0.0	0.0	0.0
Others	0	0	1	0	0.0	0.0	0.0	0.0
EUROPE-TOTAL	**1 693**	**2 882**	**2 284**	**2 810**	**37.4**	**37.0**	**54.7**	**59.0**
EC	**1 032**	**1 055**	**1 975**	**2 609**	**22.8**	**13.5**	**47.3**	**54.8**
Belgium-Luxembourg	78	100	83	108	1.7	1.3	2.0	2.3
Denmark	2	14	25	58	0.1	0.2	0.6	1.2
France	281	224	699	539	6.2	2.9	16.7	11.3
Germany	357	366	...	505	7.9	4.7	...	10.6
former FR Germany	357	341	...	...	7.9	4.4	...	...
Greece	26	13	18	92	0.6	0.2	0.4	1.9
Ireland	6	11	45	50	0.1	0.1	1.1	1.1
Italy	65	76	369	346	1.4	1.0	8.8	7.3
Netherlands	32	40	46	48	0.7	0.5	1.1	1.0
Portugal	82	107	308	456	1.8	1.4	7.4	9.6
Spain	.	.	.	.			.	
United Kingdom	104	129	382	405	2.3	1.7	9.1	8.5
EFTA	**198**	**139**	**121**	**106**	**4.4**	**1.8**	**2.9**	**2.2**
Austria	10	6	11	6	0.2	0.1	0.3	0.1
Finland	18	44	18	19	0.4	0.6	0.4	0.4
Iceland	1	3	...	...	0.0	0.0	...	...
Norway	60	31	23	25	1.3	0.4	0.6	0.5
Sweden	53	23	39	31	1.2	0.3	0.9	0.7
Switzerland	58	33	30	25	1.3	0.4	0.7	0.5
EASTERN EUROPE	**91**	**141**	**19**	**35**	**2.0**	**1.8**	**0.4**	**0.7**
Albania	2	0	...	...	0.0	0.0	...	...
Bulgaria	26	81	4	3	0.6	1.0	0.1	0.1
former Czechoslovakia	...	1	0	0	...	0.0	0.0	0.0
former GDR	...	25	...	...	...	0.3	...	...
Hungary	11	21	4	0	0.3	0.3	0.1	0.0
Poland	9	10	5	32	0.2	0.1	0.1	0.7
Romania	43	4	6	0	1.0	0.1	0.1	0.0
former USSR	**221**	**958**	**115**	**...**	**4.9**	**12.3**	**2.7**	**...**
OTHER EUROPE	**151**	**588**	**55**	**60**	**3.3**	**7.6**	**1.3**	**1.3**
Turkey	50	576	54	60	1.1	7.4	1.3	1.3
former Yugoslavia	101	13	1	...	2.2	0.2	0.0	...
OTHERS	**11**	**59**	**510**	**75**	**0.2**	**0.8**	**12.2**	**1.6**
WORLD-TOTAL	**4 533**	**7 792**	**4 179**	**4 763**	**100.0**	**100.0**	**100.0**	**100.0**

Annex table 16-8 **Destination and relative importance of steel exports by major exporting regions and countries : *United Kingdom***

Destination	Amount (1000t)				Share (%)			
	1980	1985	1990	1992	1980	1985	1990	1992
AFRICA	**180**	**191**	**111**	**143**	**6.5**	**3.9**	**1.6**	**1.7**
South Africa	15	13	19	19	0.6	0.3	0.3	0.2
Others	164	178	92	124	5.9	3.6	1.3	1.5
MIDDLE EAST	**200**	**270**	**247**	**378**	**7.2**	**5.5**	**3.5**	**4.5**
Egypt	40	46	22	20	1.4	0.9	0.3	0.2
Iran	4	38	79	112	0.1	0.8	1.1	1.3
Others	156	186	147	246	5.6	3.8	2.1	2.9
FAR EAST	**338**	**744**	**745**	**953**	**12.2**	**15.2**	**10.5**	**11.3**
China	33	255	20	52	1.2	5.2	0.3	0.6
India	126	171	133	137	4.6	3.5	1.9	1.6
Indonesia	1	5	38	23	0.0	0.1	0.5	0.3
Japan	1	1	3	29	0.0	0.0	0.0	0.3
Korea, DPR	0	0	...	...	0.0	0.0	...	...
Korea, Republic of	34	6	103	75	1.2	0.1	1.5	0.9
Others	143	307	446	637	5.2	6.3	6.3	7.5
NORTH AMERICA	**289**	**712**	**891**	**673**	**10.5**	**14.5**	**12.5**	**8.0**
Canada	74	185	121	98	2.7	3.8	1.7	1.2
United States	214	526	769	573	7.7	10.7	10.8	6.8
Others	1	1	0	2	0.0	0.0	0.0	0.0
OTHER AMERICA	**113**	**68**	**94**	**123**	**4.1**	**1.4**	**1.3**	**1.5**
Argentina	16	0	3	14	0.6	0.0	0.0	0.2
Brazil	15	4	2	5	0.6	0.1	0.0	0.1
Mexico	20	10	28	20	0.7	0.2	0.4	0.2
Venezuela	4	3	5	21	0.1	0.1	0.1	0.2
Others	57	51	57	63	2.1	1.0	0.8	0.7
OCEANIA	**24**	**44**	**47**	**29**	**0.9**	**0.9**	**0.7**	**0.3**
Australia	12	14	26	20	0.4	0.3	0.4	0.2
New Zealand	12	29	19	7	0.4	0.6	0.3	0.1
Others	0	1	2	2	0.0	0.0	0.0	0.0
EUROPE-TOTAL	**1 615**	**2 834**	**4 054**	**6 141**	**58.4**	**57.8**	**57.0**	**72.8**
EC	**1 181**	**2 070**	**3 312**	**5 182**	**42.7**	**42.2**	**46.5**	**61.4**
Belgium-Luxembourg	67	177	444	441	2.4	3.6	6.2	5.2
Denmark	77	167	180	248	2.8	3.4	2.5	2.9
France	129	231	496	616	4.7	4.7	7.0	7.3
Germany	318	587	...	1 468	11.5	12.0	...	17.4
former FR Germany	318	587	...	...	11.5	12.0	...	...
Greece	51	172	376	364	1.9	3.5	5.3	4.3
Ireland	168	184	301	364	6.1	3.8	4.2	4.3
Italy	215	216	566	623	7.8	4.4	8.0	7.4
Netherlands	112	168	371	392	4.1	3.4	5.2	4.6
Portugal	10	39	104	128	0.4	0.8	1.5	1.5
Spain	34	128	475	539	1.2	2.6	6.7	6.4
United Kingdom	.	.	.	.	.	.	.	.
EFTA	**279**	**453**	**632**	**824**	**10.1**	**9.2**	**8.9**	**9.8**
Austria	6	20	26	27	0.2	0.4	0.4	0.3
Finland	33	57	86	53	1.2	1.2	1.2	0.6
Iceland	2	4	1	10	0.1	0.1	0.0	0.1
Norway	70	96	150	349	2.5	2.0	2.1	4.1
Sweden	111	180	208	187	4.0	3.7	2.9	2.2
Switzerland	56	97	160	199	2.0	2.0	2.3	2.4
EASTERN EUROPE	**16**	**15**	**8**	**27**	**0.6**	**0.3**	**0.1**	**0.3**
Albania	0	0	2	...	0.0	0.0	0.0	...
Bulgaria	0	1	1	1	0.0	0.0	0.0	0.0
former Czechoslovakia	1	0	0	1	0.0	0.0	0.0	0.0
former GDR	0	0	...	...	0.0	0.0	...	...
Hungary	2	1	1	1	0.1	0.0	0.0	0.0
Poland	8	12	3	25	0.3	0.3	0.0	0.3
Romania	4	0	1	0	0.1	0.0	0.0	0.0
former USSR	**92**	**162**	**20**	**...**	**3.3**	**3.3**	**0.3**	**...**
OTHER EUROPE	**48**	**134**	**81**	**108**	**1.7**	**2.7**	**1.1**	**1.3**
Turkey	13	111	72	108	0.5	2.3	1.0	1.3
former Yugoslavia	35	23	10	...	1.3	0.5	0.1	...
OTHERS	**7**	**40**	**927**	**...**	**0.2**	**0.8**	**13.0**	**...**
WORLD-TOTAL	**2 766**	**4 903**	**7 116**	**8 439**	**100.0**	**100.0**	**100.0**	**100.0**

Annex table 16-9 **Destination and relative importance of steel exports by major exporting regions and countries : *EFTA***

Destination	Amount (1000t)				Share (%)			
	1980	1985	1990	1992	1980	1985	1990	1992
AFRICA	**69**	**64**	**81**	**120**	**1.0**	**0.7**	**0.9**	**1.2**
South Africa	8	7	4	3	0.1	0.1	0.0	0.0
Others	61	57	77	116	0.9	0.6	0.9	1.2
MIDDLE EAST	**147**	**218**	**91**	**129**	**2.2**	**2.3**	**1.1**	**1.3**
Egypt	12	43	12	18	0.2	0.5	0.1	0.2
Iran	59	42	34	44	0.9	0.5	0.4	0.5
Others	76	133	45	67	1.1	1.4	0.5	0.7
FAR EAST	**173**	**382**	**359**	**675**	**2.6**	**4.1**	**4.2**	**7.0**
China	4	291	7	55	0.1	3.1	0.1	0.6
India	11	36	78	73	0.2	0.4	0.9	0.8
Indonesia	1	1	7	18	0.0	0.0	0.1	0.2
Japan	90	7	35	118	1.3	0.1	0.4	1.2
Korea, DPR	0	0	0	0	0.0	0.0	0.0	0.0
Korea, Republic of	2	17	34	65	0.0	0.2	0.4	0.7
Others	64	30	198	346	1.0	0.3	2.3	3.6
NORTH AMERICA	**165**	**1 052**	**501**	**452**	**2.5**	**11.3**	**5.8**	**4.7**
Canada	24	52	36	27	0.4	0.6	0.4	0.3
United States	141	1 000	464	426	2.1	10.7	5.4	4.4
Others	0	0	1	0	0.0	0.0	0.0	0.0
OTHER AMERICA	**84**	**47**	**45**	**56**	**1.3**	**0.5**	**0.5**	**0.6**
Argentina	30	3	3	10	0.5	0.0	0.0	0.1
Brazil	11	4	3	5	0.2	0.0	0.0	0.1
Mexico	9	10	24	16	0.1	0.1	0.3	0.2
Venezuela	14	15	10	5	0.2	0.2	0.1	0.1
Others	20	16	6	19	0.3	0.2	0.1	0.2
OCEANIA	**11**	**15**	**26**	**43**	**0.2**	**0.2**	**0.3**	**0.5**
Australia	10	13	24	40	0.2	0.1	0.3	0.4
New Zealand	0	1	2	4	0.0	0.0	0.0	0.0
Others	1	...	0	0	0.0	...	0.0	0.0
EUROPE-TOTAL	**6 023**	**7 520**	**5 231**	**7 867**	**90.2**	**80.7**	**60.8**	**81.7**
EC	**3 896**	**5 086**	**3 135**	**6 109**	**58.3**	**54.6**	**36.4**	**63.5**
Belgium-Luxembourg	92	173	135	127	1.4	1.9	1.6	1.3
Denmark	407	541	497	695	6.1	5.8	5.8	7.2
France	248	366	420	454	3.7	3.9	4.9	4.7
Germany	1 558	2 555	...	2 668	23.3	27.4	...	27.7
former FR Germany	1 697	2 352	...	...	25.4	25.2	...	...
Greece	49	23	20	26	0.7	0.2	0.2	0.3
Ireland	13	29	24	29	0.2	0.3	0.3	0.3
Italy	487	654	957	1 041	7.3	7.0	11.1	10.8
Netherlands	342	362	274	354	5.1	3.9	3.2	3.7
Portugal	26	31	28	27	0.4	0.3	0.3	0.3
Spain	45	37	86	118	0.7	0.4	1.0	1.2
United Kingdom	489	519	694	570	7.3	5.6	8.1	5.9
EFTA	**1 186**	**1 100**	**1 500**	**1 555**	**17.8**	**11.8**	**17.4**	**16.2**
Austria	74	69	107	121	1.1	0.7	1.2	1.3
Finland	165	148	169	172	2.5	1.6	2.0	1.8
Iceland	22	11	6	10	0.3	0.1	0.1	0.1
Norway	332	320	327	380	5.0	3.4	3.8	3.9
Sweden	414	328	650	675	6.2	3.5	7.6	7.0
Switzerland	180	224	243	197	2.7	2.4	2.8	2.1
EASTERN EUROPE	**177**	**300**	**69**	**154**	**2.7**	**3.2**	**0.8**	**1.6**
Albania	1	5	2	0	0.0	0.0	0.0	0.0
Bulgaria	37	36	13	2	0.6	0.4	0.1	0.0
former Czechoslovakia	6	5	6	45	0.1	0.0	0.1	0.5
former GDR	25	203	...		0.4	2.2	...	
Hungary	22	19	23	50	0.3	0.2	0.3	0.5
Poland	50	30	24	53	0.7	0.3	0.3	0.5
Romania	35	3	2	5	0.5	0.0	0.0	0.0
former USSR	**561**	**825**	**352**	**0**	**8.4**	**8.8**	**4.1**	**0.0**
OTHER EUROPE	**202**	**209**	**176**	**49**	**3.0**	**2.2**	**2.0**	**0.5**
Turkey	17	78	49	49	0.3	0.8	0.6	0.5
former Yugoslavia	185	131	127	...	2.8	1.4	1.5	...
OTHERS	**5**	**25**	**2 266**	**284**	**0.1**	**0.3**	**26.3**	**2.9**
WORLD-TOTAL	**6 677**	**9 323**	**8 601**	**9 625**	**100.0**	**100.0**	**100.0**	**100.0**

Annex table 16-10 Destination and relative importance of steel exports by major exporting regions and countries : *EASTERN EUROPE*

Destination	Amount (1000t)				Share (%)			
	1980	1985	1990	1992	1980	1985	1990	1992
AFRICA	...	...	...	**398**	...	...	...	**3.6**
South Africa	...	...	...	2	...	...	...	0.0
Others	...	...	...	396	...	...	...	3.5
MIDDLE EAST	...	...	...	**1 244**	...	...	...	**11.1**
Egypt	...	...	...	193	...	...	...	1.7
Iran	...	...	...	462	...	...	...	4.1
Others	...	...	...	589	...	...	...	5.3
FAR EAST	...	...	...	**2 927**	...	...	...	**26.2**
China	...	...	...	351	...	...	...	3.1
India	...	...	...	93	...	...	...	0.8
Indonesia	...	...	...	43	...	...	...	0.4
Japan	...	...	...	257	...	...	...	2.3
Korea, DPR	...	...	...	79	...	...	...	0.7
Korea, Republic of	...	...	...	61	...	...	...	0.5
Others	...	...	...	2 043	...	...	...	18.3
NORTH AMERICA	...	...	...	**76**	...	...	...	**0.7**
Canada	...	...	...	2	...	...	...	0.0
United States	...	...	...	74	...	...	...	0.7
Others	...	...	...	1	...	...	...	0.0
OTHER AMERICA	...	...	...	**187**	...	...	...	**1.7**
Argentina	...	...	...	45	...	...	...	0.4
Brazil	...	...	...	0	...	...	...	0.0
Mexico	...	...	...	31	...	...	...	0.3
Venezuela	...	...	...	21	...	...	...	0.2
Others	...	...	...	89	...	...	...	0.8
OCEANIA	...	...	...	**2**	...	...	...	**0.0**
Australia	...	...	...	1	...	...	...	0.0
New Zealand	...	...	...	...	...	...	...	...
Others	...	...	...	1	...	...	...	0.0
EUROPE-TOTAL	...	...	...	**6 177**	...	...	...	**55.3**
EC	...	...	...	**3 500**	...	...	...	**31.3**
Belgium-Luxembourg	...	...	...	169	...	...	...	1.5
Denmark	...	...	...	54	...	...	...	0.5
France	...	...	...	259	...	...	...	2.3
Germany	...	...	...	1 882	...	...	...	16.8
former FR Germany	...	...	...	...	...	...	...	...
Greece	...	...	...	132	...	...	...	1.2
Ireland	...	...	...	6	...	...	...	0.0
Italy	...	...	...	586	...	...	...	5.2
Netherlands	...	...	...	113	...	...	...	1.0
Portugal	...	...	...	33	...	...	...	0.3
Spain	...	...	...	77	...	...	...	0.7
United Kingdom	...	...	...	190	...	...	...	1.7
EFTA	...	...	...	**831**	...	...	...	**7.4**
Austria	...	...	...	425	...	...	...	3.8
Finland	...	...	...	55	...	...	...	0.5
Iceland	...	...	...	4	...	...	...	0.0
Norway	...	...	...	30	...	...	...	0.3
Sweden	...	...	...	62	...	...	...	0.6
Switzerland	...	...	...	256	...	...	...	2.3
EASTERN EUROPE	...	...	...	**724**	...	...	...	**6.5**
Albania	...	...	...	3	...	...	...	0.0
Bulgaria	...	...	...	68	...	...	...	0.6
former Czechoslovakia	...	...	...	99	...	...	...	0.9
former GDR	...	...	...	...	...	...	...	...
Hungary	...	...	...	345	...	...	...	3.1
Poland	...	...	...	136	...	...	...	1.2
Romania	...	...	...	73	...	...	...	0.7
former USSR	...	...	...	**509**	...	...	...	**4.6**
OTHER EUROPE	...	...	...	**612**	...	...	...	**5.5**
Turkey	...	...	...	298	...	...	...	2.7
former Yugoslavia	...	...	...	314	...	...	...	2.8
OTHERS	...	...	...	**166**	...	...	...	**1.5**
WORLD-TOTAL	...	...	...	**11 176**	...	...	...	**100.0**

Annex table 16-11 Destination and relative importance of steel exports by major exporting regions and countries : *former Czechoslovakia*

Destination	Amount (1000t)				Share (%)			
	1980	1985	1990	1992	1980	1985	1990	1992
AFRICA	**59**	**6**	**97**	**138**	**1.7**	**0.2**	**2.6**	**2.7**
South Africa	...	...	...	2	...	...	...	0.0
Others	59	6	97	136	1.7	0.2	2.6	2.6
MIDDLE EAST	**428**	**442**	**247**	**448**	**12.5**	**11.4**	**6.7**	**8.7**
Egypt	117	319	132	17	3.4	8.2	3.6	0.3
Iran	110	12	40	197	3.2	0.3	1.1	3.8
Others	202	111	75	234	5.9	2.9	2.0	4.5
FAR EAST	**157**	**314**	**398**	**949**	**4.6**	**8.1**	**10.7**	**18.4**
China	69	234	101	174	2.0	6.0	2.7	3.4
India	60	62	66	42	1.8	1.6	1.8	0.8
Indonesia	...	...	...	41	...	...	...	0.8
Japan	...	...	...	57	...	...	...	1.1
Korea, DPR	...	...	...	6	...	...	...	0.1
Korea, Republic of	...	...	...	7	...	...	...	0.1
Others	29	18	231	624	0.8	0.5	6.2	12.1
NORTH AMERICA	**18**	**37**	**25**	**23**	**0.5**	**1.0**	**0.7**	**0.5**
Canada	10	11	12	1	0.3	0.3	0.3	0.0
United States	5	26	13	22	0.2	0.7	0.4	0.4
Others	2	...	...	1	0.1	...	...	0.0
OTHER AMERICA	**30**	**31**	**26**	**46**	**0.9**	**0.8**	**0.7**	**0.9**
Argentina	1	0	...	11	0.0	0.0	...	0.2
Brazil	...	...	...	0	...	...	...	0.0
Mexico	...	...	...	9	...	...	...	0.2
Venezuela	...	...	...	7	...	...	...	0.1
Others	29	31	26	19	0.8	0.8	0.7	0.4
OCEANIA	...	...	...	**1**	...	...	...	**0.0**
Australia	...	...	...	1	...	...	...	0.0
New Zealand	...	...	...	...	...	...	...	...
Others	...	...	...	1	...	...	...	0.0
EUROPE-TOTAL	**2 728**	**2 998**	**2 910**	**3 461**	**79.6**	**77.3**	**78.3**	**67.2**
EC	**867**	**834**	**1 078**	**1 838**	**25.3**	**21.5**	**29.0**	**35.7**
Belgium-Luxembourg	21	23	24	54	0.6	0.6	0.6	1.1
Denmark	18	45	26	23	0.5	1.2	0.7	0.5
France	107	118	239	235	3.1	3.0	6.4	4.6
Germany	670	680	563	1 030	19.6	17.5	15.2	20.0
former FR Germany	390	404	444	...	11.4	10.4	12.0	...
Greece	98	47	95	46	2.9	1.2	2.6	0.9
Ireland	...	0	...	...	...	0.0	...	...
Italy	154	139	159	303	4.5	3.6	4.3	5.9
Netherlands	25	30	44	40	0.7	0.8	1.2	0.8
Portugal	4	1	8	10	0.1	0.0	0.2	0.2
Spain	36	2	0	25	1.0	0.0	0.0	0.5
United Kingdom	16	25	38	72	0.5	0.6	1.0	1.4
EFTA	**182**	**265**	**291**	**489**	**5.3**	**6.8**	**7.8**	**9.5**
Austria	65	152	86	293	1.9	3.9	2.3	5.7
Finland	44	48	52	22	1.3	1.2	1.4	0.4
Iceland	2	1	3	3	0.1	0.0	0.1	0.1
Norway	23	7	19	14	0.7	0.2	0.5	0.3
Sweden	29	21	36	27	0.8	0.5	1.0	0.5
Switzerland	19	36	96	131	0.6	0.9	2.6	2.5
EASTERN EUROPE	**730**	**894**	**324**	**581**	**21.3**	**23.0**	**8.7**	**11.3**
Albania	33	26	20	1	1.0	0.7	0.5	0.0
Bulgaria	51	296	28	52	1.5	7.6	0.8	1.0
former Czechoslovakia	.	.	.	.	.	.	.	.
former GDR	281	276	119	...	8.2	7.1	3.2	...
Hungary	56	74	97	333	1.6	1.9	2.6	6.5
Poland	241	210	55	132	7.0	5.4	1.5	2.6
Romania	69	13	5	63	2.0	0.3	0.1	1.2
former USSR	**487**	**518**	**585**	**375**	**14.2**	**13.4**	**15.8**	**7.3**
OTHER EUROPE	**463**	**487**	**632**	**177**	**13.5**	**12.5**	**17.0**	**3.4**
Turkey	34	12	82	34	1.0	0.3	2.2	0.7
former Yugoslavia	429	475	550	143	12.5	12.2	14.8	2.8
OTHERS	**7**	**52**	**14**	**86**	**0.2**	**1.3**	**0.4**	**1.7**
WORLD-TOTAL	**3 426**	**3 880**	**3 715**	**5 153**	**100.0**	**100.0**	**100.0**	**100.0**

Annex table 16-12 Destination and relative importance of steel exports by major exporting regions and countries : *Poland*

Destination	Amount (1000t)				Share (%)			
	1980	1985	1990	1992	1980	1985	1990	1992
AFRICA	**9**	**1**	**2**	**227**	**0.5**	**0.0**	**0.1**	**6.7**
South Africa	...	...	0	...	...	...	0.0	...
Others	9	1	2	227	0.5	0.0	0.1	6.7
MIDDLE EAST	**188**	**138**	**241**	**456**	**9.8**	**6.5**	**6.8**	**13.4**
Egypt	41	14	6	125	2.1	0.7	0.2	3.7
Iran	119	118	...	224	6.2	5.5	...	6.6
Others	28	6	235	107	1.5	0.3	6.6	3.1
FAR EAST	**133**	**251**	**86**	**1 290**	**6.9**	**11.8**	**2.4**	**37.9**
China	91	212	38	107	4.7	10.0	1.1	3.1
India	...	...	11	34	...	...	0.3	1.0
Indonesia	0	1	2	2	0.0	0.0	0.1	0.1
Japan	...	...	...	19	...	...	...	0.6
Korea, DPR	...	...	...	69	...	...	...	2.0
Korea, Republic of	...	0	6	55	...	0.0	0.2	1.6
Others	41	38	29	1 004	2.1	1.8	0.8	29.5
NORTH AMERICA	**22**	**54**	**63**	**28**	**1.1**	**2.5**	**1.8**	**0.8**
Canada	4	12	1	1	0.2	0.5	0.0	0.0
United States	19	43	61	28	1.0	2.0	1.7	0.8
Others	...	...	...	...	...	...	...	...
OTHER AMERICA	**70**	**26**	**3**	**127**	**3.6**	**1.2**	**0.1**	**3.7**
Argentina	...	0	0	34	...	0.0	0.0	1.0
Brazil	40	...	...	...	2.1	...	...	...
Mexico	...	...	...	22	...	...	...	0.6
Venezuela	...	...	...	15	...	...	...	0.4
Others	29	26	3	57	1.5	1.2	0.1	1.7
OCEANIA	**0**	...	**0**	**0**	**0.0**	...	**0.0**	**0.0**
Australia	0	...	0	0	0.0	...	0.0	0.0
New Zealand	...	...	...	...	...	...	...	...
Others	...	...	...	...	...	...	...	...
EUROPE-TOTAL	**1 474**	**1 661**	**3 002**	**1 244**	**76.3**	**77.9**	**84.3**	**36.5**
EC	**328**	**283**	**1 393**	**911**	**17.0**	**13.3**	**39.1**	**26.7**
Belgium-Luxembourg	29	14	51	66	1.5	0.7	1.4	1.9
Denmark	8	8	24	25	0.4	0.4	0.7	0.7
France	4	15	25	13	0.2	0.7	0.7	0.4
Germany	366	293	1 043	591	18.9	13.7	29.3	17.4
former FR Germany	193	169	1 039	...	10.0	7.9	29.2	...
Greece	1	8	2	6	0.1	0.4	0.1	0.2
Ireland	...	0	5	6	...	0.0	0.1	0.2
Italy	20	0	34	62	1.1	0.0	0.9	1.8
Netherlands	9	16	53	56	0.5	0.7	1.5	1.6
Portugal	...	...	1	1	...	...	0.0	0.0
Spain	1	1	3	17	0.1	0.0	0.1	0.5
United Kingdom	62	52	156	68	3.2	2.4	4.4	2.0
EFTA	**285**	**78**	**906**	**218**	**14.8**	**3.7**	**25.4**	**6.4**
Austria	69	16	88	51	3.6	0.8	2.5	1.5
Finland	29	23	17	23	1.5	1.1	0.5	0.7
Iceland	1	...	2	1	0.0	...	0.1	0.0
Norway	43	21	12	16	2.2	1.0	0.3	0.5
Sweden	57	17	138	33	2.9	0.8	3.9	1.0
Switzerland	86	0	649	94	4.5	0.0	18.2	2.8
EASTERN EUROPE	**467**	**409**	**70**	**77**	**24.2**	**19.2**	**2.0**	**2.2**
Albania	21	21	7	...	1.1	1.0	0.2	...
Bulgaria	58	51	1	1	3.0	2.4	0.0	0.0
former Czechoslovakia	114	106	40	69	5.9	5.0	1.1	2.0
former GDR	173	124	4	...	8.9	5.8	0.1	...
Hungary	18	27	0	6	0.9	1.3	0.0	0.2
Poland	.	.	.	.	.	.	.	.
Romania	83	80	19	1	4.3	3.8	0.5	0.0
former USSR	**209**	**717**	**559**	**15**	**10.8**	**33.6**	**15.7**	**0.4**
OTHER EUROPE	**185**	**173**	**74**	**23**	**9.6**	**8.1**	**2.1**	**0.7**
Turkey	0	12	5	15	0.0	0.6	0.1	0.4
former Yugoslavia	185	161	69	8	9.6	7.5	1.9	0.2
OTHERS	**36**	**1**	**163**	**36**	**1.8**	**0.0**	**4.6**	**1.1**
WORLD-TOTAL	**1 930**	**2 133**	**3 560**	**3 408**	**100.0**	**100.0**	**100.0**	**100.0**

Annex table 16-13 Destination and relative importance of steel exports by major exporting regions and countries : *Romania*

Destination	Amount (1000t)				Share (%)			
	1980	1985	1990	1992	1980	1985	1990	1992
AFRICA	...	...	**19**	**9**	...	...	**1.0**	**1.2**
South Africa	...	...	4	...	...	...	0.2	...
Others	...	...	15	9	...	...	0.8	1.2
MIDDLE EAST	...	...	**297**	**59**	...	...	**15.8**	**7.5**
Egypt	...	...	139	0	...	...	7.4	0.0
Iran	...	...	89	8	...	...	4.7	1.1
Others	...	...	70	50	...	...	3.7	6.4
FAR EAST	...	...	**424**	**322**	...	...	**22.5**	**41.0**
China	...	...	99	60	...	...	5.3	7.6
India	...	...	21	...	...	...	1.1	...
Indonesia	...	...	3	...	...	...	0.1	...
Japan	...	...	168	132	...	...	8.9	16.8
Korea, DPR	...	...	8	...	...	...	0.4	...
Korea, Republic of	...	...	...	...	...	...	...	...
Others	...	...	124	131	...	...	6.6	16.6
NORTH AMERICA	...	...	**72**	**2**	...	...	**3.8**	**0.3**
Canada	...	...	11	0	...	...	0.6	0.0
United States	...	...	61	2	...	...	3.2	0.2
Others	...	...	...	...	...	...	...	...
OTHER AMERICA	...	...	**3**	**6**	...	...	**0.1**	**0.7**
Argentina	...	...	...	...	...	...	...	...
Brazil	...	...	...	...	...	...	...	...
Mexico	...	...	...	...	...	...	...	...
Venezuela	...	...	...	...	...	...	...	...
Others	...	...	3	6	...	...	0.1	0.7
OCEANIA	...	...	...	...	...	...	...	...
Australia	...	...	...	...	...	...	...	...
New Zealand	...	...	...	...	...	...	...	...
Others	...	...	...	...	...	...	...	...
EUROPE-TOTAL	...	...	**972**	**369**	...	...	**51.6**	**47.0**
EC	...	...	**156**	**122**	...	...	**8.3**	**15.5**
Belgium-Luxembourg	...	...	11	3	...	...	0.6	0.3
Denmark	...	...	0	...	...	...	0.0	...
France	...	...	49	8	...	...	2.6	1.1
Germany	...	...	...	17	...	...	...	2.1
former FR Germany	...	...	...	...	...	...	...	...
Greece	...	...	22	15	...	...	1.1	1.9
Ireland	...	...	...	...	...	...	...	...
Italy	...	...	45	58	...	...	2.4	7.3
Netherlands	...	...	...	0	...	...	...	0.0
Portugal	...	...	1	15	...	...	0.0	1.9
Spain	...	...	18	...	...	...	0.9	...
United Kingdom	...	...	10	6	...	...	0.5	0.8
EFTA	...	...	**52**	**32**	...	...	**2.8**	**4.1**
Austria	...	...	11	14	...	...	0.6	1.8
Finland	...	...	15	0	...	...	0.8	0.0
Iceland	...	...	...	...	...	...	...	...
Norway	...	...	15	...	...	...	0.8	...
Sweden	...	...	1	...	...	...	0.1	...
Switzerland	...	...	10	18	...	...	0.5	2.2
EASTERN EUROPE	...	...	**72**	**16**	...	...	**3.8**	**2.0**
Albania	...	...	16	0	...	...	0.8	0.0
Bulgaria	...	...	27	11	...	...	1.4	1.4
former Czechoslovakia	...	...	5	0	...	...	0.3	0.0
former GDR	...	...	...	...	...	...	...	...
Hungary	...	...	8	4	...	...	0.4	0.5
Poland	...	...	17	1	...	...	0.9	0.1
Romania	.	.	.	.	.	.	.	.
former USSR	...	...	**477**	**56**	...	...	**25.3**	**7.1**
OTHER EUROPE	...	...	**215**	**144**	...	...	**11.4**	**18.3**
Turkey	...	...	73	71	...	...	3.9	9.1
former Yugoslavia	...	...	142	72	...	...	7.5	9.2
OTHERS	...	...	**98**	**19**	...	...	**5.2**	**2.4**
WORLD-TOTAL	...	...	**1 884**	**786**	...	...	**100.0**	**100.0**

Annex table 16-14 Destination and relative importance of steel exports by major exporting regions and countries : *former USSR*

Destination	Amount (1000t)				Share (%)			
	1980	1985	1990	1992	1980	1985	1990	1992
AFRICA	...	...	**2**	...	...	...	**0.0**	...
South Africa	...	...	...	...	...	...	...	...
Others	...	...	2	...	...	...	0.0	...
MIDDLE EAST	...	...	**52**	...	...	...	**1.1**	...
Egypt	...	...	...	...	...	...	...	...
Iran	...	...	28	...	...	...	0.6	...
Others	...	...	24	...	...	...	0.5	...
FAR EAST	...	...	**811**	...	...	...	**16.5**	...
China	...	...	424	...	...	...	8.6	...
India	...	...	...	...	...	...	...	...
Indonesia	...	...	...	...	...	...	...	...
Japan	...	...	151	...	...	...	3.1	...
Korea, DPR	...	...	1	...	...	...	0.0	...
Korea, Republic of	...	...	35	...	...	...	0.7	...
Others	...	...	201	...	...	...	4.1	...
NORTH AMERICA	...	...	...	...	...	...	...	...
Canada	...	...	...	...	...	...	...	...
United States	...	...	...	...	...	...	...	...
Others	...	...	...	...	...	...	...	...
OTHER AMERICA	...	...	**437**	...	...	...	**8.9**	...
Argentina	...	...	...	...	...	...	...	...
Brazil	...	...	...	...	...	...	...	...
Mexico	...	...	44	...	...	...	0.9	...
Venezuela	...	...	...	...	...	...	...	...
Others	...	...	393	...	...	...	8.0	...
OCEANIA	...	...	...	...	...	...	...	...
Australia	...	...	...	...	...	...	...	...
New Zealand	...	...	...	...	...	...	...	...
Others	...	...	...	...	...	...	...	...
EUROPE-TOTAL	...	...	**2 347**	...	...	...	**47.6**	...
EC	...	...	**1 238**	...	...	...	**25.1**	...
Belgium-Luxembourg	...	...	6	...	...	...	0.1	...
Denmark	...	...	...	...	...	...	...	...
France	...	...	7	...	...	...	0.1	...
Germany	...	...	409	...	...	...	8.3	...
former FR Germany	...	...	409	...	...	...	8.3	...
Greece	...	...	28	...	...	...	0.6	...
Ireland	...	...	...	...	...	...	...	...
Italy	...	...	781	...	...	...	15.8	...
Netherlands	...	...	7	...	...	...	0.1	...
Portugal	...	...	...	...	...	...	...	...
Spain	...	...	...	...	...	...	...	...
United Kingdom	...	...	...	...	...	...	...	...
EFTA	...	...	**103**	...	...	...	**2.1**	...
Austria	...	...	78	...	...	...	1.6	...
Finland	...	...	2	...	...	...	0.0	...
Iceland	...	...	...	...	...	...	...	...
Norway	...	...	...	...	...	...	...	...
Sweden	...	...	1	...	...	...	0.0	...
Switzerland	...	...	23	...	...	...	0.5	...
EASTERN EUROPE	...	...	**924**	...	...	...	**18.7**	...
Albania	...	...	1	...	...	...	0.0	...
Bulgaria	...	...	531	...	...	...	10.8	...
former Czechoslovakia	...	...	2	...	...	...	0.0	...
former GDR	...	...	...	...	...	...	...	...
Hungary	...	...	94	...	...	...	1.9	...
Poland	...	...	107	...	...	...	2.2	...
Romania	...	...	190	...	...	...	3.9	...
former USSR	.	.	.	.	.	.	.	.
OTHER EUROPE	...	...	**82**	...	...	...	**1.7**	...
Turkey	...	...	56	...	...	...	1.1	...
former Yugoslavia	...	...	26	...	...	...	0.5	...
OTHERS	...	...	**1 283**	...	...	...	**26.0**	...
WORLD-TOTAL	...	...	**4 932**	...	...	...	**100.0**	...

Annex table 16-15 Destination and relative importance of steel exports by major exporting regions and countries : *Canada*

Destination	Amount (1000t)				Share (%)			
	1980	1985	1990	1992	1980	1985	1990	1992
AFRICA	**71**	**21**	**44**	**19**	**2.0**	**0.7**	**1.1**	**0.4**
South Africa	1	0	0	0	0.0	0.0	0.0	0.0
Others	71	21	43	19	2.0	0.7	1.1	0.4
MIDDLE EAST	**72**	**7**	**267**	**42**	**2.0**	**0.2**	**6.7**	**0.8**
Egypt	34	5	51	5	1.0	0.2	1.3	0.1
Iran	...	0	216	34	...	0.0	5.4	0.7
Others	38	2	0	2	1.1	0.1	0.0	0.0
FAR EAST	**547**	**72**	**401**	**379**	**15.5**	**2.5**	**10.1**	**7.6**
China	158	0	43	13	4.5	0.0	1.1	0.3
India	77	28	16	12	2.2	0.9	0.4	0.2
Indonesia	22	2	18	2	0.6	0.1	0.4	0.0
Japan	49	1	23	10	1.4	0.0	0.6	0.2
Korea, DPR	...	0	...	...	...	...	...	...
Korea, Republic of	14	2	110	22	0.4	0.1	2.8	0.5
Others	226	40	191	319	6.4	1.4	4.8	6.4
NORTH AMERICA	**2 235**	**2 653**	**2 772**	**3 951**	**63.4**	**90.7**	**69.7**	**79.6**
Canada	.	.	.	.	.	.	.	.
United States	2 234	2 652	2 772	3 951	63.4	90.7	69.7	79.6
Others	1	0	0	0	0.0	0.0	0.0	0.0
OTHER AMERICA	**317**	**93**	**127**	**183**	**9.0**	**3.2**	**3.2**	**3.7**
Argentina	29	1	1	1	0.8	0.0	0.0	0.0
Brazil	2	1	0	1	0.1	0.0	0.0	0.0
Mexico	138	75	93	155	3.9	2.6	2.3	3.1
Venezuela	27	1	13	5	0.8	0.0	0.3	0.1
Others	121	15	19	22	3.4	0.5	0.5	0.4
OCEANIA	**7**	**4**	**4**	**5**	**0.2**	**0.1**	**0.1**	**0.1**
Australia	2	1	2	4	0.0	0.0	0.1	0.1
New Zealand	5	4	1	1	0.1	0.1	0.0	0.0
Others	...	...	0	0	...	...	0.0	0.0
EUROPE-TOTAL	**274**	**75**	**339**	**364**	**7.8**	**2.6**	**8.5**	**7.3**
EC	**262**	**58**	**314**	**355**	**7.4**	**2.0**	**7.9**	**7.2**
Belgium-Luxembourg	7	11	10	0	0.2	0.4	0.2	0.0
Denmark	0	0	0	0	0.0	0.0	0.0	0.0
France	16	2	16	5	0.5	0.1	0.4	0.1
Germany	...	...	...	18	...	...	...	0.4
former FR Germany	21	16	...	...	0.6	0.5	...	...
Greece	17	0	...	...	0.5	0.0	...	...
Ireland	4	0	0	0	0.1	0.0	0.0	0.0
Italy	134	21	231	215	3.8	0.7	5.8	4.3
Netherlands	1	0	1	2	0.0	0.0	0.0	0.0
Portugal	12	...	...	...	0.4	...	...	...
Spain	8	1	40	81	0.2	0.0	1.0	1.6
United Kingdom	41	7	17	32	1.2	0.2	0.4	0.7
EFTA	**4**	**1**	**9**	**1**	**0.1**	**0.0**	**0.2**	**0.0**
Austria	0	0	...	0	0.0	0.0	...	0.0
Finland	0	0	1	0	0.0	0.0	0.0	0.0
Iceland	1	...	...	0	0.0	...	...	0.0
Norway	0	0	6	0	0.0	0.0	0.2	0.0
Sweden	1	0	1	0	0.0	0.0	0.0	0.0
Switzerland	2	0	2	0	0.1	0.0	0.0	0.0
EASTERN EUROPE	**0**	**0**	**0**	**0**	**0.0**	**0.0**	**0.0**	**0.0**
Albania	...	...	...	...	...	...	...	...
Bulgaria	...	...	...	...	...	...	...	...
former Czechoslovakia	0	...	...	...	0.0		...	...
former GDR	...	...	...	...	...	...	...	...
Hungary	...	...	...	0	...	...	...	0.0
Poland	0	0	0	0	0.0	0.0	0.0	0.0
Romania	...	...	...	0	...	...	...	0.0
former USSR	**0**	**...**	**16**	**...**	**0.0**	**...**	**0.4**	**...**
OTHER EUROPE	**7**	**17**	**0**	**8**	**0.2**	**0.6**	**0.0**	**0.2**
Turkey	7	17	0	8	0.2	0.6	0.0	0.2
former Yugoslavia	0	...	...	...	0.0	...	...	...
OTHERS	**...**	**0**	**22**	**19**	**...**	**0.0**	**0.6**	**0.4**
WORLD-TOTAL	**3 522**	**2 925**	**3 976**	**4 963**	**100.0**	**100.0**	**100.0**	**100.0**

Annex table 16-16 **Destination and relative importance of steel exports by major exporting regions and countries : *United States***

Destination	Amount (1000t)				Share (%)			
	1980	1985	1990	1992	1980	1985	1990	1992
AFRICA	**91**	**31**	**93**	**85**	**2.4**	**3.5**	**2.2**	**2.1**
South Africa	7	1	14	40	0.2	0.1	0.3	1.0
Others	84	30	79	45	2.2	3.4	1.8	1.1
MIDDLE EAST	**257**	**41**	**85**	**77**	**6.7**	**4.7**	**2.0**	**1.9**
Egypt	136	41	49	8	3.6	4.7	1.1	0.2
Iran	0	...	0	2	0.0	...	0.0	0.1
Others	120	...	36	67	3.1	...	0.8	1.7
FAR EAST	**686**	**164**	**957**	**596**	**17.9**	**18.8**	**22.3**	**14.9**
China	115	...	8	88	3.0	...	0.2	2.2
India	80	36	27	18	2.1	4.1	0.6	0.5
Indonesia	48	...	43	11	1.3	...	1.0	0.3
Japan	27	7	433	120	0.7	0.8	10.1	3.0
Korea, DPR	...	...	...	...	...	...	...	...
Korea, Republic of	33	6	281	119	0.9	0.6	6.5	3.0
Others	382	115	166	239	10.0	13.3	3.9	6.0
NORTH AMERICA	**444**	**296**	**1 833**	**1 314**	**11.6**	**34.1**	**42.7**	**32.9**
Canada	443	296	1 831	1 303	11.5	34.1	42.7	32.6
United States	.	.	.	.	.	.	.	.
Others	1	...	2	12	0.0	...	0.0	0.3
OTHER AMERICA	**1 836**	**220**	**941**	**1 667**	**47.8**	**25.3**	**21.9**	**41.7**
Argentina	28	2	19	17	0.7	0.3	0.4	0.4
Brazil	152	2	14	22	4.0	0.3	0.3	0.6
Mexico	1 189	140	645	1 307	31.0	16.1	15.0	32.7
Venezuela	70	7	30	49	1.8	0.8	0.7	1.2
Others	396	68	233	272	10.3	7.8	5.4	6.8
OCEANIA	**13**	**6**	**17**	**11**	**0.3**	**0.7**	**0.4**	**0.3**
Australia	11	5	10	10	0.3	0.5	0.2	0.2
New Zealand	1	...	3	1	0.0	...	0.1	0.0
Others	1	2	5	1	0.0	0.2	0.1	0.0
EUROPE-TOTAL	**515**	**104**	**334**	**203**	**13.4**	**12.0**	**7.8**	**5.1**
EC	**471**	**83**	**266**	**183**	**12.3**	**9.6**	**6.2**	**4.6**
Belgium-Luxembourg	14	2	12	14	0.4	0.2	0.3	0.3
Denmark	1	...	0	3	0.0	...	0.0	0.1
France	8	2	38	16	0.2	0.2	0.9	0.4
Germany			...	18	0.0	0.0	...	0.5
former FR Germany	21	11	...	...	0.5	1.3	...	...
Greece	12	...	31	1	0.3	...	0.7	0.0
Ireland	1	...	1	2	0.0	...	0.0	0.1
Italy	211	48	114	51	5.5	5.5	2.7	1.3
Netherlands	7	3	12	15	0.2	0.3	0.3	0.4
Portugal	52	...	0	0	1.4	...	0.0	0.0
Spain	6	2	15	5	0.2	0.2	0.4	0.1
United Kingdom	138	17	42	57	3.6	1.9	1.0	1.4
EFTA	**15**	**3**	**11**	**11**	**0.4**	**0.3**	**0.2**	**0.3**
Austria	0	...	0	0	0.0	...	0.0	0.0
Finland	1	...	1	1	0.0	...	0.0	0.0
Iceland	0	...	2	1	0.0	...	0.0	0.0
Norway	1	...	3	1	0.0	...	0.1	0.0
Sweden	9	3	1	1	0.2	0.3	0.0	0.0
Switzerland	3	...	3	7	0.1	...	0.1	0.2
EASTERN EUROPE	**17**	**...**	**1**	**3**	**0.5**	...	**0.0**	**0.1**
Albania	...	...	...	...	...	...	...	...
Bulgaria	...	...	...	0	...	...	...	0.0
former Czechoslovakia	0	...	0	0	0.0	...	0.0	0.0
former GDR	0	...	...	...	0.0	...	...	...
Hungary	0	...	0	0	0.0	...	0.0	0.0
Poland	0	...	0	0	0.0	...	0.0	0.0
Romania	17	...	1	3	0.5	...	0.0	0.1
former USSR	**0**	**...**	**41**	**...**	**0.0**	...	**1.0**	...
OTHER EUROPE	**11**	**18**	**16**	**6**	**0.3**	**2.0**	**0.4**	**0.2**
Turkey	7	18	15	6	0.2	2.0	0.4	0.2
former Yugoslavia	4	...	0	...	0.1	...	0.0	...
OTHERS	**0**	**8**	**32**	**46**	**0.0**	**0.9**	**0.7**	**1.1**
WORLD-TOTAL	**3 842**	**869**	**4 291**	**4 000**	**100.0**	**100.0**	**100.0**	**100.0**

Annex table 16-17 Destination and relative importance of steel exports by major exporting regions and countries : *Japan*

Destination	Amount (1000t)				Share (%)			
	1980	1985	1990	1992	1980	1985	1990	1992
AFRICA	**1 342**	**471**	**302**	**499**	**4.5**	**1.5**	**1.9**	**2.7**
South Africa	101	61	44	65	0.3	0.2	0.3	0.4
Others	1 242	411	258	434	4.2	1.3	1.6	2.4
MIDDLE EAST	**3 996**	**3 410**	**1 104**	**1 528**	**13.5**	**10.8**	**7.0**	**8.3**
Egypt	82	99	36	22	0.3	0.3	0.2	0.1
Iran	768	873	499	535	2.6	2.8	3.2	2.9
Others	3 146	2 439	569	971	10.6	7.7	3.6	5.3
FAR EAST	**12 940**	**17 902**	**9 620**	**12 357**	**43.7**	**56.8**	**60.8**	**66.9**
China	3 198	9 840	1 643	2 341	10.8	31.2	10.4	12.7
India	500	622	244	199	1.7	2.0	1.5	1.1
Indonesia	1 192	745	623	657	4.0	2.4	3.9	3.6
Japan	.	.	.	.	.	.	.	.
Korea, DPR	43	18	5	4	0.1	0.1	0.0	0.0
Korea, Republic of	1 819	1 904	1 473	1 444	6.1	6.0	9.3	7.8
Others	6 187	4 772	5 632	7 713	20.9	15.2	35.6	41.8
NORTH AMERICA	**5 268**	**5 288**	**3 114**	**2 706**	**17.8**	**16.8**	**19.7**	**14.7**
Canada	425	321	187	166	1.4	1.0	1.2	0.9
United States	4 843	4 966	2 927	2 540	16.3	15.8	18.5	13.8
Others	...	...	0	0	...	...	0.0	0.0
OTHER AMERICA	**2 548**	**962**	**437**	**372**	**8.6**	**3.1**	**2.8**	**2.0**
Argentina	306	72	9	17	1.0	0.2	0.1	0.1
Brazil	203	33	36	14	0.7	0.1	0.2	0.1
Mexico	596	167	193	...	2.0	0.5	1.2	...
Venezuela	477	46	25	14	1.6	0.1	0.2	0.1
Others	966	645	174	327	3.3	2.0	1.1	1.8
OCEANIA	**646**	**737**	**300**	**302**	**2.2**	**2.3**	**1.9**	**1.6**
Australia	363	400	225	224	1.2	1.3	1.4	1.2
New Zealand	265	315	48	53	0.9	1.0	0.3	0.3
Others	19	22	28	24	0.1	0.1	0.2	0.1
EUROPE-TOTAL	**2 889**	**2 719**	**898**	**547**	**9.7**	**8.6**	**5.7**	**3.0**
EC	**921**	**383**	**369**	**299**	**3.1**	**1.2**	**2.3**	**1.6**
Belgium-Luxembourg	67	71	48	74	0.2	0.2	0.3	0.4
Denmark	10	9	9	9	0.0	0.0	0.1	0.1
France	31	10	6	8	0.1	0.0	0.0	0.0
Germany	...	...	...	49	...	...	...	0.3
former FR Germany	150	69	...	...	0.5	0.2	...	...
Greece	160	66	1	2	0.5	0.2	0.0	0.0
Ireland	1	0	1	0	0.0	0.0	0.0	0.0
Italy	104	57	7	8	0.4	0.2	0.0	0.0
Netherlands	24	14	9	15	0.1	0.0	0.1	0.1
Portugal	91	5	1	1	0.3	0.0	0.0	0.0
Spain	53	6	8	18	0.2	0.0	0.1	0.1
United Kingdom	229	75	279	115	0.8	0.2	1.8	0.6
EFTA	**147**	**128**	**123**	**200**	**0.5**	**0.4**	**0.8**	**1.1**
Austria	10	7	3	5	0.0	0.0	0.0	0.0
Finland	7	6	2	2	0.0	0.0	0.0	0.0
Iceland	...	...	...	...	...	...	...	...
Norway	39	81	97	169	0.1	0.3	0.6	0.9
Sweden	50	15	5	13	0.2	0.0	0.0	0.1
Switzerland	42	19	16	12	0.1	0.1	0.1	0.1
EASTERN EUROPE	**80**	**25**	**10**	**6**	**0.3**	**0.1**	**0.1**	**0.0**
Albania	1	0	...	...	0.0	0.0	...	...
Bulgaria	9	3	1	0	0.0	0.0	0.0	0.0
former Czechoslovakia	0	1	0	0	0.0	0.0	0.0	0.0
former GDR	7	3	...	...	0.0	0.0	...	...
Hungary	1	3	0	0	0.0	0.0	0.0	0.0
Poland	22	8	5	4	0.1	0.0	0.0	0.0
Romania	41	8	4	1	0.1	0.0	0.0	0.0
former USSR	**1 651**	**2 114**	**355**	**...**	**5.6**	**6.7**	**2.2**	**...**
OTHER EUROPE	**90**	**69**	**42**	**42**	**0.3**	**0.2**	**0.3**	**0.2**
Turkey	67	45	33	42	0.2	0.1	0.2	0.2
former Yugoslavia	23	24	8	...	0.1	0.1	0.1	...
OTHERS	**...**	**1**	**59**	**151**	**...**	**0.0**	**0.4**	**0.8**
WORLD-TOTAL	**29 629**	**31 490**	**15 835**	**18 462**	**100.0**	**100.0**	**100.0**	**100.0**

Annex table 16-18 Destination and relative importance of steel exports by major exporting regions and countries : *Korea, Republic of*

Destination	Amount (1000t)				Share (%)			
	1980	1985	1990	1992	1980	1985	1990	1992
AFRICA	**31**	**18**	**12**	**28**	**0.7**	**0.3**	**0.2**	**0.2**
South Africa	...	...	...	3	...	...	...	0.0
Others	31	18	12	25	0.7	0.3	0.2	0.2
MIDDLE EAST	**915**	**861**	**230**	**563**	**20.4**	**15.2**	**3.2**	**4.8**
Egypt	30	18	2	2	0.7	0.3	0.0	0.0
Iran	...	...	...	127	...	...	...	1.1
Others	885	843	228	433	19.7	14.9	3.1	3.7
FAR EAST	**2 068**	**2 706**	**4 600**	**7 906**	**46.1**	**47.9**	**63.4**	**67.6**
China	...	...	...	2 072	...	...	...	17.7
India	170	97	94	255	3.8	1.7	1.3	2.2
Indonesia	252	94	258	290	5.6	1.7	3.5	2.5
Japan	903	1 623	2 986	2 805	20.1	28.7	41.1	24.0
Korea, DPR	...	...	...	...	...	...	...	...
Korea, Republic of	.	.	.	.	.	.	.	.
Others	743	891	1 263	2 484	16.6	15.8	17.4	21.2
NORTH AMERICA	**947**	**1 614**	**1 452**	**2 645**	**21.1**	**28.6**	**20.0**	**22.6**
Canada	40	82	37	40	0.9	1.5	0.5	0.3
United States	907	1 532	1 415	2 604	20.2	27.1	19.5	22.3
Others	...	...	...	...	...	...	...	...
OTHER AMERICA	**125**	**73**	**82**	**127**	**2.8**	**1.3**	**1.1**	**1.1**
Argentina	2	0	0	0	0.0	0.0	0.0	0.0
Brazil	0	...	0	0	0.0	...	0.0	0.0
Mexico	33	8	22	...	0.7	0.1	0.3	...
Venezuela	0	1	3	0	0.0	0.0	0.0	0.0
Others	90	64	57	126	2.0	1.1	0.8	1.1
OCEANIA	**62**	**106**	**146**	**179**	**1.4**	**1.9**	**2.0**	**1.5**
Australia	61	98	82	146	1.4	1.7	1.1	1.2
New Zealand	1	6	19	19	0.0	0.1	0.3	0.2
Others	0	1	46	14	0.0	0.0	0.6	0.1
EUROPE-TOTAL	**388**	**55**	**182**	**240**	**8.7**	**1.0**	**2.5**	**2.1**
EC	**386**	**55**	**134**	**207**	**8.6**	**1.0**	**1.8**	**1.8**
Belgium-Luxembourg	26	5	34	42	0.6	0.1	0.5	0.4
Denmark	13	...	8	2	0.3	...	0.1	0.0
France	0	0	0	0	0.0	0.0	0.0	0.0
Germany	...	...	...	17	...	...	...	0.1
former FR Germany	1	8	...	...	0.0	0.1	...	...
Greece	102	0	...	0	2.3	0.0	...	0.0
Ireland	0	0	0	0	0.0	0.0	0.0	0.0
Italy	214	...	36	49	4.8	...	0.5	0.4
Netherlands	7	10	16	31	0.2	0.2	0.2	0.3
Portugal	0	0	0	0	0.0	0.0	0.0	0.0
Spain	14	0	13	38	0.3	0.0	0.2	0.3
United Kingdom	7	31	27	28	0.2	0.5	0.4	0.2
EFTA	**1**	**0**	**9**	**6**	**0.0**	**0.0**	**0.1**	**0.1**
Austria	1	0	0	0	0.0	0.0	0.0	0.0
Finland	0	0	0	4	0.0	0.0	0.0	0.0
Iceland	...	...	...	...	...	...	...	...
Norway	0	0	0	0	0.0	0.0	0.0	0.0
Sweden	...	0	3	0	...	0.0	0.0	0.0
Switzerland	0	0	5	2	0.0	0.0	0.1	0.0
EASTERN EUROPE	**0**	**...**	**...**	**0**	**0.0**	**...**	**...**	**0.0**
Albania	...	...	...	0	...	...	...	0.0
Bulgaria	...	...	...	...	...	...	...	...
former Czechoslovakia	...	...	...	...	...	...	...	...
former GDR	...	...	...	...	...	...	...	...
Hungary	...	...	...	...	...	...	...	...
Poland	...	...	...	0	...	...	...	0.0
Romania	0	...	...	...	0.0	...	...	...
former USSR	**...**	**...**	**...**	**...**	**...**	**...**	**...**	**...**
OTHER EUROPE	**...**	**0**	**39**	**27**	**...**	**0.0**	**0.5**	**0.2**
Turkey	...	0	39	27	...	0.0	0.5	0.2
former Yugoslavia	...	...	...	...	...	...	...	...
OTHERS	**- 53**	**217**	**553**	**10**	**...**	**3.8**	**7.6**	**0.1**
WORLD-TOTAL	**4 482**	**5 650**	**7 258**	**11 697**	**100.0**	**100.0**	**100.0**	**100.0**

Annex table 16-19 Destination and relative importance of steel exports by major exporting regions and countries : *Brazil*

Destination	Amount (1000t)				Share (%)			
	1980	1985	1990	1992	1980	1985	1990	1992
AFRICA	**101**	**206**	**335**	**228**	**6.7**	**2.9**	**3.8**	**2.0**
South Africa	4	0	11	12	0.3	0.0	0.1	0.1
Others	97	206	324	216	6.5	2.9	3.6	1.9
MIDDLE EAST	**51**	**552**	**687**	**439**	**3.4**	**7.8**	**7.7**	**3.8**
Egypt	0	20	1	7	0.0	0.3	0.0	0.1
Iran	0	145	614	346	0.0	2.0	6.9	3.0
Others	50	387	72	86	3.4	5.4	0.8	0.7
FAR EAST	**73**	**3 052**	**4 996**	**5 726**	**4.8**	**42.9**	**55.9**	**49.4**
China	53	1 830	178	400	3.5	25.7	2.0	3.5
India	11	119	138	135	0.7	1.7	1.5	1.2
Indonesia	2	27	107	118	0.1	0.4	1.2	1.0
Japan	6	581	687	444	0.4	8.2	7.7	3.8
Korea, DPR	...	...	...	...	...	...	...	...
Korea, Republic of	...	42	855	747	...	0.6	9.6	6.4
Others	1	454	3 031	3 882	0.0	6.4	33.9	33.5
NORTH AMERICA	**571**	**1 480**	**1 528**	**1 488**	**38.0**	**20.8**	**17.1**	**12.8**
Canada	30	103	134	81	2.0	1.5	1.5	0.7
United States	540	1 377	1 394	1 406	35.9	19.4	15.6	12.1
Others	...	...	0	1	...	...	0.0	0.0
OTHER AMERICA	**449**	**491**	**651**	**2 458**	**29.9**	**6.9**	**7.3**	**21.2**
Argentina	168	118	107	811	11.2	1.7	1.2	7.0
Brazil	.	.	.	.	.	.	.	.
Mexico	83	48	67	386	5.5	0.7	0.7	3.3
Venezuela	33	17	77	155	2.2	0.2	0.9	1.3
Others	165	309	400	1 107	11.0	4.3	4.5	9.6
OCEANIA	**5**	**61**	**58**	**74**	**0.3**	**0.9**	**0.6**	**0.6**
Australia	5	58	55	67	0.3	0.8	0.6	0.6
New Zealand	0	4	3	7	0.0	0.1	0.0	0.1
Others	...	...	...	...	...	...	...	...
EUROPE-TOTAL	**296**	**1 268**	**876**	**1 170**	**19.7**	**17.8**	**9.8**	**10.1**
EC	**290**	**791**	**537**	**666**	**19.3**	**11.1**	**6.0**	**5.7**
Belgium-Luxembourg	12	18	21	41	0.8	0.3	0.2	0.4
Denmark	0	0	1	6	0.0	0.0	0.0	0.1
France	1	1	8	8	0.1	0.0	0.1	0.1
Germany	...	...	...	102	...	...	...	0.9
former FR Germany	17	53	...	...	1.1	0.7	...	...
Greece	104	21	86	249	6.9	0.3	1.0	2.2
Ireland	...	...	0	0	...	...	0.0	0.0
Italy	132	482	263	139	8.8	6.8	2.9	1.2
Netherlands	12	7	8	2	0.8	0.1	0.1	0.0
Portugal	1	5	17	0	0.1	0.1	0.2	0.0
Spain	0	186	75	98	0.0	2.6	0.8	0.8
United Kingdom	10	18	56	20	0.6	0.3	0.6	0.2
EFTA	**2**	**9**	**29**	**9**	**0.1**	**0.1**	**0.3**	**0.1**
Austria	0	0	0	0	0.0	0.0	0.0	0.0
Finland	1	2	3	3	0.1	0.0	0.0	0.0
Iceland	...	...	...	...	...	...	...	...
Norway	...	4	7	0	...	0.1	0.1	0.0
Sweden	0	0	10	5	0.0	0.0	0.1	0.0
Switzerland	0	3	9	1	0.0	0.0	0.1	0.0
EASTERN EUROPE	**...**	**2**	**2**	**7**	**...**	**0.0**	**0.0**	**0.1**
Albania	...	...	...	...	...	...	...	...
Bulgaria	...	...	0	...	...	...	0.0	...
former Czechoslovakia	...	0	...	...	...	0.0	...	...
former GDR	...	...	...	...	...	...	...	...
Hungary	...	1	...	7	...	0.0	...	0.1
Poland	...	0	0	0	...	0.0	0.0	0.0
Romania	...	1	2	...	...	0.0	0.0	...
former USSR	**0**	**126**	**6**	**...**	**0.0**	**1.8**	**0.1**	**...**
OTHER EUROPE	**4**	**341**	**303**	**488**	**0.2**	**4.8**	**3.4**	**4.2**
Turkey	4	341	297	488	0.2	4.8	3.3	4.2
former Yugoslavia	...	0	6	...	...	0.0	0.1	...
OTHERS	**- 42**	**- 2**	**- 192**	**...**	**...**	**...**	**...**	**...**
WORLD-TOTAL	**1 502**	**7 109**	**8 940**	**11 584**	**100.0**	**100.0**	**100.0**	**100.0**

Annex table 17-1 Origin and relative importance of steel imports by major importing regions and countries : *Africa*

Origin of imports	Amount (1000t)				Share (%)			
	1980	1985	1990	1992	1980	1985	1990	1992
EC	4 387	3 244	2 361	3 178	73.3	63.9	56.7	71.8
Belgium-Luxembourg	680	268	235	231	11.4	5.3	5.6	5.2
France	819	681	482	359	13.7	13.4	11.6	8.1
Germany	721	722	469	505	12.0	14.2	11.3	11.4
Italy	1 160	460	633	1 048	19.4	9.1	15.2	23.7
Netherlands	149	139	99	95	2.5	2.7	2.4	2.1
Spain	596	715	272	725	10.0	14.1	6.5	16.4
United Kingdom	180	191	111	143	3.0	3.8	2.7	3.2
EFTA	69	64	81	120	1.2	1.3	1.9	2.7
EASTERN EUROPE	...	...	...	398	...	...	...	9.0
former Czechoslovakia	59	6	97	138	1.0	0.1	2.3	3.1
Poland	9	1	2	227	0.2	0.0	0.0	5.1
Romania	...	...	19	9	...	...	0.5	0.2
former USSR	...	...	2	...	...	...	0.0	...
Canada	71	21	44	19	1.2	0.4	1.1	0.4
United States	91	31	93	85	1.5	0.6	2.2	1.9
Japan	1 342	471	302	499	22.4	9.3	7.3	11.3
Korea, Republic of	31	18	12	28	0.5	0.4	0.3	0.6
Brazil	101	206	335	228	1.7	4.1	8.1	5.2
Others	- 105	1 024	931	- 126	-1.8	20.2	22.4	-2.8
WORLD TOTAL	5 987	5 079	4 161	4 429	100.0	100.0	100.0	100.0

Annex table 17-2 Origin and relative importance of steel imports by major importing regions and countries : *Middle East*

Origin of imports	Amount (1000t)				Share (%)			
	1980	1985	1990	1992	1980	1985	1990	1992
EC	4 540	4 826	2 651	3 046	38.6	31.0	21.8	22.9
Belgium-Luxembourg	469	290	372	318	4.0	1.9	3.1	2.4
France	467	486	348	327	4.0	3.1	2.9	2.5
Germany	1 180	1 060	796	852	10.0	6.8	6.5	6.4
Italy	796	719	560	712	6.8	4.6	4.6	5.4
Netherlands	99	158	87	146	0.8	1.0	0.7	1.1
Spain	1 109	1 703	172	187	9.4	10.9	1.4	1.4
United Kingdom	200	270	247	378	1.7	1.7	2.0	2.8
EFTA	147	218	91	129	1.2	1.4	0.7	1.0
EASTERN EUROPE	...	...	...	1 244	...	...	...	9.4
former Czechoslovakia	428	442	247	448	3.6	2.8	2.0	3.4
Poland	188	138	241	456	1.6	0.9	2.0	3.4
Romania	...	...	297	59	...	...	2.4	0.4
former USSR	...	...	52	...	...	...	0.4	...
Canada	72	7	267	42	0.6	0.0	2.2	0.3
United States	257	41	85	77	2.2	0.3	0.7	0.6
Japan	3 996	3 410	1 104	1 528	34.0	21.9	9.1	11.5
Korea, Republic of	915	861	230	563	7.8	5.5	1.9	4.2
Brazil	51	552	687	439	0.4	3.5	5.6	3.3
Others	1 785	5 673	7 009	6 218	15.2	36.4	57.6	46.8
WORLD TOTAL	11 761	15 587	12 176	13 285	100.0	100.0	100.0	100.0

Annex table 17-3 Origin and relative importance of steel imports by major importing regions and countries : *Iran*

Origin of imports	Amount (1000t)				Share (%)			
	1980	1985	1990	1992	1980	1985	1990	1992
EC	1 165	1 009	1 152	965	46.3	22.3	23.5	16.1
Belgium-Luxembourg	80	87	183	106	3.2	1.9	3.7	1.8
France	25	48	92	64	1.0	1.1	1.9	1.1
Germany	504	341	373	374	20.0	7.5	7.6	6.2
Italy	47	89	343	211	1.9	2.0	7.0	3.5
Netherlands	0	32	27	7	0.0	0.7	0.6	0.1
Spain	477	369	56	81	19.0	8.1	1.1	1.4
United Kingdom	4	38	79	112	0.2	0.8	1.6	1.9
EFTA	59	42	34	44	2.3	0.9	0.7	0.7
EASTERN EUROPE	...	...	...	462	...	...	...	7.7
former Czechoslovakia	110	12	40	197	4.4	0.3	0.8	3.3
Poland	119	118	...	224	4.7	2.6	...	3.7
Romania	...	...	89	8	...	...	1.8	0.1
former USSR	...	...	28	...	...	...	0.6	...
Canada	...	0	216	34	...	0.0	4.4	0.6
United States	0	...	0	2	0.0	...	0.0	0.0
Japan	768	873	499	535	30.5	19.3	10.2	8.9
Korea, Republic of	...	...	...	127	...	...	...	2.1
Brazil	0	145	614	346	0.0	3.2	12.5	5.8
Others	523	2 463	2 357	3 484	20.8	54.3	48.1	58.1
WORLD TOTAL	2 515	4 532	4 900	6 000	100.0	100.0	100.0	100.0

Annex table 17-4 Origin and relative importance of steel imports by major importing regions and countries : *Far East*

Origin of imports	Amount (1000t)				Share (%)			
	1980	1985	1990	1992	1980	1985	1990	1992
EC	2 485	6 792	3 950	5 243	10.5	17.1	10.0	9.7
Belgium-Luxembourg	204	489	721	613	0.9	1.2	1.8	1.1
France	405	910	427	670	1.7	2.3	1.1	1.2
Germany	924	1 971	1 140	1 579	3.9	5.0	2.9	2.9
Italy	215	488	414	617	0.9	1.2	1.0	1.1
Netherlands	145	349	102	265	0.6	0.9	0.3	0.5
Spain	251	1 686	381	507	1.1	4.3	1.0	0.9
United Kingdom	338	744	745	953	1.4	1.9	1.9	1.8
EFTA	173	382	359	675	0.7	1.0	0.9	1.2
EASTERN EUROPE	...	...	...	2 927	...	...	...	5.4
former Czechoslovakia	157	314	398	949	0.7	0.8	1.0	1.8
Poland	133	251	86	1 290	0.6	0.6	0.2	2.4
Romania	...	...	424	322	...	...	1.1	0.6
former USSR	...	...	811	...	...	...	2.0	...
Canada	547	72	401	379	2.3	0.2	1.0	0.7
United States	686	164	957	596	2.9	0.4	2.4	1.1
Japan	12 940	17 902	9 620	12 357	54.8	45.2	24.3	22.8
Korea, Republic of	2 068	2 706	4 600	7 906	8.8	6.8	11.6	14.6
Brazil	73	3 052	4 996	5 726	0.3	7.7	12.6	10.6
Others	4 628	8 573	13 891	18 403	19.6	21.6	35.1	33.9
WORLD TOTAL	23 599	39 643	39 585	54 212	100.0	100.0	100.0	100.0

Annex table 17-5 Origin and relative importance of steel imports by major importing regions and countries : *China*

Origin of imports	Amount (1000t)				Share (%)			
	1980	1985	1990	1992	1980	1985	1990	1992
EC	755	3 977	302	1 112	15.1	20.3	8.2	13.8
Belgium-Luxembourg	78	174	33	68	1.6	0.9	0.9	0.8
France	148	523	16	212	3.0	2.7	0.4	2.6
Germany	255	1 123	150	482	5.1	5.7	4.1	6.0
Italy	99	429	31	194	2.0	2.2	0.8	2.4
Netherlands	71	145	1	42	1.4	0.7	0.0	0.5
Spain	72	1 194	48	42	1.4	6.1	1.3	0.5
United Kingdom	33	255	20	52	0.7	1.3	0.5	0.6
EFTA	4	291	7	55	0.1	1.5	0.2	0.7
EASTERN EUROPE	...	...	...	351	...	...	...	4.4
former Czechoslovakia	69	234	101	174	1.4	1.2	2.7	2.2
Poland	91	212	38	107	1.8	1.1	1.0	1.3
Romania	...	...	99	60	...	...	2.7	0.7
former USSR	...	...	424	...	...	...	11.5	...
Canada	158	0	43	13	3.2	0.0	1.2	0.2
United States	115	...	8	88	2.3	...	0.2	1.1
Japan	3 198	9 840	1 643	2 341	63.9	50.1	44.6	29.1
Korea, Republic of	...	...	...	2 072	...	...	...	25.7
Brazil	53	1 830	178	400	1.1	9.3	4.8	5.0
Others	723	3 697	1 078	1 620	14.4	18.8	29.3	20.1
WORLD TOTAL	5 006	19 635	3 683	8 052	100.0	100.0	100.0	100.0

Annex table 17-6 Origin and relative importance of steel imports by major importing regions and countries : *Japan*

Origin of imports	Amount (1000t)				Share (%)			
	1980	1985	1990	1992	1980	1985	1990	1992
EC	38	118	277	185	3.4	4.1	3.9	3.0
Belgium-Luxembourg	...	1	117	14	...	0.0	1.6	0.2
France	6	3	14	5	0.5	0.1	0.2	0.1
Germany	4	4	39	60	0.4	0.1	0.5	1.0
Italy	0	2	2	3	0.0	0.1	0.0	0.0
Netherlands	0	0	40	0	0.0	0.0	0.6	0.0
Spain	27	106	62	74	2.4	3.7	0.9	1.2
United Kingdom	1	1	3	29	0.1	0.0	0.0	0.5
EFTA	90	7	35	118	8.1	0.2	0.5	1.9
EASTERN EUROPE	...	...	...	257	...	...	...	4.2
former Czechoslovakia	...	...	...	57	...	...	...	0.9
Poland	...	...	...	19	...	...	...	0.3
Romania	...	...	168	132	...	...	2.4	2.1
former USSR	...	...	151	...	...	...	2.1	...
Canada	49	1	23	10	4.4	0.0	0.3	0.2
United States	27	7	433	120	2.4	0.2	6.1	2.0
Japan	.	.	.	.	.	.	.	.
Korea, Republic of	903	1 623	2 986	2 805	81.4	55.9	41.9	45.7
Brazil	6	581	687	444	0.6	20.0	9.6	7.2
Others	- 4	566	2 529	2 204	-0.4	19.5	35.5	35.9
WORLD TOTAL	1 109	2 903	7 121	6 143	100.0	100.0	100.0	100.0

Annex table 17-7 Origin and relative importance of steel imports by major importing regions and countries : *Korea, Republic of*

Origin of imports	Amount (1000t)				Share (%)			
	1980	1985	1990	1992	1980	1985	1990	1992
EC	94	119	326	255	4.4	4.5	5.9	4.2
Belgium-Luxembourg	1	65	53	18	0.0	2.5	1.0	0.3
France	54	6	55	30	2.5	0.2	1.0	0.5
Germany	4	27	26	48	0.2	1.0	0.5	0.8
Italy	1	3	4	23	0.0	0.1	0.1	0.4
Netherlands	0	11	1	1	0.0	0.4	0.0	0.0
Spain	...	2	84	59	...	0.1	1.5	1.0
United Kingdom	34	6	103	75	1.6	0.2	1.9	1.2
EFTA	2	17	34	65	0.1	0.6	0.6	1.1
EASTERN EUROPE	...	...	...	61	...	...	...	1.0
former Czechoslovakia	...	...	...	7	...	...	...	0.1
Poland	...	0	6	55	...	0.0	0.1	0.9
Romania	...	...	...	...	...	...	...	...
former USSR	...	...	35	...	...	...	0.6	...
Canada	14	2	110	22	0.7	0.1	2.0	0.4
United States	33	6	281	119	1.5	0.2	5.1	2.0
Japan	1 819	1 904	1 473	1 444	85.1	72.3	26.5	23.8
Korea, Republic of	.	.	.	.	.	.	.	.
Brazil	...	42	855	747	...	1.6	15.4	12.3
Others	175	542	2 449	3 365	8.2	20.6	44.0	55.4
WORLD TOTAL	2 137	2 632	5 563	6 078	100.0	100.0	100.0	100.0

Annex table 17-8 Origin and relative importance of steel imports by major importing regions and countries : *North America*

Origin of imports	Amount (1000t)				Share (%)			
	1980	1985	1990	1992	1980	1985	1990	1992
EC	4 148	7 166	5 963	4 826	28.4	30.3	31.8	27.0
Belgium-Luxembourg	825	773	788	520	5.7	3.3	4.2	2.9
France	883	1 500	1 261	910	6.0	6.3	6.7	5.1
Germany	1 184	2 299	1 614	1 501	8.1	9.7	8.6	8.4
Italy	192	599	493	278	1.3	2.5	2.6	1.6
Netherlands	249	450	478	656	1.7	1.9	2.6	3.7
Spain	474	557	349	200	3.2	2.4	1.9	1.1
United Kingdom	289	712	891	673	2.0	3.0	4.8	3.8
EFTA	165	1 052	501	452	1.1	4.4	2.7	2.5
EASTERN EUROPE	...	...	...	76	...	...	...	0.4
former Czechoslovakia	18	37	25	23	0.1	0.2	0.1	0.1
Poland	22	54	63	28	0.2	0.2	0.3	0.2
Romania	...	...	72	2	...	...	0.4	0.0
former USSR	...	...	...	...	...	...	...	...
Canada	2 235	2 653	2 772	3 951	15.3	11.2	14.8	22.1
United States	444	296	1 833	1 314	3.0	1.3	9.8	7.4
Japan	5 268	5 288	3 114	2 706	36.1	22.3	16.6	15.2
Korea, Republic of	947	1 614	1 452	2 645	6.5	6.8	7.8	14.8
Brazil	571	1 480	1 528	1 488	3.9	6.3	8.2	8.3
Others	823	4 121	1 563	385	5.6	17.4	8.3	2.2
WORLD TOTAL	14 601	23 670	18 726	17 842	100.0	100.0	100.0	100.0

Annex table 17-9 Origin and relative importance of steel imports by major importing regions and countries : *United States*

Origin of imports	Amount (1000t)				Share (%)			
	1980	1985	1990	1992	1980	1985	1990	1992
EC	3 832	5 846	5 318	4 238	28.1	26.9	33.5	27.1
Belgium-Luxembourg	777	667	714	435	5.7	3.1	4.5	2.8
France	823	1 301	1 110	826	6.0	6.0	7.0	5.3
Germany	1 107	2 094	1 512	1 337	8.1	9.6	9.5	8.6
Italy	188	552	376	231	1.4	2.5	2.4	1.5
Netherlands	248	436	437	560	1.8	2.0	2.8	3.6
Spain	427	458	321	190	3.1	2.1	2.0	1.2
United Kingdom	214	526	769	573	1.6	2.4	4.8	3.7
EFTA	141	1 000	464	426	1.0	4.6	2.9	2.7
EASTERN EUROPE	...	...	...	74	...	...	...	0.5
former Czechoslovakia	5	26	13	22	0.0	0.1	0.1	0.1
Poland	19	43	61	28	0.1	0.2	0.4	0.2
Romania	...	...	61	2	...	...	0.4	0.0
former USSR	...	...	...	...	...	...	...	...
Canada	2 234	2 652	2 772	3 951	16.4	12.2	17.5	25.3
United States								
Japan	4 843	4 966	2 927	2 540	35.5	22.9	18.4	16.3
Korea, Republic of	907	1 532	1 415	2 604	6.6	7.1	8.9	16.7
Brazil	540	1 377	1 394	1 406	4.0	6.3	8.8	9.0
Others	1 159	4 354	1 588	373	8.5	20.0	10.0	2.4
WORLD TOTAL	13 656	21 727	15 878	15 613	100.0	100.0	100.0	100.0

Annex table 17-10 Origin and relative importance of steel imports by major importing regions and countries : *Other America*

Origin of imports	Amount (1000t)				Share (%)			
	1980	1985	1990	1992	1980	1985	1990	1992
EC	2 510	1 299	1 204	1 262	30.9	31.7	26.0	17.2
Belgium-Luxembourg	202	116	90	135	2.5	2.8	1.9	1.8
France	658	358	256	168	8.1	8.7	5.5	2.3
Germany	670	415	416	326	8.3	10.1	9.0	4.5
Italy	247	79	72	157	3.0	1.9	1.6	2.1
Netherlands	216	74	68	95	2.7	1.8	1.5	1.3
Spain	398	188	207	255	4.9	4.6	4.5	3.5
United Kingdom	113	68	94	123	1.4	1.7	2.0	1.7
EFTA	84	47	45	56	1.0	1.1	1.0	0.8
EASTERN EUROPE	...	...	...	187	...	...	...	2.6
former Czechoslovakia	30	31	26	46	0.4	0.8	0.6	0.6
Poland	70	26	3	127	0.9	0.6	0.1	1.7
Romania	...	...	3	6	...	...	0.1	0.1
former USSR	...	...	437	...	...	...	9.4	...
Canada	317	93	127	183	3.9	2.3	2.7	2.5
United States	1 836	220	941	1 667	22.6	5.4	20.3	22.8
Japan	2 548	962	437	372	31.4	23.5	9.4	5.1
Korea, Republic of	125	73	82	127	1.5	1.8	1.8	1.7
Brazil	449	491	651	2 458	5.5	12.0	14.1	33.6
Others	249	908	703	1 007	3.1	22.2	15.2	13.8
WORLD TOTAL	8 118	4 093	4 627	7 319	100.0	100.0	100.0	100.0

Annex table 17-11 Origin and relative importance of steel imports by major importing regions and countries : *Oceania*

Origin of imports	Amount (1000t)				Share (%)			
	1980	1985	1990	1992	1980	1985	1990	1992
EC	68	94	133	100	7.1	7.5	14.0	12.0
Belgium-Luxembourg	6	7	11	10	0.6	0.6	1.2	1.2
France	19	19	36	36	2.0	1.5	3.8	4.3
Germany	10	13	24	13	1.0	1.0	2.5	1.6
Italy	5	6	6	6	0.5	0.5	0.6	0.7
Netherlands	1	2	5	2	0.1	0.2	0.5	0.2
Spain	2	2	4	5	0.2	0.2	0.4	0.6
United Kingdom	24	44	47	29	2.5	3.5	4.9	3.5
EFTA	11	15	26	43	1.1	1.2	2.7	5.2
EASTERN EUROPE	...	...	...	2	...	...	...	0.2
former Czechoslovakia	...	...	...	1	...	...	...	0.1
Poland	0	...	0	0	0.0	...	0.0	0.0
Romania	...	...	...	...	...	...	...	...
former USSR	...	...	...	...	...	...	...	...
Canada	7	4	4	5	0.7	0.3	0.4	0.6
United States	13	6	17	11	1.4	0.5	1.8	1.3
Japan	646	737	300	302	67.2	59.0	31.5	36.2
Korea, Republic of	62	106	146	179	6.4	8.4	15.4	21.5
Brazil	5	61	58	74	0.5	4.9	6.1	8.8
Others	150	227	267	118	15.6	18.2	28.0	14.2
WORLD TOTAL	962	1 250	951	834	100.0	100.0	100.0	100.0

Annex table 17-12 Origin and relative importance of steel imports by major importing regions and countries : *Europe*

Origin of imports	Amount (1000t)				Share (%)			
	1980	1985	1990	1992	1980	1985	1990	1992
EC	45 110	46 182	52 239	52 857	63.0	63.3	61.8	68.9
Belgium-Luxembourg	11 251	9 287	12 997	11 873	15.7	12.7	15.4	15.5
France	7 447	6 882	8 696	6 770	10.4	9.4	10.3	8.8
Germany	14 337	13 555	13 725	13 357	20.0	18.6	16.2	17.4
Italy	4 104	5 611	4 291	5 945	5.7	7.7	5.1	7.7
Netherlands	3 755	4 015	4 755	4 601	5.2	5.5	5.6	6.0
Spain	1 693	2 882	2 284	2 810	2.4	4.0	2.7	3.7
United Kingdom	1 615	2 834	4 054	6 141	2.3	3.9	4.8	8.0
EFTA	6 023	7 520	5 231	7,867	8.4	10.3	6.2	10.3
EASTERN EUROPE	...	...	...	6 177	...	...	...	8.1
former Czechoslovakia	2 728	2 998	2 910	3 461	3.8	4.1	3.4	4.5
Poland	1 474	1 661	3 002	1 244	2.1	2.3	3.6	1.6
Romania	...	...	972	369	...	...	1.2	0.5
former USSR	...	...	2 347	...	...	...	2.8	...
Canada	274	75	339	364	0.4	0.1	0.4	0.5
United States	515	104	334	203	0.7	0.1	0.4	0.3
Japan	2 889	2 719	898	547	4.0	3.7	1.1	0.7
Korea, Republic of	388	55	182	240	0.5	0.1	0.2	0.3
Brazil	296	1 268	876	1 170	0.4	1.7	1.0	1.5
Others	16 152	14 984	22 019	7 285	22.5	20.6	26.1	9.5
WORLD TOTAL	71 647	72 908	84 465	76 711	100.0	100.0	100.0	100.0

Annex table 17-13 Origin and relative importance of steel imports by major importing regions and countries : *EC*

Origin of imports	Amount (1000t)				Share (%)			
	1980	1985	1990	1992	1980	1985	1990	1992
EC	33 276	31 942	43 422	45 379	77.4	77.8	72.1	73.4
Belgium-Luxembourg	9 887	7 842	11 840	10 877	23.0	19.1	19.7	17.6
France	5 912	5 390	7 751	6 008	13.7	13.1	12.9	9.7
Germany	8 684	8 211	10 275	10 847	20.2	20.0	17.1	17.5
Italy	2 877	3 229	3 028	4 860	6.7	7.9	5.0	7.9
Netherlands	3 180	3 414	4 211	3 943	7.4	8.3	7.0	6.4
Spain	1 032	1 055	1 975	2 609	2.4	2.6	3.3	4.2
United Kingdom	1 181	2 070	3 312	5 182	2.7	5.0	5.5	8.4
EFTA	3 896	5 086	3 135	6,109	9.1	12.4	5.2	9.9
EASTERN EUROPE	...	...	...	3 500	...	...	...	5.7
former Czechoslovakia	867	834	1 078	1 838	2.0	2.0	1.8	3.0
Poland	328	283	1 393	911	0.8	0.7	2.3	1.5
Romania	...	...	156	122	...	...	0.3	0.2
former USSR	...	...	1 238	...	...	...	2.1	...
Canada	262	58	314	355	0.6	0.1	0.5	0.6
United States	471	83	266	183	1.1	0.2	0.4	0.3
Japan	921	383	369	299	2.1	0.9	0.6	0.5
Korea, Republic of	386	55	134	207	0.9	0.1	0.2	0.3
Brazil	290	791	537	666	0.7	1.9	0.9	1.1
Others	3 512	2 640	10 818	5 150	8.2	6.4	18.0	8.3
WORLD TOTAL	43 014	41 038	60 232	61 848	100.0	100.0	100.0	100.0

Annex table 17-14 Origin and relative importance of steel imports by major importing regions and countries : *Belgium-Luxembourg*

Origin of imports	Amount (1000t)				Share (%)			
	1980	1985	1990	1992	1980	1985	1990	1992
EC	2 715	3 083	4 536	4 688	91.7	90.7	98.1	99.8
Belgium-Luxembourg	.	.	.	.	.	.	.	.
France	869	877	1 511	1 584	29.3	25.8	32.7	33.7
Germany	811	1 020	1 236	1 380	27.4	30.0	26.7	29.4
Italy	58	143	190	173	2.0	4.2	4.1	3.7
Netherlands	817	1 706	1 021	928	27.6	20.8	22.1	19.8
Spain	78	100	83	108	2.6	2.9	1.8	2.3
United Kingdom	67	177	444	441	2.3	5.2	9.6	9.4
EFTA	92	173	135	127	3.1	5.1	2.9	2.7
EASTERN EUROPE	...	...	...	169	...	...	...	3.6
former Czechoslovakia	21	23	24	54	0.7	0.7	0.5	1.1
Poland	29	14	51	66	1.0	0.4	1.1	1.4
Romania	...	...	11	3	...	...	0.2	0.1
former USSR	...	...	6	...	...	...	0.1	...
Canada	7	11	10	0	0.2	0.3	0.2	0.0
United States	14	2	12	14	0.5	0.1	0.3	0.3
Japan	67	71	48	74	2.3	2.1	1.0	1.6
Korea, Republic of	26	5	34	42	0.9	0.1	0.7	0.9
Brazil	12	18	21	41	0.4	0.5	0.5	0.9
Others	29	36	- 178	- 457	1.0	1.0	-3.8	-9.7
WORLD TOTAL	2 962	3 398	4 624	4 698	100.0	100.0	100.0	100.0

Annex table 17-15 Origin and relative importance of steel imports by major importing regions and countries : *France*

Origin of imports	Amount (1000t)				Share (%)			
	1980	1985	1990	1992	1980	1985	1990	1992
EC	7 881	6 647	9 689	9 273	103.5	101.0	92.1	92.1
Belgium-Luxembourg	3 567	2 796	4 199	3 931	46.9	42.5	39.9	39.1
France	.	.	.	.	.	.	.	.
Germany	2 521	1 955	2 297	2 421	33.1	29.7	21.8	24.1
Italy	1 145	1 183	1 576	1 395	15.0	18.0	15.0	13.9
Netherlands	216	219	355	316	2.8	3.3	3.4	3.1
Spain	281	224	699	539	3.7	3.4	6.6	5.4
United Kingdom	129	231	496	616	1.7	3.5	4.7	6.1
EFTA	248	366	420	454	3.3	5.6	4.0	4.5
EASTERN EUROPE	...	...	...	259	...	...	...	2.6
former Czechoslovakia	107	118	239	235	1.4	1.8	2.3	2.3
Poland	4	15	25	13	0.1	0.2	0.2	0.1
Romania	...	...	49	8	...	...	0.5	0.1
former USSR	...	...	7	...	...	...	0.1	...
Canada	16	2	16	5	0.2	0.0	0.2	0.0
United States	8	2	38	16	0.1	0.0	0.4	0.2
Japan	31	10	6	8	0.4	0.2	0.1	0.1
Korea, Republic of	0	0	0	0	0.0	0.0	0.0	0.0
Brazil	1	1	8	8	0.0	0.0	0.1	0.1
Others	- 573	- 449	333	41	-7.5	-6.8	3.2	0.4
WORLD TOTAL	7 613	6 579	10 517	10 065	100.0	100.0	100.0	100.0

Annex table 17-16 Origin and relative importance of steel imports by major importing regions and countries : *Germany*

Origin of imports	Amount (1000t)				Share (%)			
	1980	1985	1990	1992	1980	1985	1990	1992
EC	7 859	7 477	7 375	8 748	68.6	64.7	49.6	50.3
Belgium-Luxembourg	3 135	2 447	3 710	3 405	27.4	21.2	25.0	19.6
France	1 633	1 508	2 134	...	14.3	13.0	14.4	...
Germany	.	.	.	.	.	.	.	.
Italy	1 120	1 264	...	1 910	9.8	10.9	...	11.0
Netherlands	1 075	1 061	1 349	1 355	9.4	9.2	9.1	7.8
Spain	357	341	...	505	3.1	2.9	...	2.9
United Kingdom	318	587	...	1 468	2.8	5.1	...	8.4
EFTA	1 697	2 352	...	2,668	14.8	20.3	...	15.3
EASTERN EUROPE	...	...	...	1 882	...	...	...	10.8
former Czechoslovakia	390	404	444	1 030	3.4	3.5	3.0	5.9
Poland	193	169	1 039	591	1.7	1.5	7.0	3.4
Romania	...	...	...	17	...	...	...	0.1
former USSR	...	...	409	...	...	...	2.8	...
Canada	21	16	...	18	0.2	0.1	...	0.1
United States	21	11	...	18	0.2	0.1	...	0.1
Japan	150	69	...	49	1.3	0.6	...	0.3
Korea, Republic of	1	8	...	17	0.0	0.1	...	0.1
Brazil	17	53	...	102	0.1	0.5	...	0.6
Others	1 688	1 576	7 072	3 889	14.7	13.6	47.6	22.4
WORLD TOTAL	11 454	11 562	14 856	17 391	100.0	100.0	100.0	100.0

Annex table 17-17 Origin and relative importance of steel imports by major importing regions and countries : *Netherlands*

Origin of imports	Amount (1000t)				Share (%)			
	1980	1985	1990	1992	1980	1985	1990	1992
EC	3 142	3 161	4 595	4 331	81.0	78.0	88.7	88.4
Belgium-Luxembourg	1 058	847	1 259	1 259	27.3	20.9	24.3	25.7
France	325	292	498	379	8.4	7.2	9.6	7.7
Germany	1 528	1 621	2 102	2 001	39.4	40.0	40.6	40.8
Italy	70	140	252	153	1.8	3.5	4.9	3.1
Netherlands	.	.	.	.	.	.	.	.
Spain	32	40	46	48	0.8	1.0	0.9	1.0
United Kingdom	112	168	371	392	2.9	4.1	7.2	8.0
EFTA	342	362	274	354	8.8	8.9	5.3	7.2
EASTERN EUROPE	...	...	...	113	...	...	...	2.3
former Czechoslovakia	25	30	44	40	0.6	0.7	0.8	0.8
Poland	9	16	53	56	0.2	0.4	1.0	1.1
Romania	...	...	...	0	...	...	...	0.0
former USSR	...	...	7	...	...	...	0.1	...
Canada	1	0	1	2	0.0	0.0	0.0	0.0
United States	7	3	12	15	0.2	0.1	0.2	0.3
Japan	24	14	9	15	0.6	0.3	0.2	0.3
Korea, Republic of	7	10	16	31	0.2	0.3	0.3	0.6
Brazil	12	7	8	2	0.3	0.2	0.1	0.0
Others	343	493	259	37	8.8	12.2	5.0	0.8
WORLD TOTAL	3 878	4 050	5 180	4 900	100.0	100.0	100.0	100.0

Annex table 17-18 Origin and relative importance of steel imports by major importing regions and countries : *United Kingdom*

Origin of imports	Amount (1000t)				Share (%)			
	1980	1985	1990	1992	1980	1985	1990	1992
EC	3 249	2 932	4 087	4 310	69.0	75.9	78.2	82.7
Belgium-Luxembourg	597	435	654	585	12.7	11.3	12.5	11.2
France	479	373	564	808	10.2	9.7	10.8	15.5
Germany	1 125	1 060	1 285	1 473	23.9	27.5	24.6	28.3
Italy	250	268	277	306	5.3	6.9	5.3	5.9
Netherlands	548	456	695	528	11.6	11.8	13.3	10.1
Spain	104	129	382	405	2.2	3.3	7.3	7.8
United Kingdom	.	.	.	.	.	.	.	.
EFTA	489	519	694	570	10.4	13.4	13.3	10.9
EASTERN EUROPE	...	...	...	190	...	...	...	3.6
former Czechoslovakia	16	25	38	72	0.3	0.6	0.7	1.4
Poland	62	52	156	68	1.3	1.3	3.0	1.3
Romania	...	...	10	6	...	...	0.2	0.1
former USSR	...	...	...	...	...	...	...	...
Canada	41	7	17	32	0.9	0.2	0.3	0.6
United States	138	17	42	57	2.9	0.4	0.8	1.1
Japan	229	75	279	115	4.9	1.9	5.3	2.2
Korea, Republic of	7	31	27	28	0.2	0.8	0.5	0.5
Brazil	10	18	56	20	0.2	0.5	1.1	0.4
Others	543	262	26	- 109	11.5	6.8	0.5	-2.1
WORLD TOTAL	4 706	3 861	5 228	5 213	100.0	100.0	100.0	100.0

Annex table 17-19 Origin and relative importance of steel imports by major importing regions and countries : *EFTA*

Origin of imports	Amount (1000t)				Share (%)			
	1980	1985	1990	1992	1980	1985	1990	1992
EC	5 013	4 873	5 954	6 034	73.1	72.7	73.5	79.7
Belgium-Luxembourg	691	628	785	852	10.1	9.4	9.7	11.3
France	653	617	659	626	9.5	9.2	8.1	8.3
Germany	2 183	1 971	2 230	2 083	31.8	29.4	27.5	27.5
Italy	344	431	780	725	5.0	6.4	9.6	9.6
Netherlands	411	394	485	569	6.0	5.9	6.0	7.5
Spain	198	139	121	106	2.9	2.1	1.5	1.4
United Kingdom	279	453	632	824	4.1	6.8	7.8	10.9
EFTA	1 186	1 100	1 500	1,555	17.3	16.4	18.5	20.5
EASTERN EUROPE	...	...	...	831	...	...	...	11.0
former Czechoslovakia	182	265	291	489	2.7	4.0	3.6	6.5
Poland	285	78	906	218	4.2	1.2	11.2	2.9
Romania	...	...	52	32	...	...	0.6	0.4
former USSR	...	...	103	...	...	...	1.3	...
Canada	4	1	9	1	0.1	0.0	0.1	0.0
United States	15	3	11	11	0.2	0.0	0.1	0.1
Japan	147	128	123	200	2.1	1.9	1.5	2.6
Korea, Republic of	1	0	9	6	0.0	0.0	0.1	0.1
Brazil	2	9	29	9	0.0	0.1	0.4	0.1
Others	488	585	362	-1 080	7.1	8.7	4.5	-14.3
WORLD TOTAL	6 856	6 699	8 099	7 567	100.0	100.0	100.0	100.0

Annex table 17-20 Origin and relative importance of steel imports by major importing regions and countries : *Eastern Europe*

Origin of imports	Amount (1000t)				Share (%)			
	1980	1985	1990	1992	1980	1985	1990	1992
EC	1 195	716	400	498	11.9	7.4	7.2	26.1
Belgium-Luxembourg	147	18	30	11	1.5	0.2	0.5	0.6
France	168	73	26	23	1.7	0.8	0.5	1.2
Germany	549	294	228	280	5.5	3.0	4.1	14.7
Italy	146	154	63	92	1.5	1.6	1.1	4.8
Netherlands	18	7	7	5	0.2	0.1	0.1	0.3
Spain	91	141	19	35	0.9	1.5	0.3	1.8
United Kingdom	16	15	8	27	0.2	0.2	0.1	1.4
EFTA	177	300	69	154	1.8	3.1	1.2	8.1
EASTERN EUROPE	...	...	...	724	...	...	...	37.9
former Czechoslovakia	730	894	324	581	7.3	9.2	5.8	30.4
Poland	467	409	70	77	4.6	4.2	1.3	4.0
Romania	...	...	72	16	...	...	1.3	0.8
former USSR	...	...	924	...	...	...	16.6	...
Canada	0	0	0	0	0.0	0.0	0.0	0.0
United States	17	...	1	3	0.2	...	0.0	0.2
Japan	80	25	10	6	0.8	0.3	0.2	0.3
Korea, Republic of	0	...	...	0	0.0	...	...	0.0
Brazil	...	2	2	7	...	0.0	0.0	0.4
Others	8 581	8 670	4 155	518	85.4	89.3	74.7	27.1
WORLD TOTAL	10 050	9 713	5 561	1 910	100.0	100.0	100.0	100.0

Annex table 17-21 Origin and relative importance of steel imports by major importing regions and countries : *former USSR*

Origin of imports	Amount (1000t)				Share (%)			
	1980	1985	1990	1992	1980	1985	1990	1992
EC	4 638	6 537	1 324	8	51.2	61.0	18.8	0.4
Belgium-Luxembourg	444	632	194	...	4.9	5.9	2.7	...
France	631	573	121	...	7.0	5.3	1.7	...
Germany	2 552	2 804	630	...	28.2	26.2	8.9	...
Italy	579	1 354	148	...	6.4	12.6	2.1	...
Netherlands	117	28	3	1	1.3	0.3	0.0	0.1
Spain	221	958	115	...	2.4	8.9	1.6	...
United Kingdom	92	162	20	...	1.0	1.5	0.3	...
EFTA	561	825	352	...	6.2	7.7	5.0	...
EASTERN EUROPE	...	...	...	509	...	...	...	25.5
former Czechoslovakia	487	518	585	375	5.4	4.8	8.3	18.8
Poland	209	717	559	15	2.3	6.7	7.9	0.8
Romania	...	...	477	56	...	...	6.8	2.8
former USSR	.	.	.	.	.	.	.	
Canada	0	...	16	...	0.0	...	0.2	...
United States	0	...	41	...	0.0	...	0.6	...
Japan	1 651	2 114	355	...	18.2	19.7	5.0	...
Korea, Republic of	...	...	...	...	...	...	...	...
Brazil	0	126	6	...	0.0	1.2	0.1	...
Others	2 214	1 111	4 966	1 483	24.4	10.4	70.3	74.2
WORLD TOTAL	9 064	10 713	7 060	2 000	100.0	100.0	100.0	100.0

Annex table 17-22 Origin and relative importance of steel imports by major importing regions and countries : *Other Europe*

Origin of imports	Amount (1000t)				Share (%)			
	1980	1985	1990	1992	1980	1985	1990	1992
EC	989	2 113	1 138	938	37.1	44.5	32.4	27.7
Belgium-Luxembourg	82	167	148	133	3.1	3.5	4.2	3.9
France	83	229	139	114	3.1	4.8	4.0	3.4
Germany	369	275	362	147	13.9	5.8	10.3	4.3
Italy	158	443	272	268	5.9	9.3	7.7	7.9
Netherlands	30	172	49	82	1.1	3.6	1.4	2.4
Spain	151	588	55	60	5.7	12.4	1.6	1.8
United Kingdom	48	134	81	108	1.8	2.8	2.3	3.2
EFTA	202	209	176	49	7.6	4.4	5.0	1.4
EASTERN EUROPE	...	...	...	612	...	...	...	18.1
former Czechoslovakia	463	487	632	177	17.4	10.3	18.0	5.2
Poland	185	173	74	23	6.9	3.6	2.1	0.7
Romania	...	...	215	144	...	...	6.1	4.3
former USSR	...	...	82	...	...	...	2.3	...
Canada	7	17	0	8	0.3	0.4	0.0	0.2
United States	11	18	16	6	0.4	0.4	0.5	0.2
Japan	90	69	42	42	3.4	1.5	1.2	1.2
Korea, Republic of	...	0	39	27	...	0.0	1.1	0.8
Brazil	4	341	303	488	0.1	7.2	8.6	14.4
Others	1 360	1 978	1 717	1 216	51.1	41.7	48.9	35.9
WORLD TOTAL	2 663	4 745	3 513	3 387	100.0	100.0	100.0	100.0

Annex table 18-1 Steel exports by products : EC

Products	Amount (1000t)				Share (%)			
	1980	1985	1990	1992	1980	1985	1990	1992
Ingots and semis	4 358	7 110	6 035	5 901	6.9	10.0	8.4	7.9
Long products	20 464	21 276	20 186	21 762	32.3	29.9	28.3	29.2
Flat products	30 840	32 901	38 117	39 452	48.7	46.3	53.3	53.0
Tubes and fittings	7 665	9 825	7 110	7 262	12.1	13.8	10.0	9.8
Total	63 327	71 112	71 448	74 447	100.0	100.0	100.0	100.0

Annex table 18-2 Steel exports by products : Belgium-Luxembourg

Products	Amount (1000t)				Share (%)			
	1980	1985	1990	1992	1980	1985	1990	1992
Ingots and semis	704	589	835	487	5.2	4.7	5.5	3.6
Long products	4 623	3 541	3 931	3 698	33.9	28.1	26.1	27.0
Flat products	7 966	7 866	9 836	9 018	58.4	62.4	65.2	65.8
Tubes and fittings	359	615	479	503	2.6	4.9	3.2	3.7
Total	13 652	12 611	15 081	13 706	100.0	100.0	100.0	100.0

Annex table 18-3 Steel exports by products : France

Products	Amount (1000t)				Share (%)			
	1980	1985	1990	1992	1980	1985	1990	1992
Ingots and semis	360	976	1 072	1 000	3.4	9.0	9.4	8.5
Long products	3 075	3 017	2 877	3 013	28.7	27.8	25.1	25.5
Flat products	6 087	5 680	6 451	6 789	56.9	52.4	56.4	57.6
Tubes and fittings	1 185	1 169	1 044	995	11.1	10.8	9.1	8.4
Total	10 707	10 842	11 444	11 797	100.0	100.0	100.0	100.0

Annex table 18-4 Steel exports by products : Germany

Products	Amount (1000t)				Share (%)			
	1980	1985	1990	1992	1980	1985	1990	1992
Ingots and semis	1 487	1 950	1 043	1 448	7.8	9.7	5.8	7.7
Long products	4 317	3 922	3 639	4 132	22.7	19.6	20.1	21.9
Flat products	10 199	10 037	10 899	11 104	53.6	50.1	60.2	59.0
Tubes and fittings	3 031	4 136	2 513	2 152	15.9	20.6	13.9	11.4
Total	19 034	20 045	18 094	18 836	100.0	100.0	100.0	100.0

Annex table 18-5 Steel exports by products : Italy

Products	Amount (1000t)				Share (%)			
	1980	1985	1990	1992	1980	1985	1990	1992
Ingots and semis	189	350	485	379	2.8	4.4	5.9	4.2
Long products	3 151	2 757	3 669	3 997	46.7	34.6	44.7	44.6
Flat products	1 800	2 774	2 609	2 871	26.7	34.8	31.8	32.0
Tubes and fittings	1 606	2 090	1 450	1 645	23.8	26.2	17.7	18.4
Total	6 746	7 971	8 213	8 962	100.0	100.0	100.0	100.0

Annex table 18-6 Steel exports by products : *Netherlands*

Products	Amount (1000t)				Share (%)			
	1980	1985	1990	1992	1980	1985	1990	1992
Ingots and semis	878	1 284	1 153	1 068	19.0	24.7	20.6	17.8
Long products	650	617	703	834	14.1	11.9	12.6	13.9
Flat products	2 600	2 799	3 204	3 367	56.3	53.9	57.3	56.3
Tubes and fittings	488	489	533	715	10.6	9.4	9.5	11.9
Total	4 616	5 189	5 593	5 984	100.0	100.0	100.0	100.0

Annex table 18-7 Steel exports by products : *Spain*

Products	Amount (1000t)				Share (%)			
	1980	1985	1990	1992	1980	1985	1990	1992
Ingots and semis	545	1 189	81	257	12.0	15.3	1.9	5.4
Long products	2 846	4 681	1 951	2 245	62.8	60.1	46.7	47.1
Flat products	675	1 308	1 827	1 968	14.9	16.8	43.7	41.3
Tubes and fittings	467	614	320	293	10.3	7.9	7.7	6.2
Total	4 533	7 792	4 179	4 763	100.0	100.0	100.0	100.0

Annex table 18-8 Steel exports by products : *United Kingdom*

Products	Amount (1000t)				Share (%)			
	1980	1985	1990	1992	1980	1985	1990	1992
Ingots and semis	185	742	1 310	1 221	6.7	15.1	18.4	14.5
Long products	1 429	1 994	2 870	3 280	51.7	40.7	40.3	38.9
Flat products	793	1 660	2 436	3 187	28.7	33.9	34.2	37.8
Tubes and fittings	359	507	500	751	13.0	10.3	7.0	8.9
Total	2 766	4 903	7 116	8 439	100.0	100.0	100.0	100.0

Annex table 18-9 Steel exports by products : *EFTA*

Products	Amount (1000t)				Share (%)			
	1980	1985	1990	1992	1980	1985	1990	1992
Ingots and semis	776	1 043	396	527	11.6	11.2	4.6	5.5
Long products	2 041	2 694	2 471	2 564	30.6	28.9	28.7	26.6
Flat products	3 170	4 475	4 629	5 595	47.5	48.0	53.8	58.1
Tubes and fittings	690	1 111	1 095	940	10.3	11.9	12.7	9.8
Total	6 677	9 323	8 601	9 626	100.0	100.0	100.0	100.0

Annex table 18-10 Steel exports by products : *EASTERN EUROPE*

Products	Amount (1000t)				Share (%)			
	1980	1985	1990	1992	1980	1985	1990	1992
Ingots and semis	1 066	1 225	1 970	2 443	9.1	8.4	13.7	21.9
Long products	5 311	7 229	6 026	3 763	45.2	49.6	42.0	33.7
Flat products	4 089	4 752	5 209	4 175	34.8	32.6	36.3	37.4
Tubes and fittings	1 276	1 378	1 132	786	10.9	9.4	7.9	7.0
Total	11 742	14 584	14 337	11 177	100.0	100.0	100.0	100.0

Annex table 18-11 Steel exports by products : *former Czechoslovakia*

Products	Amount (1000t)				Share (%)			
	1980	1985	1990	1992	1980	1985	1990	1992
Ingots and semis	341	570	311	628	10.0	14.7	8.4	12.2
Long products	1 369	1 551	1 507	1 675	40.0	40.0	40.6	32.5
Flat products	1 191	1 165	1 354	2 351	34.8	30.0	36.4	45.6
Tubes and fittings	525	594	543	489	15.3	15.3	14.6	9.5
Total	3 426	3 880	3 715	5 153	100.0	100.0	100.0	100.0

Annex table 18-12 Steel exports by products : *Poland*

Products	Amount (1000t)				Share (%)			
	1980	1985	1990	1992	1980	1985	1990	1992
Ingots and semis	87	82	1 423	1 642	4.5	3.8	40.0	48.2
Long products	1 240	1 526	1 753	1 341	64.2	71.5	49.2	39.3
Flat products	555	491	346	318	28.8	23.0	9.7	9.3
Tubes and fittings	48	34	38	107	2.5	1.6	1.1	3.1
Total	1 930	2 133	3 560	3 408	100.0	100.0	100.0	100.0

Annex table 18-13 Steel exports by products : *Romania*

Products	Amount (1000t)				Share (%)			
	1980	1985	1990	1992	1980	1985	1990	1992
Ingots and semis	34	55	0	10	1.7	1.9	0.0	1.3
Long products	810	1 595	735	304	40.8	54.4	39.0	38.7
Flat products	707	872	918	362	35.6	29.7	48.7	46.1
Tubes and fittings	433	410	231	110	21.8	14.0	12.2	14.0
Total	1 984	2 932	1 884	786	100.0	100.0	100.0	100.0

Annex table 18-14 Steel exports by products : *former USSR*

Products	Amount (1000t)				Share (%)			
	1980	1985	1990	1992	1980	1985	1990	1992
Ingots and semis	...	...	3 852	...	...	...	78.1	...
Long products	...	...	706	...	...	...	14.3	...
Flat products	...	...	117	...	...	...	2.4	...
Tubes and fittings	...	...	256	...	...	...	5.2	...
Total	...	8 738	4 932	4 052	100.0	100.0	100.0	100.0

Annex table 18-15 Steel exports by products : *Canada*

Products	Amount (1000t)				Share (%)			
	1980	1985	1990	1992	1980	1985	1990	1992
Ingots and semis	327	77	327	269	9.3	2.6	8.2	5.4
Long products	1 492	924	1 311	1 588	42.4	31.6	33.0	32.0
Flat products	1 314	1 483	1 941	2 664	37.3	50.7	48.8	53.7
Tubes and fittings	389	441	397	442	11.0	15.1	10.0	8.9
Total	3 522	2 925	3 976	4 963	100.0	100.0	100.0	100.0

Annex table 18-16 Steel exports by products : *United States*

Products	Amount (1000t)				Share (%)			
	1980	1985	1990	1992	1980	1985	1990	1992
Ingots and semis	830	81	451	384	21.6	9.3	10.5	9.6
Long products	953	165	1 258	976	24.8	19.0	29.3	24.4
Flat products	1 571	416	2 120	1 984	40.9	47.9	49.4	49.6
Tubes and fittings	488	207	462	656	12.7	23.8	10.8	16.4
Total	3 842	869	4 291	4 000	100.0	100.0	100.0	100.0

Annex table 18-17 Steel exports by products : *Japan*

Products	Amount (1000t)				Share (%)			
	1980	1985	1990	1992	1980	1985	1990	1992
Ingots and semis	189	324	43	325	0.6	1.0	0.3	1.8
Long products	8 032	8 524	2 469	2 944	27.1	27.1	15.6	15.9
Flat products	14 957	15 993	10 141	12 212	50.5	50.8	64.0	66.1
Tubes and fittings	6 453	6 649	3 182	2 981	21.8	21.1	20.1	16.1
Total	29 631	31 490	15 835	18 462	100.0	100.0	100.0	100.0

Annex table 18-18 Steel exports by products : *Korea, Republic of*

Products	Amount (1000t)				Share (%)			
	1980	1985	1990	1992	1980	1985	1990	1992
Ingots and semis 1/	1 209	1 517	331	470	25.9	26.7	4.6	4.0
Long products	1 108	1 704	953	1 893	23.8	30.0	13.1	16.2
Flat products	1 551	1 478	5 123	7 516	33.3	26.0	70.6	64.3
Tubes and fittings	792	977	851	1 817	17.0	17.2	11.7	15.5
Total	4 660	5 676	7 258	11 696	100.0	100.0	100.0	100.0

1/ "Ingots and semis" for 1980 and 1985 include hot-rolled coil for rerolling (hot band), which is normally included under "flat products".

Annex table 18-19 Steel exports by products : *Brazil*

Products	Amount (1000t)				Share (%)			
	1980	1985	1990	1992	1980	1985	1990	1992
Ingots and semis 1/	285	2 493	3 522	4 640	18.5	35.1	39.4	40.1
Long products	232	2 045	1 880	2 113	15.0	28.8	21.0	18.2
Flat products	771	2 239	3 225	4 648	49.9	31.5	36.1	40.1
Tubes and fittings	256	335	313	183	16.6	4.7	3.5	1.6
Total	1 544	7 112	8 940	11 584	100.0	100.0	100.0	100.0

1/ "Ingots and semis" for 1980 and 1985 include hot-rolled coil for rerolling (hot band), which is normally included under "flat products".

Annex table 19-1 Steel imports by products : *EC*

Products	Amount (1000t)				Share (%)			
	1980	1985	1990	1992	1980	1985	1990	1992
Ingots and semis	5 041	11 776	5 976	5 518	11.7	28.7	9.9	8.9
Long products	12 427	11 535	17 709	17 368	28.9	28.1	29.4	28.1
Flat products	22 731	14 671	31 158	33 687	52.8	35.8	51.7	54.5
Tubes and fittings	2 816	3 055	5 391	5 274	6.5	7.4	9.0	8.5
Total	43 015	41 037	60 234	61 847	100.0	100.0	100.0	100.0

Annex table 19-2 Steel imports by products : *Belgium-Luxembourg*

Products	Amount (1000t)				Share (%)			
	1980	1985	1990	1992	1980	1985	1990	1992
Ingots and semis	628	520	909	1 020	21.2	15.3	19.7	21.7
Long products	991	989	1 276	1 350	33.5	29.1	27.6	28.7
Flat products	1 100	1 580	2 109	1 998	37.1	46.5	45.6	42.5
Tubes and fittings	243	310	330	330	8.2	9.1	7.1	7.0
Total	2 962	3 399	4 624	4 698	100.0	100.0	100.0	100.0

Annex table 19-3 Steel imports by products : *France*

Products	Amount (1000t)				Share (%)			
	1980	1985	1990	1992	1980	1985	1990	1992
Ingots and semis	980	630	477	357	12.9	9.6	4.5	3.5
Long products	2 260	1 865	3 341	2 972	29.7	28.4	31.8	29.5
Flat products	4 373	4 083	5 854	5 927	57.4	62.1	55.7	58.9
Tubes and fittings	...	...	845	808	...	...	8.0	8.0
Total	7 613	6 578	10 517	10 064	100.0	100.0	100.0	100.0

Annex table 19-4 Steel imports by products : *Germany*

Products	Amount (1000t)				Share (%)			
	1980	1985	1990	1992	1980	1985	1990	1992
Ingots and semis	1 943	1 964	1 175	1 205	17.0	17.0	7.9	6.9
Long products	4 378	4 270	5 604	6 152	38.2	36.9	37.7	35.4
Flat products	4 387	4 426	6 723	8 389	38.3	38.3	45.3	48.2
Tubes and fittings	748	902	1 355	1 646	6.5	7.8	9.1	9.5
Total	11 456	11 562	14 857	17 392	100.0	100.0	100.0	100.0

Annex table 19-5 Steel imports by products : *Italy*

Products	Amount (1000t)				Share (%)			
	1980	1985	1990	1992	1980	1985	1990	1992
Ingots and semis	1 063	1 090	1 788	1 298	14.9	17.6	16.5	12.4
Long products	1 186	1 141	1 992	1 659	16.7	18.4	18.4	15.8
Flat products	4 578	3 677	6 433	6 845	64.3	59.3	59.5	65.4
Tubes and fittings	289	288	601	668	4.1	4.6	5.6	6.4
Total	7 116	6 196	10 814	10 470	100.0	100.0	100.0	100.0

Annex table 19-6 Steel imports by products : *Netherlands*

Products	Amount (1000t)				Share (%)			
	1980	1985	1990	1992	1980	1985	1990	1992
Ingots and semis	182	158	78	42	4.7	3.9	1.5	0.9
Long products	1 537	1 462	1 892	1 893	39.6	36.1	36.5	38.6
Flat products	1 426	1 654	2 155	2 246	36.8	40.8	41.6	45.8
Tubes and fittings	732	776	1 055	719	18.9	19.2	20.4	14.7
Total	3 877	4 050	5 180	4 900	100.0	100.0	100.0	100.0

Annex table 19-7 Steel imports by products : *United Kingdom*

Products	Amount (1000t)				Share (%)			
	1980	1985	1990	1992	1980	1985	1990	1992
Ingots and semis	351	253	325	222	7.5	6.6	6.2	4.3
Long products	996	857	1 198	1 152	21.2	22.2	22.9	22.1
Flat products	2 896	2 397	3 093	3 255	61.5	62.1	59.2	62.4
Tubes and fittings	464	354	612	584	9.9	9.2	11.7	11.2
Total	4 707	3 861	5 228	5 213	100.0	100.0	100.0	100.0

Annex table 19-8 Steel imports by products : *EFTA*

Products	Amount (1000t)				Share (%)			
	1980	1985	1990	1992	1980	1985	1990	1992
Ingots and semis	386	260	894	768	5.6	3.9	11.0	10.1
Long products	2 140	2 056	2 618	2 398	31.2	30.7	32.3	31.7
Flat products	3 491	3 418	3 552	3 352	50.9	51.0	43.9	44.3
Tubes and fittings	840	962	1 035	1 049	12.3	14.4	12.8	13.9
Total	6 857	6 696	8 099	7 567	100.0	100.0	100.0	100.0

Annex table 19-9 Steel imports by products : *United States*

Products	Amount (1000t)				Share (%)			
	1980	1985	1990	1992	1980	1985	1990	1992
Ingots and semis	141	2 121	2 146	2 129	1.0	9.8	13.5	13.6
Long products	3 792	5 558	3 414	3 178	27.8	25.6	21.5	20.4
Flat products	6 275	9 874	7 805	8 795	46.0	45.4	49.2	56.3
Tubes and fittings	3 448	4 174	2 513	1 512	25.2	19.2	15.8	9.7
Total	13 656	21 727	15 878	15 614	100.0	100.0	100.0	100.0

Annex table 19-10 Steel imports by products : *Japan*

Products	Amount (1000t)				Share (%)			
	1980	1985	1990	1992	1980	1985	1990	1992
Ingots and semis	449	353	1 023	615	40.6	12.2	14.4	10.0
Long products	46	81	958	712	4.2	2.8	13.5	11.6
Flat products	585	2 426	4 790	4 520	52.8	83.5	67.3	73.6
Tubes and fittings	27	44	350	296	2.4	1.5	4.9	4.8
Total	1 107	2 904	7 121	6 143	100.0	100.0	100.0	100.0

Annex table 20-1 Steel trade balance by products : *EC*

Products	Amount (1000t)				Share (%)			
	1980	1985	1990	1992	1980	1985	1990	1992
Ingots and semis	- 683	- 4 666	59	383	-3.4	-15.5	0.5	3.0
Long products	8 037	9 741	2 477	4 394	39.6	32.4	22.1	34.9
Flat products	8 109	18 230	6 959	5 765	39.9	60.6	62.1	45.8
Tubes and fittings	4 849	6 770	1 719	1 988	23.9	22.5	15.3	15.8
Total	20 312	30 075	11 214	12 600	100.0	100.0	100.0	100.0

Annex table 20-2 Steel trade balance by products : *Belgium-Luxembourg*

Products	Amount (1000t)				Share (%)			
	1980	1985	1990	1992	1980	1985	1990	1992
Ingots and semis	76	69	- 74	- 533	0.7	0.7	-0.7	-5.9
Long products	3 632	2 552	2 655	2 348	34.0	27.7	25.4	26.1
Flat products	6 866	6 286	7 727	7 020	64.2	68.2	73.9	77.9
Tubes and fittings	116	305	149	173	1.1	3.3	1.4	1.9
Total	10 690	9 212	10 457	9 008	100.0	100.0	100.0	100.0

Annex table 20-3 Steel trade balance by products : *France*

Products	Amount (1000t)				Share (%)			
	1980	1985	1990	1992	1980	1985	1990	1992
Ingots and semis	- 620	346	595	643	-20.0	8.1	64.2	37.1
Long products	815	1 152	- 464	41	26.3	27.0	-50.0	2.4
Flat products	1 714	1 597	597	862	55.4	37.5	64.4	49.8
Tubes and fittings	1 185	1 169	199	187	38.3	27.4	21.5	10.8
Total	3 094	4 264	927	1 733	100.0	100.0	100.0	100.0

Annex table 20-4 Steel trade balance by products : *Germany*

Products	Amount (1000t)				Share (%)			
	1980	1985	1990	1992	1980	1985	1990	1992
Ingots and semis	- 456	- 14	- 132	243	-6.0	-0.2	-4.1	16.8
Long products	- 61	- 348	- 1 965	- 2 020	-0.8	-4.1	-60.7	-139.9
Flat products	5 812	5 611	4 176	2 715	76.7	66.1	129.0	188.0
Tubes and fittings	2 283	3 234	1 158	506	30.1	38.1	35.8	35.0
Total	7 578	8 483	3 237	1 444	100.0	100.0	100.0	100.0

Annex table 20-5 Steel trade balance by products : *Italy*

Products	Amount (1000t)				Share (%)			
	1980	1985	1990	1992	1980	1985	1990	1992
Ingots and semis	- 874	- 740	- 1 304	- 919	-236.2	-41.7	-50.1	-60.9
Long products	1 965	1 616	1 677	2 338	531	91.0	64.5	155.1
Flat products	- 2 778	- 903	- 3 825	- 3 974	-751	-50.9	-147.0	-263.5
Tubes and fittings	1 317	1 802	849	977	355.9	101.5	32.7	64.8
Total	- 370	1 775	- 2 601	- 1 508	-100.0	100.0	-100.0	-100.0

Annex table 20-6 Steel trade balance by products : *Netherlands*

Products	Amount (1000t)				Share (%)			
	1980	1985	1990	1992	1980	1985	1990	1992
Ingots and semis	696	1 126	1 075	1 026	94.2	98.9	260.3	94.6
Long products	- 887	- 845	- 1 189	- 1 059	-120.0	-74.2	-287.9	-97.7
Flat products	1 174	1 145	1 049	1 121	158.9	100.5	254.0	103.4
Tubes and fittings	- 244	- 287	- 522	4	-33.0	-25.2	-126.4	-0.4
Total	739	1 139	413	1 084	100.0	100.0	100.0	100.0

Annex table 20-7 Steel trade balance by products : *United Kingdom*

Products	Amount (1000t)				Share (%)			
	1980	1985	1990	1992	1980	1985	1990	1992
Ingots and semis	- 166	489	985	999	-8.6	46.9	52.2	31.0
Long products	433	1 137	1 672	2 128	22.3	109.1	88.6	66.0
Flat products	- 2 103	- 737	- 657	- 68	-108.3	-70.7	-34.8	-2.1
Tubes and fittings	- 105	153	- 112	167	-5.4	14.7	-5.9	5.2
Total	- 1 941	1 042	1 888	3 226	-100.0	100.0	100.0	100.0

Annex table 20-8 Steel trade balance by products : *EFTA*

Products	Amount (1000t)				Share (%)			
	1980	1985	1990	1992	1980	1985	1990	1992
Ingots and semis	390	783	- 498	- 241	216.7	29.8	-99.2	-11.7
Long products	- 99	638	- 147	166	-55.0	24.3	-29.3	8.1
Flat products	- 321	1 057	1 077	2 243	-178.3	40.2	214.5	108.9
Tubes and fittings	- 150	149	60	- 109	-83.3	5.7	12.0	-5.3
Total	- 180	2 627	502	2 059	-100.0	100.0	100.0	100.0

Annex table 20-9 Steel trade balance by products : *United States*

Products	Amount (1000t)				Share (%)			
	1980	1985	1990	1992	1980	1985	1990	1992
Ingots and semis	689	- 2 040	- 1 695	- 1 745	7.0	-9.8	-14.6	-15.0
Long products	- 2 839	- 5 393	- 2 156	- 2 202	-28.9	-25.9	-18.6	-19.0
Flat products	- 4 704	- 9 458	- 5 685	- 6 811	-47.9	-45.3	-49.1	-58.6
Tubes and fittings	- 2 960	- 3 967	- 2 051	- 856	-30.2	-19.0	-17.7	-7.4
Total	- 9 814	- 20 858	- 11 587	- 11 614	-100.0	-100.0	-100.0	-100.0

Annex table 20-10 Steel trade balance by products : *Japan*

Products	Amount (1000t)				Share (%)			
	1980	1985	1990	1992	1980	1985	1990	1992
Ingots and semis	- 260	- 29	- 980	- 290	-0.9	-0.1	-11.2	-2.4
Long products	7 986	8 443	1 511	2 232	28.0	29.5	17.3	18.1
Flat products	14 372	13 567	5 351	7 692	50.4	47.5	61.4	62.4
Tubes and fittings	6 426	6 605	2 832	2 685	22.5	23.1	32.5	21.8
Total	28 524	28 586	8 714	12 319	100.0	100.0	100.0	100.0

Annex table 21-1 Consumption of steel by major steel using sectors : <u>EC</u> 1/

Item		1980	1981	1982	1983	1984	1985	1986	1987	1988	1989	1990	1991
Steel consumption (1000t)													
Total	2/	87 874	78 525	77 562	74 855	79 945	86 243	89 462	89 490	102 256	108 807	108 462	105 899
Manufacturing	3/	52 956	49 358	48 059	46 188	46 879	50 233	52 133	52 661	57 271	60 790	60 758	61 913
Metal	4/	20 065	19 123	18 217	18 251	19 091	19 615	19 817	19 871	21 518	23 030	23 510	25 498
Machinery	5/	12 134	11 142	10 988	10 021	10 051	10 874	11 472	11 341	12 318	13 312	13 257	13 573
Electronics	6/	3 967	3 547	3 551	3 451	3 554	3 822	4 058	4 112	4 233	4 562	4 682	4 797
Transport	7/	16 790	15 546	15 303	14 465	14 183	15 922	16 786	17 337	19 202	19 886	19 309	18 045
Construction	8/	14 530	13 445	13 679	12 607	12 250	14 329	15 354	17 083	18 672	21 318	18 662	20 883
Mfg.+Const.	9/	67 486	62 803	61 738	58 795	59 129	64 562	67 487	69 744	75 943	82 108	79 420	82 796
Share of steel consumption (%)													
Total		100.0	100.0	100.0	100.0	100.0	100.0	100.0	100.0	100.0	100.0	100.0	100.0
Manufacturing		60.3	62.9	62.0	61.7	58.6	58.2	58.3	58.8	56.0	55.9	56.0	58.5
Metal		22.8	24.4	23.5	24.4	23.9	22.7	22.2	22.2	21.0	21.2	21.7	24.1
Machinery		13.8	14.2	14.2	13.4	12.6	12.6	12.8	12.7	12.0	12.2	12.2	12.8
Electronics		4.5	4.5	4.6	4.6	4.4	4.4	4.5	4.6	4.1	4.2	4.3	4.5
Transport		19.1	19.8	19.7	19.3	17.7	18.5	18.8	19.4	18.8	18.3	17.8	17.0
Construction		16.5	17.1	17.6	16.8	15.3	16.6	17.2	19.1	18.3	19.6	17.2	19.7
Mfg.+Const.		76.8	80.0	79.6	78.5	74.0	74.9	75.4	77.9	74.3	75.5	73.2	78.2
Production of steel consuming sectors													
Total GDP	10/	2 888	2 895	2 914	2 956	3 029	3 356	3 450	3 547	3 678	3 803	3 913	3 957
Manufacturing	10/	747	737	734	746	762	840	849	854	895	924	865	...
Metal		...	...	...	...	...	...	...	...	...	...	...	...
Machinery		...	...	...	...	...	...	...	...	...	...	...	...
Electronics		...	...	...	...	...	...	...	...	...	...	...	...
Transport		...	...	...	...	...	...	...	...	...	...	...	...
Construction	10/	193	181	178	178	177	194	200	204	217	228	204	...
Mfg.+Const.	10/	940	919	912	924	939	1 035	1 049	1 059	1 112	1 152	1 068	...
Steel intensity													
Total	11/	30.4	27.1	26.6	25.3	26.4	25.7	25.9	25.2	27.8	28.6	27.7	26.8
Manufacturing	11/	70.9	66.9	65.5	61.9	61.5	59.8	61.4	61.6	64.0	65.8	70.3	...
Metal		...	...	...	...	...	...	...	...	...	...	...	...
Machinery		...	...	...	...	...	...	...	...	...	...	...	...
Electronics		...	...	...	...	...	...	...	...	...	...	...	...
Transport		...	...	...	...	...	...	...	...	...	...	...	...
Construction	11/	75.4	74.1	76.7	70.7	69.1	73.8	76.7	83.6	85.9	93.3	91.6	...
Mfg.+Const.	11/	71.8	68.4	67.7	63.6	62.9	62.4	64.3	65.9	68.3	71.3	74.3	...

1/ EC includes 10 member countries before 1985 and 12 after 1985.
2/ Apparent finished steel consumption.
3/ Total manufacturing (ISIC 380): the sum of ISIC 381, 382, 383 and 384.
4/ Fabricated metals (ISIC 381).
5/ Machinery except electrical (ISIC 382).
6/ Electrical machinery (ISIC 383).
7/ Transport equipment (ISIC 384).
8/ Construction (ISIC 500).
9/ Manufacturing plus construction (ISIC 380 plus ISIC 500).
10/ Billions of constant 1980 US dollars.
11/ Grams of steel consumed per 1980 US dollar.

Annex table 21-2 Consumption of steel by major steel using sectors : *France*

Item		1980	1981	1982	1983	1984	1985	1986	1987	1988	1989	1990	1991	1992
Steel consumption (1000t)														
Total		16 049	15 098	14 226	13 085	12 689	12 626	12 669	13 115	14 464	15 047	15 226	14 499	13 941
Manufacturing	1/	10 283	9 468	9 183	8 507	8 256	8 091	7 941	8 104	8 740	9 081	8 923	8 358	8 525
Fabricated metals	2/	1 409	1 367	1 385	1 296	1 332	1 325	1 279	1 302	1 389	1 367	1 404	1 391	1 400
Machinery	3/	3 526	3 314	3 175	2 901	2 865	2 833	2 652	2 647	2 798	2 963	3 033	2 744	2 545
Electrical mach.	4/	948	936	905	821	810	801	810	812	859	884	915	906	890
Transport	5/	4 400	3 851	3 718	3 489	3 249	3 132	3 200	3 343	3 694	3 867	3 571	3 317	3 690
Construction	6/	4 235	4 198	3 802	3 527	3 364	3 440	3 595	3 814	4 302	4 581	4 890	4 722	4 437
Mnfg.+Const.	7/	14 518	13 666	12 985	12 034	11 620	11 531	11 536	11 918	13 042	13 662	13 813	13 080	12 962
Share of steel consumption (%)														
Total		100.0	100.0	100.0	100.0	100.0	100.0	100.0	100.0	100.0	100.0	100.0	100.0	100.0
Manufacturing		64.1	62.7	64.6	65.0	65.1	64.1	62.7	61.8	60.4	60.4	58.6	57.6	61.2
Fabricated metals		8.8	9.1	9.7	9.9	10.5	10.5	10.1	9.9	9.6	9.1	9.2	9.6	10.0
Machinery		22.0	21.9	22.3	22.2	22.6	22.4	20.9	20.2	19.3	19.7	19.9	18.9	18.3
Electrical mach.		5.9	6.2	6.4	6.3	6.4	6.3	6.4	6.2	5.9	5.9	6.0	6.2	6.4
Transport		27.4	25.5	26.1	26.7	25.6	24.8	25.3	25.5	25.5	25.7	23.5	22.9	26.5
Construction		26.4	27.8	26.7	27.0	26.5	27.2	28.4	29.1	29.7	30.4	32.1	32.6	31.8
Mnfg.+Const.		90.5	90.5	91.3	92.0	91.6	91.3	91.1	90.9	90.2	90.8	90.7	90.2	93.0
Production of steel consuming sectors														
Total GDP	8/	665.5	672.1	687.7	691.9	701.8	714.8	733.1	749.7	779.2	807.7	828.7	...	...
Manufacturing	8/	161.0	159.9	161.3	162.0	159.0	158.4	158.1	156.7	164.7	171.2	174.3	...	...
Fabricated metals	9/	...	...	...	...	2 178	2 157	2 101	2 135	2 289	2 365	2 474	2 408	2 360
Machinery	9/	...	...	...	...	1 856	1 796	1 592	1 618	1 724	1 865	1 908	1 808	1 650
Electrical mach.	10/	...	...	...	...	77.4	79.5	80.8	82.0	86.0	90.0	96.8	96.5	97.1
Transport	9/	...	...	...	...	2 724	2 606	2 836	3 109	3 371	3 534	3 340	3 191	3 254
Construction	8/	45.9	45.7	45.7	44.4	43.4	42.8	44.0	44.5	46.6	47.6	48.1	...	...
Mnfg.+Const.	8/	206.9	205.6	207.0	206.4	202.4	201.2	202.1	201.2	211.3	218.8	222.4	...	...
Steel intensity														
Total	11/	24.1	22.5	20.7	18.9	18.1	17.7	17.3	17.5	18.6	18.6	18.4	...	...
Manufacturing	11/	63.9	59.2	56.9	52.5	51.9	51.1	50.2	51.7	53.1	53.0	51.2	...	...
Fabricated metals	12/	...	...	...	...	612	614	609	610	607	578	568	578	593
Machinery	12/	...	...	...	...	1 544	1 577	1 666	1 636	1 623	1 589	1 590	1 518	1 542
Electrical mach.	13/	...	...	...	...	1 047	1 008	1 002	990	999	982	945	939	917
Transport	12/	...	...	...	...	1 193	1 202	1 128	1 075	1 096	1 094	1 069	1 039	1 134
Construction	11/	92.3	91.9	83.2	79.4	77.5	80.4	81.7	85.7	92.3	96.2	101.7	...	...
Mnfg.+Const.	11/	70.2	66.5	62.7	58.3	57.4	57.3	57.1	59.2	61.7	62.4	62.1	...	...

1/ Total manufacturing (ISIC 380): the sum of ISIC 381, 382, 383 and 384.
2/ Fabricated metals (ISIC 381).
3/ Machinery except electrical (ISIC 382).
4/ Electrical machinery (ISIC 383).
5/ Transport equipment (ISIC 384).
6/ Construction (ISIC 500).
7/ Manufacturing plus construction (ISIC 380 plus ISIC 500).
8/ Billions of constant 1980 US dollars.
9/ Thousands of tonnes.
10/ Billions of 1970 francs.
11/ Grams of steel consumed per 1980 US dollar.
12/ Kilograms per tonne of production.
13/ Grams of steel consumed per hundred 1970 francs.

Annex table 21-3 Consumption of steel by major steel using sectors : *Netherlands*

Item		1980	1981	1982	1983	1984	1985	1986	1987	1988	1989	1990	1991
Steel consumption (1000t)													
Total	1/	3 215	2 692	2 874	2 624	3 270	3 396	3 312	2 904	3 758	4 179	4 431	3 592
Manufacturing	2/	1 575	1 464	1 461	1 377	1 452	1 525	1 503	1 629	2 011	2 327	2 397	2 315
Fabricated metals	3/	787	740	729	733	777	876	881	1 004	1 229	1 391	1 465	1 410
Machinery	4/	312	268	267	245	251	277	272	298	324	382	377	394
Electrical mach.	5/	105	91	86	81	95	82	77	73	117	114	109	113
Transport	6/	371	365	379	318	329	290	273	254	341	440	446	398
Construction		...	...	...	...	...	...	...	...	...	...	...	...
Mnfg.+Const.		...	...	...	...	...	...	...	...	...	...	...	...
Share of steel consumption (%)													
Total		100.0	100.0	100.0	100.0	100.0	100.0	100.0	100.0	100.0	100.0	100.0	100.0
Manufacturing		49.0	54.4	50.8	52.5	44.4	44.9	45.4	56.1	53.5	55.7	54.1	64.4
Fabricated metals		24.5	27.5	25.4	27.9	23.8	25.8	26.6	34.6	32.7	33.3	33.1	39.3
Machinery		9.7	10.0	9.3	9.3	7.7	8.2	8.2	10.3	8.6	9.1	8.5	11.0
Electrical mach.		3.3	3.4	3.0	3.1	2.9	2.4	2.3	2.5	3.1	2.7	2.5	3.1
Transport		11.5	13.6	13.2	12.1	10.1	8.5	8.2	8.7	9.1	10.5	10.1	11.1
Construction		...	...	...	...	...	...	...	...	...	...	...	...
Mnfg.+Const.		...	...	...	...	...	...	...	...	...	...	...	...
Production of steel consuming sectors													
Total GDP	7/	169.6	168.4	166.1	168.4	173.2	177.6	181.2	182.5	187.1	194.6	202.1	206.0
Manufacturing	7/	30.4	29.8	28.4	29.5	30.6	31.7	31.7	31.7	33.2	34.6	36.1	...
Fabricated metals	8/	9.4	8.8	8.4	8.3	8.8	9.4	9.7	8.4	9.2	9.8	10.8	11.4
Machinery	8/	10.0	9.8	9.3	9.8	10.2	10.5	10.7	10.7	11.1	12.0	12.8	12.6
Electrical mach.	8/	18.6	18.2	18.3	18.0	21.6	22.8	23.7	21.6	18.4	18.4	19.4	19.4
Transport	8/	9.4	9.6	10.6	10.4	9.4	9.8	9.9	9.4	11.4	12.2	13.0	13.4
Construction	7/	12.0	11.0	10.4	10.1	10.4	10.5	10.7	10.8	12.0	12.4	12.9	...
Mnfg.+Const.	7/	42.4	40.9	38.8	39.6	41.0	42.2	42.5	42.4	45.3	47.1	49.0	...
Steel intensity													
Total	9/	19.0	16.0	17.3	15.6	18.9	19.1	18.3	15.9	20.1	21.5	21.9	17.4
Manufacturing	9/	51.8	49.1	51.4	46.7	47.5	48.1	47.3	51.4	60.5	67.2	66.5	...
Fabricated metals	10/	83.3	84.5	86.5	88.7	88.7	93.2	90.6	119.4	134.3	141.9	136.1	123.8
Machinery	10/	31.2	27.4	28.6	25.1	24.6	26.3	25.4	27.9	29.1	31.9	29.4	31.2
Electrical mach.	10/	5.6	5.0	4.7	4.5	4.4	3.6	3.2	3.4	6.4	6.2	5.6	5.8
Transport	10/	39.5	38.0	35.7	30.4	35.0	29.6	27.6	26.9	29.9	36.0	34.4	29.6
Construction		...	...	...	...	...	...	...	...	...	...	...	...
Mnfg.+Const.		...	...	...	...	...	...	...	...	...	...	...	...

1/ Apparent finished steel consumption.
2/ Total manufacturing (ISIC 380): the sum of ISIC 381, 382, 383 and 384.
3/ Fabricated metals (ISIC 381).
4/ Machinery except electrical (ISIC 382).
5/ Electrical machinery (ISIC 383).
6/ Transport equipment (ISIC 384).
7/ Billions of constant 1980 US dollars.
8/ Billions of guilders.
9/ Grams of steel consumed per 1980 US dollar.
10/ Grams of steel consumed per guilder.

Annex table 21-4 Consumption of steel by major steel using sectors : *United Kingdom*

Item		1973	1980	1981	1982	1983	1984	1985	1986	1987	1988	1989	1990	1991
Steel consumption (1000t)														
Total	1/	19 091	11 813	11 451	11 542	11 207	11 241	11 615	11 938	12 449	14 967	14 660	13 974	12 049
Manufacturing	2/	13 448	9 522	8 451	8 763	8 273	8 275	8 914	8 816	9 234	9 992	10 296	9 901	8 332
Fabricated metals	3/	6 949	4 536	4 100	4 334	4 203	4 374	4 443	4 436	4 576	4 831	4 898	4 839	3 910
Machinery	4/	2 409	2 432	2 049	2 083	1 906	1 803	2 079	2 144	2 237	2 595	2 749	2 516	2 229
Electrical mach.	5/	870	648	495	519	460	509	626	604	639	652	721	676	601
Transport	6/	3 220	1 906	1 807	1 827	1 704	1 589	1 766	1 632	1 782	1 914	1 928	1 870	1 592
Construction	7/	2 282	1 041	1 064	1 137	1 026	1 000	1 069	1 146	1 449	2 180	2 289	2 151	1 839
Mnfg.+Const.	8/	15 730	10 563	9 515	9 900	9 299	9 275	9 983	9 962	10 683	12 172	12 585	12 052	10 171
Share of steel consumption (%)														
Total		100.0	100.0	100.0	100.0	100.0	100.0	100.0	100.0	100.0	100.0	100.0	100.0	100.0
Manufacturing		70.4	80.6	73.8	75.9	73.8	73.6	76.7	73.8	74.2	66.8	70.2	70.9	69.2
Fabricated metals		36.4	38.4	35.8	37.5	37.5	38.9	38.3	37.2	36.8	32.3	33.4	34.6	32.5
Machinery		12.6	20.6	17.9	18.0	17.0	16.0	17.9	18.0	18.0	17.3	18.8	18.0	18.5
Electrical mach.		4.6	5.5	4.3	4.5	4.1	4.5	5.4.	5.1	5.1	4.4	4.9	4.8	5.0
Transport		16.9	16.1	15.8	15.8	15.2	14.1	15.2	13.7	14.3	12.8	13.2	13.4	13.2
Construction		12.0	8.8	9.3	9.9	9.2	8.9	9.2	9.6	11.6	14.6	15.6	15.4	15.3
Mnfg.+Const.		82.4	89.4	83.1	85.8	83.0	82.5	85.9	83.4	85.8	81.3	85.8	86.2	84.4
Production of steel consuming sectors														
Total GDP	9/	447.6	485.9	479.8	490.1	506.8	520.8	539.1	557.4	582.8	609.8	623.8	627.6	611.9
Manufacturing	9/	144.8	124.4	122.6	125.1	129.6	133.3	136.9	140.2	144.8	150.0	150.4	149.6	145.2
Fabricated metals	10/	...	4 731	4 385	4 480	4 381	4 388	4 562	4 486	4 775	5 017	5 295	5 236	4 482
Machinery	10/	...	2 825	2 393	2 389	2 241	2 286	2 434	2 476	2 497	2 578	2 716	2 743	2 439
Electrical mach.	10/	...	2 213	1 942	1 908	1 824	1 797	1 779	1 817	1 929	2 117	2 295	2 275	2 165
Transport	10/	...	3 439	2 902	2 839	2 818	2 447	2 859	2 435	2 751	2 976	3 078	3 039	2 853
Construction	9/	33.0	28.0	23.3	25.2	26.8	27.8	28.2	29.5	31.2	34.0	37.5	37.8	34.5
Mnfg.+Const.	9/	177.8	152.4	145.9	150.3	156.4	161.1	165.1	169.7	176.0	184.0	187.9	187.4	179.7
Steel intensity														
Total	11/	42.7	24.3	23.9	23.6	22.1	21.6	21.5	21.4	21.4	24.5	23.5	22.3	19.7
Manufacturing	11/	92.9	76.5	68.9	70.0	63.8	62.1	65.1	62.9	63.8	66.6	68.5	66.2	57.4
Fabricated metals	12/	...	958.8	935.0	967.4	959.4	996.8	973.9	988.9	958.3	962.9	925.0	924.2	872.4
Machinery	12/	...	860.9	856.2	871.9	850.5	788.7	854.1	865.9	895.9	1,006.6	1,012.2	917.2	913.9
Electrical mach.	12/	...	292.8	254.9	272.0	252.2	283.2	351.9	332.4	331.3	308.0	314.2	297.1	277.6
Transport	12/	...	554.2	622.7	643.5	604.7	649.4	617.7	670.2	647.8	643.1	626.4	615.3	558.0
Construction	11/	69.2	37.2	45.7	45.1	38.3	36.0	37.9	38.8	46.4	64.1	61.0	56.9	53.3
Mnfg.+Const.	11/	88.5	69.3	65.2	65.9	59.5	57.6	60.5	58.7	60.7	66.2	67.0	64.3	56.6

1/ Apparent finished steel consumption.
2/ Total manufacturing (ISIC 380): the sum of ISIC 381, 382, 383 and 384.
3/ Fabricated metals (ISIC 381).
4/ Machinery except electrical (ISIC 382).
5/ Electrical machinery (ISIC 383).
6/ Transport equipment (ISIC 384).
7/ Construction (ISIC 500).
8/ Manufacturing plus construction (ISIC 380 plus ISIC 500).
9/ Billions of constant 1980 US dollars.
10/ Thousands of tonnes.
11/ Grams of steel consumed per 1980 US dollars.
12/ Kilograms of steel consumed per tonne of material.

Annex table 21-5 Consumption of steel by major steel using sectors : *Japan*

Item		1980	1981	1982	1983	1984	1985	1986	1987	1988	1989	1990	1991	1992
Steel consumption (1000t)														
Total	1/	71 307	63 272	63 601	61 168	69 097	69 861	67 702	72 891	80 961	88 306	92 807	93 132	79 029
Manufacturing	2/	30 818	30 207	27 493	28 225	30 784	31 480	29 276	30 356	33 443	35 673	37 888	36 676	34 164
Fabricated metals	3/	7 042	6 626	6 400	6 433	7 007	6 959	7 134	7 399	7 972	8 088	8 524	8 392	7 684
Machinery	4/	5 306	5 337	4 954	4 768	5 127	5 364	4 890	5 206	6 009	6 695	7 124	6 462	5 805
Electric machine	5/	3 445	3 449	3 297	3 617	4 172	4 182	4 162	4 396	4 829	4 995	5 344	5 323	4 814
Transport	6/	15 025	14 795	12 842	13 407	14 478	14 975	13 090	13 355	14 633	15 895	16 896	16 499	15 861
Construction	7/	28 199	26 818	25 383	24 729	25 857	26 216	28 047	31 662	34 953	38 429	41 515	38 995	36 212
Mnfg.+Const.	8/	59 017	57 025	52 876	52 954	56 641	57 696	57 323	62 018	68 396	74 102	79 403	75 671	70 376
Share of steel consumption (%)														
Total		100.0	100.0	100.0	100.0	100.0	100.0	100.0	100.0	100.0	100.0	100.0	100.0	100.0
Manufacturing		43.2	47.7	43.2	46.1	44.6	45.1	43.2	41.6	41.3	40.4	40.8	39.4	43.2
Fabricated metals		9.9	10.5	10.1	10.5	10.1	10.0	10.5	10.2	9.8	9.2	9.2	9.0	9.7
Machinery		7.4	8.4	7.8	7.8	7.4	7.7	7.2	7.1	7.4	7.6	7.7	6.9	7.3
Electric machine		4.8	5.5	5.2	5.9	6.0	6.0	6.1	6.0	6.0	5.7	5.8	5.7	6.1
Transport		21.1	23.4	20.2	21.9	21.0	21.4	19.3	18.3	18.1	18.0	18.2	17.7	20.1
Construction		39.5	42.4	39.9	40.4	37.4	37.5	41.4	43.4	43.2	43.5	44.7	41.9	45.8
Mnfg.+Const.		82.8	90.1	83.1	86.6	82.0	82.6	84.7	85.1	84.5	83.9	85.6	81.3	89.1
Production of steel consuming sectors														
Total GDP	9/	1 064.0	1 104.2	1 138.7	1 169.6	1 221.2	1 284.2	1 316.4	1 369.9	1 455.0	1 521.9	1 609.4	...	...
Manufacturing	9/	311.1	325.2	344.0	371.6	414.5	443.5	445.7	477.8	516.0	557.2	598.3	...	...
Fabricated metals	10/	10 250	10 808	11 134	11 191	11 766	13 307	14 352	15 088	16 945	18 102	19 865	21 660	...
Machinery	10/	17 257	19 147	19 639	20 221	22 188	24 583	25 526	24 866	29 066	31 966	35 535	38 380	...
Electric machine	10/	22 028	25 363	27 158	31 510	38 848	41 506	45 107	46 814	52 387	55 180	58 320	62 767	...
Transport	10/	24 749	27 691	28 155	29 516	32 099	36 767	38 326	39 427	41 849	45 669	50 116	52 420	...
Construction	9/	99.7	102.0	98.2	87.7	85.6	85.4	89.4	97.9	107.8	113.0	118.3	...	...
Mnfg.+Const.	9/	410.8	427.2	442.2	459.3	500.1	528.9	535.1	575.7	623.8	670.2	716.6	...	...
Steel intensity														
Total	11/	67.0	57.3	55.9	52.3	56.6	54.4	51.4	53.2	55.6	58.0	57.7	...	...
Manufacturing	11/	99.1	92.9	79.9	76.0	74.3	71.0	65.7	63.5	64.8	64.0	63.3	...	...
Fabricated metals	12/	687.1	613.1	574.8	574.9	595.5	523.0	497.1	490.4	470.5	446.8	429.1	387.4	...
Machinery	12/	307.5	278.7	252.2	235.8	231.1	218.2	191.6	209.4	206.7	209.4	200.5	168.4	...
Electric machine	12/	156.4	136.0	121.4	114.8	107.4	100.8	92.3	93.9	92.2	90.5	91.6	84.8	...
Transport	12/	607.1	534.3	456.1	454.2	451.0	407.3	341.5	338.7	349.7	348.0	337.1	314.7	...
Construction	11/	282.8	262.9	258.5	282.0	302.1	307.0	313.7	323.4	324.2	340.1	350.9	...	...
Mnfg.+Const.	11/	143.7	133.5	119.6	115.3	113.3	109.1	107.1	107.7	109.6	110.6	110.8	...	...

1/ Apparent finished steel consumption.
2/ Total manufacturing (ISIC 380): the sum of ISIC 381, 382, 383 and 384.
3/ Fabricated metals (ISIC 381).
4/ Machinery except electrical (ISIC 382).
5/ Electrical machinery (ISIC 383).
6/ Transport equipment (ISIC 384).
7/ Construction (ISIC 500).
8/ Manufacturing plus construction (ISIC 380 plus ISIC 500).
9/ Billions of constant 1980 US dollars.
10/ Billions of constant 1980 Japanese yen.
11/ Grams of steel consumed per 1980 US dollar.
12/ Grams of steel consumed per thousand 1980 Japanese yen.

Annex table 21-6 Consumption of steel by major steel using sectors : *Finland*

Item		1970	1980	1981	1982	1983	1984	1985	1986	1987	1988	1989	1990	1991
Steel consumption (1000t)														
Total	1/	1 502	1 900	1 672	1 868	1 621	1 732	1 577	1 743	1 784	1 769	2 036	1 809	1 308
Manufacturing	2/	756	1 302	1 270	1 508	1 508	1 510	1 504	1 495	1 597	1 653	1 816	1 649	1 382
Metal	3/	301	544	564	646	719	723	707	715	782	840	900	860	700
Machinery	4/	242	383	379	430	400	436	466	421	488	463	590	448	395
Electronics	5/	41	76	71	68	67	65	73	69	66	65	89	92	87
Transport	6/	172	299	256	364	322	286	258	290	261	285	237	249	200
Construction		...	...	...	...	...	...	...	...	...	...	...	...	...
Mfg.+Const.		...	...	...	...	...	...	...	...	...	...	...	...	...
Share of steel consumption (%)														
Total		100.0	100.0	100.0	100.0	100.0	100.0	100.0	100.0	100.0	100.0	100.0	100.0	100.0
Manufacturing		50.3	68.5	76.0	80.7	93.0	87.2	95.4	85.8	89.5	93.4	89.2	91.2	105.7
Metal		20.0	28.6	33.7	34.6	44.4	41.7	44.8	41.0	43.8	47.5	44.2	47.5	53.5
Machinery		16.1	20.2	22.7	23.0	24.7	25.2	29.5	24.2	27.3	26.2	29.0	24.8	30.2
Electronics		2.7	4.0	4.2	3.6	4.1	3.8	4.6	4.0	3.7	3.7	4.4	5.1	6.7
Transport		11.5	15.7	15.3	19.5	19.9	16.5	16.4	16.6	14.6	16.1	11.6	13.8	15.3
Construction		...	...	...	...	...	...	...	...	...	...	...	...	...
Mfg.+Const.		...	...	...	...	...	...	...	...	...	...	...	...	...
Production of steel consuming sectors														
Total GDP	7/	31.2	46.4	47.1	48.7	50.3	51.3	52.9	53.8	55.7	58.0	61.1	61.7	...
Manufacturing	7/	8.2	13.0	13.4	13.6	14.1	14.6	15.2	15.4	16.2	16.8	17.4	17.2	...
Metal	8/	...	5 898	7 382	7 775	8 314	8 688	9 136	9 401	9 766	10 251	10 945	11 420	11 603
Machinery	8/	...	13 462	14 726	15 876	16 932	17 875	18 855	19 722	20 646	22 230	24 059	25 643	27 340
Electronics	8/	...	7 320	7 642	7 991	8 617	8 994	9 196	9 316	9 481	9 628	9 941	10 115	10 115
Transport	8/	...	8 795	9 606	10 454	11 215	11 744	12 284	12 751	15 402	13 991	14 864	15 477	15 846
Construction	7/	2.7	3.7	3.6	3.8	4.0	3.8	3.9	3.9	3.9	4.2	4.7	4.6	...
Mfg.+Const.	7/	10.9	16.7	17.0	17.4	18.1	18.4	19.1	19.3	20.1	21.0	22.1	21.8	...
Steel intensity														
Total	9/	48.1	41.0	35.5	38.4	32.2	33.8	29.8	32.4	32.0	30.5	33.3	29.3	...
Manufacturing	9/	92.2	100.2	94.8	110.9	107.0	103.4	98.9	97.1	98.6	98.4	104.4	95.9	...
Metal	10/	...	92.2	76.4	83.1	86.5	83.2	77.4	76.1	80.1	81.9	82.2	75.3	60.3
Machinery	10/	...	28.5	25.7	27.1	23.6	24.4	24.7	21.3	23.6	20.8	24.5	17.5	14.4
Electronics	10/	...	10.4	9.3	8.5	7.8	7.2	7.9	7.4	7.0	6.8	9.0	9.1	8.6
Transport	10/	...	34.0	26.7	34.8	28.7	24.4	21.0	22.7	16.9	20.4	15.9	16.1	12.6
Construction		...	...	...	...	...	...	...	...	...	...	...	...	...
Mfg.+Const.		...	...	...	...	...	...	...	...	...	...	...	...	...

1/ Apparent finished steel consumption.
2/ Total manufacturing (ISIC 380): the sum of ISIC 381, 382, 383 and 384.
3/ Fabricated metals (ISIC 381).
4/ Machinery except electrical (ISIC 382).
5/ Electrical machinery (ISIC 383).
6/ Transport equipment (ISIC 384).
7/ Billions of constant 1980 US dollars.
8/ Millions of constant 1980 markkaa.
9/ Grams of steel consumed per 1980 US dollar.
10/ Grams of steel consumed per markkaa.

Annex table 21-7 Consumption of steel by major steel using sectors : *Sweden*

Item		1980	1981	1982	1983	1984	1985	1986	1987	1988	1989	1990	1991	1992
Steel consumption (1000t)														
Total	1/	2 991	2 775	2 692	2 688	3 027	2 826	2 964.	3 015	3 384	3 013	3 265	2 462	2 422
Manufacturing	2/	2 576	2 459	2 390	2 461	2 494	2 479	2 404	...	...	...	...	...	...
Fabricated metals	3/	1 400	1 257	1 241	1 304	1 397	1 388	1 338	...	...	...	...	...	...
Machinery	4/	520	466	480	502	518	516	505	...	...	...	...	...	...
Electrical mach.	5/	90	83	81	80	81	85	93	...	...	...	...	...	...
Transport	6/	566	653	588	575	498	490	468	...	...	...	...	...	...
Construction		...	...	...	...	...	...	...	...	...	...	...	...	...
Mnfg.+Const.		...	...	...	...	...	...	...	...	...	...	...	...	...
Share of steel consumption (%)														
Total		100.0	100.0	100.0	100.0	100.0	100.0	100.0	...	...	...	...	...	...
Manufacturing		86.1	88.6	88.8	91.6	82.4	87.7	81.1	...	...	...	...	...	...
Fabricated metals		46.8	45.3	46.1	48.5	46.2	49.1	45.1	...	...	...	...	...	...
Machinery		17.4	16.8	17.8	18.7	17.1	18.3	17.0	...	...	...	...	...	...
Electrical mach.		3.0	3.0	3.0	3.0	2.7	3.0	3.1	...	...	...	...	...	...
Transport		18.9	23.5	21.8	21.4	16.5	17.3	15.8	...	...	...	...	...	...
Construction		...	...	...	...	...	...	...	...	...	...	...	...	...
Mnfg.+Const.		...	...	...	...	...	...	...	...	...	...	...	...	...
Production of steel consuming sectors														
Total GDP	7/	114.5	115.0	116.9	120.0	124.7	127.2	130.1	134.1	137.2	140.8	142.2	206.0	...
Manufacturing	7/	26.4	25.6	25.6	27.2	29.2	29.8	30.0	30.8	31.6	32.1	31.4	...	...
Fabricated metals	8/	37.1	36.2	35.9	37.9	41.1	42.5	42.4.	45.4	48.6	49.8	49.6	46.9	46.7
Machinery	8/	58.2	57.3	54.8	59.1	64.2	69.7	68.8	68.4	69.8	75.1	74.8	65.5	60.0
Electrical mach.	8/	31.0	33.0	33.2	34.7	38.3	38.9	40.3	41.6	46.8	48.0	50.6	49.5	48.0
Transport	8/	50.4	51.1	54.0	55.5	60.6	63.0	70.0	74.4	75.0	75.7	71.3	65.2	66.5
Construction	7/	8.3	8.0	8.4	8.2	8.5	8.7	8.9	9.2	9.3	9.8	9.8	...	...
Mnfg.+Const.	7/	34.7	33.6	34.0	35.3	37.8	38.5	39.0	40.1	40.9	42.0	41.3	...	...
Steel intensity														
Total	9/	26.1	24.1	23.0	22.4	24.3	22.2	22.8	22.5	24.7	21.4	23.0	12.0	...
Manufacturing	9/	97.7	96.2	93.3	90.5	85.3	83.1	80.0	...	...	...	...	...	...
Fabricated metals	10/	37.8	34.7	34.5	34.4	34.0	32.6	31.6	...	...	...	...	...	...
Machinery	10/	8.9	8.1	8.8	8.5	8.1	7.4	7.3	...	...	...	...	...	...
Electrical mach.	10/	2.9	2.5	2.4	2.3	2.1	2.2	2.3	...	...	...	...	...	...
Transport	10/	11.2	12.8	10.9	10.4	8.2	7.8	6.7	...	...	...	...	...	...
Construction		...	...	...	...	...	...	...	...	...	...	...	...	...
Mnfg.+Const.		...	...	...	...	...	...	...	...	...	...	...	...	...

1/ Apparent finished steel consumption.
2/ Total manufacturing (ISIC 380): the sum of ISIC 381, 382, 383 and 384.
3/ Fabricated metals (ISIC 381).
4/ Machinery except electrical (ISIC 382).
5/ Electrical machinery (ISIC 383).
6/ Transport equipment (ISIC 384).
7/ Billions of constant 1980 US dollars.
8/ Billions of kronor.
9/ Grams of steel consumed per 1980 US dollar.
10/ Grams of seel consumed per kronor.

Annex table 21-8 Consumption of steel by major steel using sectors : *Korea, Republic of*

Item		1980	1981	1982	1983	1984	1985	1986	1987	1988	1989	1990	1991	1992
Steel consumption (1000t)														
Total	1/	5 115	6 314	6 455	7 801	9 408	10 020	10 934	13 642	14 519	16 946	20 054	24 454	21 820
Manufacturing	2/	...	2 455	2 533	3 033	4 140	4 581	5 137	6 679	7 566	8 587	9 187	11 555	10 644
Metal	3/	...	1 381	1 342	1 453	2 100	2 182	2 401	2 959	3 110	3 111	3 104	3 896	3 240
Machinery	4/	...	233	302	414	627	676	817	1 093	1 333	1 510	1 769	2 157	1 964
Electronics	5/	...	210	209	295	398	415	551	859	1 098	1 357	1 444	1 871	1 728
Transport	6/	...	631	680	871	1 016	1 308	1 368	1 768	2 025	2 608	2 870	3 631	3 712
Construction	7/	...	3 228	3 277	3 988	4 327	4 437	4 704	5 599	5 501	6 665	8 862	10 454	8 994
Mfg.+Const.	8/	...	5 683	5 810	7 021	8 467	9 018	9 841	12 278	13 067	15 251	18 049	22 009	19 638
Share of steel consumption (%)														
Total		...	100.0	100.0	100.0	100.0	100.0	100.0	100.0	100.0	100.0	100.0	100.0	100.0
Manufacturing		...	38.9	39.2	38.9	44.0	45.7	47.0	49.0	52.1	50.7	45.8	47.3	48.8
Metal		...	21.9	20.8	18.6	22.3	21.8	22.0	21.7	21.4	18.4	15.5	15.9	14.9
Machinery		...	3.7	4.7	5.3	6.7	6.8	7.5	8.0	9.2	8.9	8.8	8.8	9.0
Electronics		...	3.3	3.2	3.8	4.2	4.1	5.0	6.3	7.6	8.0	7.2	7.7	7.9
Transport		...	10.0	10.5	11.2	10.8	13.1	12.5	13.0	14.0	15.4	14.3	14.9	17.0
Construction		...	51.1	50.8	51.1	46.0	44.3	43.0	41.0	37.9	39.3	44.2	42.8	41.2
Mfg.+Const.		...	90.0	90.0	90.0	90.0	90.0	90.0	90.0	90.0	90.0	90.0	90.0	90.0
Production of steel consuming sectors														
Total GDP	9/	63.1	67.5	72.2	81.0	88.7	95.3	107.2	119.8	133.5	141.1	153.6	165.9	...
Manufacturing	9/	18.6	20.5	21.9	25.2	30.0	31.7	37.5	44.4	50.4	52.2	57.0	61.8	...
Metal		...	...	...	...	...	...	...	...	...	...	...	...	...
Machinery		...	...	...	...	...	...	...	...	...	...	...	...	...
Electronics		...	...	...	...	...	...	...	...	...	...	...	...	...
Transport		...	...	...	...	...	...	...	...	...	...	...	...	...
Construction	9/	5.3	5.0	5.9	7.2	7.6	7.9	8.3	9.3	10.2	11.5	13.7	15.1	...
Mfg.+Const.	9/	23.9	25.5	27.8	32.4	37.6	39.6	45.8	53.7	60.6	63.7	70.7	76.9	...
Steel intensity														
Total	10/	81.1	93.5	89.4	96.3	106.1	105.1	102.0	113.9	108.8	120.1	130.6	147.4	...
Manufacturing	10/	...	119.8	115.7	120.4	138.0	144.5	137.0	150.4	150.1	164.5	161.2	187.0	...
Metal		...	...	...	...	...	...	...	...	...	...	...	...	...
Machinery		...	...	...	...	...	...	...	...	...	...	...	...	...
Electronics		...	...	...	...	...	...	...	...	...	...	...	...	...
Transport		...	...	...	...	...	...	...	...	...	...	...	...	...
Construction	10/	...	645.5	555.3	553.9	569.3	561.6	566.7	602.0	539.3	579.6	646.9	692.3	...
Mfg.+Const.	10/	...	222.8	209.0	216.7	225.2	227.7	214.9	228.6	215.6	239.4	255.3	286.2	...

1/ Apparent finished steel consumption.
2/ Total manufacturing (ISIC 380): t he sum of ISIC 381, 382, 383 and 384.
3/ Fabricated metals (ISIC 381).
4/ Machinery except electrical (ISIC 382).
5/ Electrical machinery (ISIC 383).
6/ Transport equipment (ISIC 384).
7/ Construction (ISIC 500).
8/ Manufacturing plus construction (ISIC 380 plus ISIC 500).
9/ Billions of constant 1980 US dollars.
10/ Grams of steel consumed per 1980 US dollar.

Annex table 21-9 Consumption of steel by major steel using sectors : *Brazil*

Item		1972	1975	1980	1982	1983	1984	1985	1986	1987	1988	1989	1990	1991	1992
Steel consumption (1000t)															
Total	1/	5 432	8 353	11 648	8 747	7 157	8 803	9 932	12 077	12 435	9 988	12 466	8 693	8 415	8 868
Manufacturing	2/	3 780	5 874	8 353	...	...	...	...	...	...	...	...	5 760	6 077	5 410
Fabricated metals	3/	1 792	2 734	4 048	...	...	...	...	...	...	...	...	3 637	3 959	3 314
Machinery	4/	480	804	1 144	...	...	...	...	...	...	...	...	400	333	333
Electric machine	5/	231	334	455	...	...	...	...	...	...	...	...	593	589	527
Transport	6/	1 277	2 002	2 706	...	...	...	...	...	...	...	...	1 130	1 196	1 236
Construction	7/	1 530	2 363	3 475	...	...	...	...	...	...	...	...	790	772	798
Mnfg.+Const.	8/	5 310	8 237	11 828	...	...	...	...	...	...	...	...	6 550	6 849	6 208
Share of steel consumption (%)															
Total		100.0	100.0	100.0	100.0	100.0	100.0	100.0	100.0	100.0	100.0	100.0	100.0	100.0	100.0
Manufacturing		69.6	70.3	71.7	...	...	...	...	...	...	...	...	66.3	72.2	61.0
Fabricated metals		33.0	32.7	34.8	...	...	...	...	...	...	...	...	41.8	47.0	37.4
Machinery		8.8	9.6	9.8	...	...	...	...	...	...	...	...	4.6	4.0	3.8
Electric machine		4.3	4.0	3.9	...	...	...	...	...	...	...	...	6.8	7.0	5.9
Transport		23.5	24.0	23.2	...	...	...	...	...	...	...	...	13.0	14.2	13.9
Construction		28.2	28.3	29.8	...	...	...	...	...	...	...	...	9.1	9.2	9.0
Mnfg.+Const.		97.8	98.6	101.5	...	...	...	...	...	...	...	...	75.4	81.4	70.0
Production of steel consuming sectors															
Total GDP	9/	137.3	179.5	252.8	248.0	244.3	259.1	278.5	296.3	307.8	310.3	324.0	311.4	315.1	...
Manufacturing	9/	38.4	50.4	71.6	64.1	60.3	64.0	69.4	77.2	78.0	75.3	76.2	70.8	67.9	...
Fabricated metals		...	...	...	...	...	...	...	...	...	...	...	...	...	...
Machinery		...	...	...	...	...	...	...	...	...	...	...	...	...	...
Electric machine		...	...	...	...	...	...	...	...	...	...	...	...	...	...
Transport		...	...	...	...	...	...	...	...	...	...	...	...	...	...
Construction	9/	7.9	11.5	16.3	15.0	14.0	13.0	13.8	16.3	16.5	16.0	16.5	15.1	14.5	...
Mnfg.+Const.	9/	46.3	61.9	87.9	79.1	74.3	77.0	83.2	93.5	94.5	91.3	92.7	85.9	82.4	...
Steel intensity															
Total	10/	39.6	46.5	46.1	35.3	29.3	34.0	35.7	40.8	40.4	32.2	38.5	27.9	26.7	...
Manufacturing	10/	98.4	116.5	116.7	...	...	...	...	...	...	...	...	81.4	89.5	...
Fabricated metals		...	...	...	...	...	...	...	...	...	...	...	...	...	...
Machinery		...	...	...	...	...	...	...	...	...	...	...	...	...	...
Electric machine		...	...	...	...	...	...	...	...	...	...	...	...	...	...
Transport		...	...	...	...	...	...	...	...	...	...	...	...	...	...
Construction	10/	193.7	205.5	213.2	...	...	...	...	...	...	...	...	52.3	53.2	...
Mnfg.+Const.	10/	114.7	133.1	134.6	...	...	...	...	...	...	...	...	76.3	83.1	...

1/ Apparent finished steel consumption.
2/ Total manufacturing (ISIC 380): the sum of ISIC 381, 382, 383 and 384.
3/ Fabricated metals (ISIC 381).
4/ Machinery except electrical (ISIC 382).
5/ Electrical machinery (ISIC 383).
6/ Transport equipment (ISIC 384).
7/ Construction (ISIC 500).
8/ Manufacturing plus construction (ISIC 380 plus ISIC 500).
9/ Billions of constant 1980 US dollars.
10/ Grams of steel consumed per 1980 US dollar.

Annex table 21-10 Consumption of steel by major steel using sectors : *Hungary*

Item		1980	1981	1982	1983	1984	1985	1986	1987	1988	1989	1990	1991	1992	1993
Steel consumption (1000t)															
Total	1/	2 955	2 996	3 050	2 942	2 768	2 868	2 946	2 847	2 678	2 394	2 027	1 196	904	1 073
Manufacturing	2/	2 180	2 105	2 150	2 055	1 950	2 010	2 010	1 970	1 870	1 690	1 135	840	720	890
Fabricated metals	3/	850	840	850	840	810	800	800	780	750	700	460	350	300	330
Machinery	4/	920	887	900	847	788	850	860	862	808	724	445	295	270	385
Electrical mach.	5/	60	58	60	58	52	50	50	48	42	36	30	25	20	25
Transport	6/	350	320	340	310	300	310	300	280	270	230	200	170	130	150
Construction	7/	250	245	250	245	200	190	190	180	180	160	140	120	100	120
Mnfg.+Const.	8/	2 430	2 350	2 400	2 300	2 150	2 200	2 200	2 150	2 050	1 850	1 275	960	820	1 010
Share of steel consumption (%)															
Total		100.0	100.0	100.0	100.0	100.0	100.0	100.0	100.0	100.0	100.0	100.0	100.0	100.0	100.0
Manufacturing		73.8	70.3	70.5	69.9	70.4	70.1	68.2	69.2	69.8	70.6	56.0	70.2	79.6	82.9
Fabricated metals		28.8	28.0	27.9	28.6	29.3	27.9	27.2	27.4	28.0	29.2	22.7	29.3	33.2	30.8
Machinery		31.1	29.6	29.5	28.8	28.5	29.6	29.2	30.3	30.2	30.2	22.0	24.7	29.9	35.9
Electrical mach.		2.0	1.9	2.0	2.0	1.9	1.7	1.7	1.7	1.6	1.5	1.5	2.1	2.2	2.3
Transport		11.8	10.7	11.1	10.5	10.8	10.8	10.2	9.8	10.1	9.6	9.9	14.2	14.4	14.0
Construction		8.5	8.2	8.2	8.3	7.2	6.6	6.4	6.3	6.7	6.7	6.9	10.0	11.1	11.2
Mnfg.+Const.		82.2	78.4	78.7	78.2	77.7	76.7	74.7	75.5	76.5	77.3	62.9	80.3	90.7	94.1
Production of steel consuming sectors															
Total GDP	9/	48.2	50.2	51.3	52.0	53.6	53.7	54.2	56.5	56.3	56.3	54.2	...	...	...
Manufacturing		...	...	...	...	...	...	...	...	...	...	...	...	...	...
Fabricated metals		...	...	...	...	...	...	...	...	...	...	...	...	...	...
Machinery		...	...	...	...	...	...	...	...	...	...	...	...	...	...
Electrical mach.		...	...	...	...	...	...	...	...	...	...	...	...	...	...
Transport		...	...	...	...	...	...	...	...	...	...	...	...	...	...
Construction		...	...	...	...	...	...	...	...	...	...	...	...	...	...
Mnfg.+Const.		...	...	...	...	...	...	...	...	...	...	...	...	...	...
Steel intensity															
Total	10/	61.3	59.7	59.5	56.6	51.6	53.4	54.4	50.4	47.6	42.5	37.4	...	...	...
Manufacturing		...	...	...	...	...	...	...	...	...	...	...	...	...	...
Fabricated metals		...	...	...	...	...	...	...	...	...	...	...	...	...	...
Machinery		...	...	...	...	...	...	...	...	...	...	...	...	...	...
Electrical mach.		...	...	...	...	...	...	...	...	...	...	...	...	...	...
Transport		...	...	...	...	...	...	...	...	...	...	...	...	...	...
Construction		...	...	...	...	...	...	...	...	...	...	...	...	...	...
Mnfg.+Const.		...	...	...	...	...	...	...	...	...	...	...	...	...	...

1/ Apparent finished steel consumption.
2/ Total manufacturing (ISIC 380): the sum of ISIC 381, 382, 383 and 384.
3/ Fabricated metals (ISIC 381).
4/ Machinery except electrical (ISIC 382).
5/ Electrical machinery (ISIC 383).
6/ Transport equipment (ISIC 384).
7/ Construction (ISIC 500).
8/ Manufacturing plus construction (ISIC 380 plus ISIC 500).
9/ Grams of steel consumed per 1980 US dollar.

Annex table 21-11 Consumption of steel by major steel using sectors : *The former Yugoslavia* 1/

Item		1977	1980	1981	1982	1983	1984	1985	1986	1987	1988	1989	1991	1992
Steel consumption (1000t)														
Total	2/	4 144	4 720	5 053	4 363	4 433	4 315	4 281	4 769	3 458	3 251	4 137	1 310	815
Manufacturing	3/	1 676	2 211	2 604	2 591	2 505	2 559	2 565	2 842	2 875	2 807	2 786	512	403
Metal	4/	977	1 171	1 439	1 539	1 416	1 414	1 426	1 540	1 581	1 466	1 446	246	211
Machinery	5/	249	319	407	351	412	371	380	486	477	476	482	85	64
Electronics	6/	163	174	171	157	154	189	196	186	177	178	162	32	28
Transport	7/	287	547	587	544	523	585	563	630	640	687	696	149	100
Construction	8/	...	687	769	693	662	555	782	397	591	444	816	200	169
Mfg.+Const.	9/	...	2 898	3 373	3 284	3 167	3 114	3 347	3 239	3 466	3 251	3 602	712	572
Share of steel consumption (%)														
Total		100.0	100.0	100.0	100.0	100.0	100.0	100.0	100.0	100.0	100.0	100.0	100.0	100.0
Manufacturing		40.4	46.8	51.5	59.4	56.5	59.3	59.9	59.6	83.1	86.3	67.3	39.1	49.4
Metal		23.6	24.8	28.5	35.3	31.9	32.8	33.3	32.3	45.7	45.1	35.0	18.8	25.9
Machinery		6.0	6.8	8.1	8.0	9.3	8.6	8.9	10.2	13.8	14.6	11.7	6.5	7.9
Electronics		3.9	3.7	3.4	3.6	3.5	4.4	4.6	3.9	5.1	5.5	3.9	2.4	3.4
Transport		6.9	11.6	11.6	12.5	11.8	13.6	13.2	13.2	18.5	21.1	16.8	11.4	12.3
Construction		...	14.6	15.2	15.9	14.9	12.9	18.3	8.3	17.1	13.7	19.7	15.3	20.7
Mfg.+Const.		...	61.4	66.8	75.3	71.4	72.2	78.2	67.9	100.2	100.0	87.1	54.4	70.2
Production of steel consuming sectors														
Total GDP	10/	54.5	64.7	65.3	66.2	64.8	66.2	68.2	70.5	71.8	70.7	71.2	...	...
Manufacturing	10/	16.2	19.8	20.7	20.6	20.9	22.0	22.6	23.5	24.0	23.8	24.1	...	...
Metal	11/	137.6	143.2	141.0	136.3	139.6	141.8	143.2	140.8	143.9	136.3	...	34.6	...
Machinery	11/	94.7	101.4	103.5	104.1	107.0	114.1	118.9	115.5	111.0	116.4	...	25.3	...
Electronics	11/	87.4	91.7	92.3	92.9	99.6	109.2	116.0	115.0	110.9	114.8	...	18.5	...
Transport	11/	96.4	98.0	98.5	99.1	107.7	110.3	113.7	121.3	125.9	125.5	...	42.3	...
Construction	10/	...	419.3	400.0	369.9	321.6	306.4	303.2	299.0	296.4	277.1	268.1	85.7	...
Mfg.+Const.	10/	21.8	26.7	27.3	26.7	26.2	27.1	27.6	28.4	29.0	28.8	29.1	...	...
Steel intensity														
Total	12/	76.0	72.9	77.4	65.9	68.4	65.2	62.8	67.6	48.2	46.0	58.1	...	...
Manufacturing	12/	103.5	111.7	125.8	125.8	119.9	116.3	113.5	120.9	119.8	117.9	115.6	...	...
Metal	13/	7.10	8.18	10.21	11.29	10.14	9.97	9.96	10.94	10.99	10.76	...	7.11	...
Machinery	13/	2.63	3.15	3.93	3.37	3.85	3.25	3.20	4.21	4.30	4.09	...	3.36	...
Electronics	13/	1.86	1.90	1.85	1.69	1.55	1.73	1.69	1.62	1.60	1.55	...	1.73	...
Transport	13/	2.98	5.58	5.96	5.49	4.86	5.30	4.95	5.19	5.08	5.47	...	3.52	...
Construction	12/	...	1.64	1.92	1.87	2.06	1.81	2.58	1.33	1.99	1.60	3.04	2.33	...
Mfg.+Const.	12/	...	108.5	123.6	123.0	120.9	114.9	121.3	114.0	119.5	112.9	123.8	...	...

1/ Data for 1991 and 1992 are for the Federal Republic of Yugoslavia.
2/ Apparent finished steel consumption.
3/ Total manufacturing (ISIC 380): the sum of ISIC 381, 382, 383 and 384.
4/ Fabricated metals (ISIC 381).
5/ Machinery except electrical (ISIC 382).
6/ Electrical machinery (ISIC 383).
7/ Transport equipment (ISIC 384).
8/ Construction (ISIC 500).
9/ Manufacturing plus construction (ISIC 380 plus ISIC 500).
10/ Billions of constant 1980 US dollars.
11/ Thousands of constant 1972 dinars.
12/ Grams of steel consumed per 1980 US dollar.
13/ Grams of steel consumed per 1972 dinars.

Annex table 21-12 Consumption of steel by major steel using sectors : *Croatia*

Item		1980	1981	1982	1983	1984	1985	1986	1987	1988	1989	1990	1991	1992
Steel consumption (1000t)														
Total	1/	856	890	875	851	843	844	840	768	784	753	614	395	316
Manufacturing	2/	690	731	693	735	689	660	673	630	650	629	500	324	277
Fabricated metals	3/	370	398	376	345	358	357	352	313	292	296	220	142	94
Machinery	4/	108	109	106	104	94	89	102	99	100	95	81	37	22
Electric machine	5/	38	42	40	39	45	43	38	33	35	33	31	19	17
Transport	6/	174	182	171	247	192	171	181	185	223	205	168	126	144
Construction	7/	166	159	182	116	154	184	167	138	134	124	114	71	39
Mnfg.+Const.	8/	856	890	875	851	843	844	840	768	784	753	614	395	316
Share of steel consumption (%)														
Total		100.0	100.0	100.0	100.0	100.0	100.0	100.0	100.0	100.0	100.0	100.0	100.0	100.0
Manufacturing		80.6	82.1	79.2	86.4	81.7	78.2	80.1	82.0	82.9	83.5	81.4	82.0	87.7
Fabricated metals		43.2	44.7	43.0	40.5	42.5	42.3	41.9	40.8	37.2	39.3	35.8	35.9	29.7
Machinery		12.6	12.2	12.1	12.2	11.2	10.5	12.1	12.9	12.8	12.6	13.2	9.4	7.0
Electric machine		4.4	4.7	4.6	4.6	5.3	5.1	4.5	4.3	4.5	4.4	5.0	4.8	5.4
Transport		20.3	20.4	19.5	29.0	22.8	20.3	21.5	24.1	28.4	27.2	27.4	31.9	45.6
Construction		19.4	17.9	20.8	13.6	18.3	21.8	19.9	18.0	17.1	16.5	18.6	18.0	12.3
Mnfg.+Const.		100.0	100.0	100.0	100.0	100.0	100.0	100.0	100.0	100.0	100.0	100.0	100.0	100.0
Production of steel consuming sectors														
Total GDP		...	...	...	...	...	...	...	...	...	...	...	...	...
Manufacturing		...	...	...	...	...	...	...	...	...	...	...	...	...
Fabricated metals	9/	393	417	362	324	329	316	331	309	294	294	238	160	126
Machinery	9/	114	120	120	123	116	114	122	118	108	111	84	44	27
Electric machine	9/	116	121	117	108	115	125	130	128	121	115	95	83	65
Transport		...	...	...	...	...	...	...	...	...	...	...	...	...
Construction		...	...	...	...	...	...	...	...	...	...	...	...	...
Mnfg.+Const.		...	...	...	...	...	...	...	...	...	...	...	...	...
Steel intensity														
Total		...	...	...	...	...	...	...	...	...	...	...	...	...
Manufacturing		...	...	...	...	...	...	...	...	...	...	...	...	...
Fabricated metals	10/	941.5	954.4	1 038.7	1 064.8	1 088.1	1 129.7	1 063.4	1 012.9	993.2	1 006.8	924.4	887.5	746.0
Machinery	10/	947.4	908.3	883.3	845.5	810.3	780.7	836.1	839.0	925.9	855.9	964.3	840.9	814.8
Electric machine	10/	327.6	347.1	341.9	361.1	391.3	344.0	292.3	257.8	289.3	287.0	326.3	228.9	261.5
Transport		...	...	...	...	...	...	...	...	...	...	...	...	...
Construction		...	...	...	...	...	...	...	...	...	...	...	...	...
Mnfg.+Const.		...	...	...	...	...	...	...	...	...	...	...	...	...

1/ Apparent finished steel consumption.
2/ Total manufacturing (ISIC 380): the sum of ISIC 381, 382, 383 and 384.
3/ Fabricated metals (ISIC 381).
4/ Machinery except electrical (ISIC 382).
5/ Electrical machinery (ISIC 383).
6/ Transport equipment (ISIC 384).
7/ Construction (ISIC 500).
8/ Manufacturing plus construction (ISIC 380 plus ISIC 500).
9/ Thousands of tonnes.
10/ Kilograms of steel consumed per tonne of material.

Annex table 21-13 Consumption of steel by major steel using sectors : *Slovenia*

Item		1980	1981	1982	1983	1984	1985	1986	1987	1988	1989	1990	1991	1992
Steel consumption (1000t)														
Total	1/	...	...	...	...	...	...	...	684	663	691	592	429	376
Manufacturing	2/	...	...	...	...	...	...	...	573	571	582	495	398	351
Fabricated metals	3/	...	...	...	...	...	...	...	285	277	293	283	233	193
Machinery	4/	...	...	...	...	...	...	...	179	184	192	126	90	77
Electric machine	5/	...	...	...	...	...	...	...	35	41	34	24	23	20
Transport	6/	...	...	...	...	...	...	...	74	69	63	62	52	61
Construction	7/	138	119	93	85	137	107	118	95	78	81	83	23	18
Mnfg.+Const.	8/	...	...	...	...	...	...	...	668	649	663	578	421	369
Share of steel consumption (%)														
Total		...	...	...	...	...	...	...	100.0	100.0	100.0	100.0	100.0	100.0
Manufacturing		...	...	...	...	...	...	...	83.8	86.1	84.2	83.6	92.8	93.4
Fabricated metals		...	...	...	...	...	...	...	41.7	41.8	42.4	47.8	54.3	51.3
Machinery		...	...	...	...	...	...	...	26.2	27.8	27.8	21.3	21.0	20.5
Electric machine		...	...	...	...	...	...	...	5.1	6.2	4.9	4.1	5.4	5.3
Transport		...	...	...	...	...	...	...	10.8	10.4	9.1	10.5	12.1	16.2
Construction		...	...	...	...	...	...	...	13.9	11.8	11.7	14.0	5.4	4.8
Mnfg.+Const.		...	...	...	...	...	...	...	97.7	97.9	95.9	97.6	98.1	98.1
Production of steel consuming sectors														
Total GDP		...	...	...	...	...	...	...	...	...	...	...	...	...
Manufacturing		...	...	...	...	...	...	...	...	...	...	...	...	...
Fabricated metals	9/	...	98.1	98.1	96.9	98.9	100.0	99.5	99.0	94.4	97.1	82.4	77.7	65.6
Machinery	9/	...	83.0	85.4	94.7	96.0	100.0	99.4	103.2	98.7	108.8	100.0	79.4	64.0
Electric machine	9/	...	75.8	82.9	89.6	97.5	100.0	105.3	95.9	93.6	98.3	74.1	62.2	50.7
Transport	9/	...	103.8	107.1	103.9	97.8	100.0	105.6	101.8	91.0	90.7	84.7	69.7	57.2
Construction	9/	134.0	120.4	105.9	101.3	102.9	100.0	101.0	100.5	103.4	114.2	97.5	72.5	64.3
Mnfg.+Const.		...	...	...	...	...	...	...	...	...	...	...	...	...
Steel intensity														
Total		...	...	...	...	...	...	...	...	...	...	...	...	...
Manufacturing		...	...	...	...	...	...	...	...	...	...	...	...	...
Fabricated metals	10/	...	...	...	...	...	...	...	2.879	2.934	3.018	3.434	2.999	2.942
Machinery	10/	...	...	...	...	...	...	...	1.734	1.864	1.765	1.260	1.134	1.203
Electric machine	10/	...	...	...	...	...	...	...	365.0	438.0	345.9	323.9	369.8	394.5
Transport	10/	...	...	...	...	...	...	...	0.727	0.758	0.695	0.732	0.746	1.066
Construction	10/	1.030	0.988	0.878	0.839	1.331	1.070	1.168	0.945	0.754	0.709	0.851	0.317	0.280
Mnfg.+Const.		...	...	...	...	...	...	...	...	...	...	...	...	...

1/ Apparent finished steel consumption.
2/ Total manufacturing (ISIC 380): the sum of ISIC 381, 382, 383 and 384.
3/ Fabricated metals (ISIC 381).
4/ Machinery except electrical (ISIC 382).
5/ Electrical machinery (ISIC 383).
6/ Transport equipment (ISIC 384).
7/ Construction (ISIC 500).
8/ Manufacturing plus construction (ISIC 380 plus ISIC 500).
9/ Index (1985=100).
10/ Thousands of tonnes of steel consumed per index.

Annex table 21-14 Consumption of steel by major steel using sectors : *Russian Federation*

Item		1980	1981	1982	1983	1984	1985	1986	1987	1988	1989	1990	1991	1992
Steel consumption (Million tonnes)														
Total	1/	62.4	...	...	...	...	66.8	...	...	...	...	67.8	52.6	41.9
Manufacturing		...	...	...	...	...	...	...	...	...	...	...	...	...
Fabricated metals	2/	29.5	...	...	...	...	32.0	...	...	...	...	34.7	26.4	21.4
Machinery	3/	...	...	...	...	...	...	...	...	...	...	...	...	...
Electric machine	4/	1.9	...	...	...	...	1.7	...	...	...	...	1.9	1.6	1.3
Transport	5/	1.9	...	...	...	...	2.0	...	...	...	...	1.9	1.7	1.5
Construction	6/	19.9	...	...	...	...	21.0	...	...	...	...	18.5	13.4	9.8
Mnfg.+Const.		...	...	...	...	...	...	...	...	...	...	...	...	...
Share of steel consumption (%)														
Total		100.0	...	...	...	...	100.0	...	...	...	...	100.0	100.0	100.0
Manufacturing		...	...	...	...	...	...	...	...	...	...	...	...	...
Fabricated metals		47.3	...	...	...	...	47.9	...	...	...	...	51.2	50.2	51.1
Machinery		...	...	...	...	...	...	...	...	...	...	...	...	...
Electric machine		3.0	...	...	...	...	2.5	...	...	...	...	2.8	3.0	3.1
Transport		3.0	...	...	...	...	3.0	...	...	...	...	2.8	3.2	3.6
Construction		31.9	...	...	...	...	31.4	...	...	...	...	27.3	25.5	23.4
Mnfg.+Const.		...	...	...	...	...	...	...	...	...	...	...	...	...
Production of steel consuming sectors														
Total GDP		...	...	...	...	...	...	...	...	...	...	...	...	...
Manufacturing		...	...	...	...	...	...	...	...	...	...	...	...	...
Fabricated metals		...	...	...	...	...	...	...	...	...	...	...	...	...
Machinery		...	...	...	...	...	...	...	...	...	...	...	...	...
Electric machine		...	...	...	...	...	...	...	...	...	...	...	...	...
Transport		...	...	...	...	...	...	...	...	...	...	...	...	...
Construction		...	...	...	...	...	...	...	...	...	...	...	...	...
Mnfg.+Const.		...	...	...	...	...	...	...	...	...	...	...	...	...
Steel intensity														
Total		...	...	...	...	...	...	...	...	...	...	...	...	...
Manufacturing		...	...	...	...	...	...	...	...	...	...	...	...	...
Fabricated metals		...	...	...	...	...	...	...	...	...	...	...	...	...
Machinery		...	...	...	...	...	...	...	...	...	...	...	...	...
Electric machine		...	...	...	...	...	...	...	...	...	...	...	...	...
Transport		...	...	...	...	...	...	...	...	...	...	...	...	...
Construction		...	...	...	...	...	...	...	...	...	...	...	...	...
Mnfg.+Const.		...	...	...	...	...	...	...	...	...	...	...	...	...

1/ Apparent finished steel consumption.
2/ Fabricated metals (ISIC 381). Includes machinery except electrical (ISIC 382).
3/ Machinery except electrical (ISIC 382). Includes fabricated metals (ISIC 381).
4/ Electrical machinery (ISIC 383).
5/ Transport equipment (ISIC 384).
6/ Construction (ISIC 500).